T0318761

TROPICAL EXTREMES: NATURAL VARIABILITY AND TRENDS

TROPICAL EXTREMES: NATURAL VARIABILITY AND TRENDS

Edited by

V. VENUGOPAL

Centre for Atmospheric and Oceanic Sciences & Divecha Centre for Climate Change, Indian Institute of Science, Bangalore, India

JAI SUKHATME

Centre for Atmospheric and Oceanic Sciences & Divecha Centre for Climate Change, Indian Institute of Science, Bangalore, India

RAGHU MURTUGUDDE

Earth System Science Interdisciplinary Centre, University of Maryland, College Park, MD, United States

RÉMY ROCA

CNRS at LEGOS, Obervatoire Midi-Pyrénées (OMP), Toulouse, France

Elsevier
Radarweg 29, PO Box 211, 1000 AE Amsterdam, Netherlands
The Boulevard, Langford Lane, Kidlington, Oxford OX5 1GB, United Kingdom
50 Hampshire Street, 5th Floor, Cambridge, MA 02139, United States

Notices
Knowledge and best practice in this field are constantly changing. As new research and experience broaden our understanding, changes in research methods, professional practices, or medical treatment may become necessary.

Practitioners and researchers must always rely on their own experience and knowledge in evaluating and using any information, methods, compounds, or experiments described herein. In using such information or methods they should be mindful of their own safety and the safety of others, including parties for whom they have a professional responsibility.

To the fullest extent of the law, neither the Publisher nor the authors, contributors, or editors, assume any liability for any injury and/or damage to persons or property as a matter of products liability, negligence or otherwise, or from any use or operation of any methods, products, instructions, or ideas contained in the material herein.

Library of Congress Cataloging-in-Publication Data
A catalog record for this book is available from the Library of Congress

British Library Cataloguing-in-Publication Data
A catalogue record for this book is available from the British Library

ISBN: 978-0-12-809248-4

Publisher: Candice Janco
Acquisition Editor: Laura Kelleher
Editorial Project Manager: Tasha Frank
Production Project Manager: Nilesh Kumar Shah
Cover Designer: Christian Bilbow

Typeset by TNQ Technologies

CONTENTS

LIST OF CONTRIBUTORS

Richard P. Allan
Department of Meteorology, University of Reading, Reading, United Kingdom

Mélanie Becker
LIENSs/CNRS, UMR 7266, ULR/CNRS, La Rochelle, France

Tobias Becker
Max Planck Institute for Meteorology, Hamburg, Germany

Florent Beucher
Centre National de Recherches Météorologiques, Météo-France & CNRS, Toulouse, France

Maria Budiarti
Badan Meteorologi Klimatologi dan Geofisika, Jakarta Pusat, Indonesia

R. Chattopadhyay
Indian Institute of Tropical Meteorology, Pune, India

Boris Dewitte
Centro de Estudios Avanzado en Zonas Áridas (CEAZA), Coquimbo, Chile; Universidad Católica del Norte, Coquimbo, Chile; Millennium Nucleus for Ecology and Sustainable Management of Oceanic Islands (ESMOI), Coquimbo, Chile; Laboratoire d'Etudes en Géophysique et Océanographie Spatiales, Toulouse, France

Gabor Drotos
Hungarian Academy of Sciences and Eötvös Loránd University, Budapest, Hungary

B.N. Goswami
Cotton University, Guwahati, India

Alice M. Grimm
Federal University of Parana, Curitiba, Brazil

Mikhail Karpytchev
LIENSs/CNRS, UMR 7266, ULR/CNRS, La Rochelle, France

Thierry Lebel
Univ. Grenoble Alpes, IRD, CNRS, IGE, Grenoble, France

Chunlei Liu
Department of Meteorology, University of Reading, Reading, United Kingdom

Brian E. Mapes
Rosenstiel School of Marine and Atmospheric Science (RSMAS), University of Miami, Miami, Florida, USA

Brian C. Matilla
Rosenstiel School of Marine and Atmospheric Science (RSMAS), University of Miami, Miami, Florida, USA

Thorsten Mauritsen
Max Planck Institute for Meteorology, Hamburg, Germany

Gérémy Panthou
Univ. Grenoble Alpes, IRD, CNRS, IGE, Grenoble, France

Fabrice Papa
LEGOS/IRD, UMR 5566, CNES/CNRS/IRD/UPS, Toulouse, France; Indo-French Cell for Water Sciences, IRD-IISc-NIO-IITM, Indian Institute of Science, Bangalore, India

Philippe Peyrillé
Centre National de Recherches Météorologiques, Météo-France & CNRS, Toulouse, France

Guillaume Quantin
Univ. Grenoble Alpes, IRD, CNRS, IGE, Grenoble, France

Ivan J. Ramírez
Department of Geography and Environmental Sciences, University of Colorado Denver, Denver, CO, United States; Consortium for Capacity Building/INSTAAR, University of Colorado Boulder, Boulder, CO, United States

Romain Roehrig
Centre National de Recherches Météorologiques, Météo-France & CNRS, Toulouse, France

Bjorn Stevens
Max Planck Institute for Meteorology, Hamburg, Germany

Ken Takahashi
Servicio Nacional de Meteorología e Hidrología, Lima, Peru

V. Venugopal
Centre for Atmospheric and Oceanic Sciences & Divecha Centre for Climate Change, Indian Institute of Science, Bangalore, India

Théo Vischel
Univ. Grenoble Alpes, IRD, CNRS, IGE, Grenoble, France

Catherine Wilcox
Univ. Grenoble Alpes, IRD, CNRS, IGE, Grenoble, France

INTRODUCTION

V. Venugopal[1], Jai Sukhatme[1], Raghu Murtugudde[2], Rémy Roca[3]
[1]Centre for Atmospheric and Oceanic Sciences & Divecha Centre for Climate Change, Indian Institute of Science, Bangalore, India; [2]Earth System Science Interdisciplinary Centre, University of Maryland, College Park, Maryland, USA; [3]CNRS at LEGOS, Obervatoire Midi-Pyrénées (OMP), Toulouse, France

Tropics immediately bring two things to mind. The first is the hot and humid climate. And the second is what is generally encapsulated in the phrase, "Global South"—developing countries with high population density. Global warming raises the specter of climate extremes, further exposing the dark underbelly of the climate vulnerability of the Global South. The floods and droughts associated with the cyclones and the El Niño—La Niña teleconnections have served well as templates for what can be expected under climate extremes (McPhaden et al., 2006; Emanuel, 2005). The evidence for increase in extremes of temperature is incontrovertible (IPCC, 2013; Alexander, 2016), whereas that for precipitation extremes is beginning to emerge beyond internal variability in many regions, especially in the tropics in terms of their intensity, spatial variability, and extent (Goswami et al., 2006; Ghosh et al., 2012; Roxy et al., 2017). The complexity of precipitation response to anthropogenic activities is complicated by the impact of aerosols on global to microphysical scale processes, which may be affecting the reliability of future projections (Yang et al., 2014). This underscores the need to understand the processes that determine the climate extremes in a warming world.

Weather-related disasters account for over 90% of the natural disasters worldwide with tens of thousands of deaths every year and billions of injuries (UNISDR Report, 2015). On a global map of top 10 countries most affected by weather-related disasters over the past two to three decades, the tropics stand out. This becomes even more crucial when you consider the fact that the Global South constitutes nearly 50% when stratified by income groups, while experiencing nearly 60% of the weather-related deaths over the same period. A particularly poignant plea (Webster, 2013) for investments in improving flood forecasting, highlighted the fact that while only 5% of the cyclones occur in the North Indian Ocean, they account for 95% of the casualties. It also points out that an investment of a few million dollars is all that is needed to save these lives and reduce the economic loss of billions of dollars. Significant work is needed in the impact of climate extremes on health and society, as well as the market mechanisms, which are supposed to protect against disasters (Ebi and Bowen, 2016; Hoeppe, 2016). The vulnerability of tropical countries extends even to potential impediments to their economic development, for instance, via the susceptibility of their port cities to extreme events (Nicholls et al., 2008).

It is in this broad context that the process and predictive understanding of climate extremes over the tropics has to be advanced. Cicero famously said that just being human saddles us with duties. In the context of global warming, attribution of climate extremes is an incredibly interesting and complex scientific challenge. But how to attribute responsibility for climate change and compensate for the loss of life and property borne by the Global South is a much more onerous duty for humanity as a whole.

Any such discussion ought to be based on a sound scientific rationale, and this book offers a detailed summary of the current state of understanding on the natural and anthropogenic-induced past and future changes in the extremes of importance to the denizens of the Global South.

Short-duration extreme rainfall events form an important part of the hydrologic cycle and have been extensively studied because of their societal and economic impacts (Easterling et al., 2000). The intensification of these types of extremes is often thought to be a direct consequence of warming, by way of an increased moisture-holding capacity of the atmosphere (Trenberth, 1999; Allen and Ingram, 2002). This first-order effect has mostly been validated by ground-based observations, especially in the tropical belt (Groisman et al., 2005). Additionally, a much discussed aspect of tropical rainfall is the so-called "wet-wetter, dry-drier" paradigm (Held and Soden, 2006). The expectation that, as a whole, wet regions will become wetter and dry regions will experience a further scarcity of rain in a warming scenario, has received some support from auxiliary observations and long-term simulations (Allan et al., 2010; Chou et al., 2013; Lau et al., 2013). However, the dynamical response to warming complicates the issue as reported by many recent studies (Greve et al., 2014).

It is also important to note that warming occurs in the presence of natural climate variability. The strongest mode of natural variability in the tropics is the El Niño Southern Oscillation (ENSO) with its warm (El Niño) and cold (La Niña) phases. To this end, anomalously wet or dry conditions experienced by various tropical regions have been associated with phases of ENSO (Dai and Wigley, 2000). In fact, fluctuations of episodic heavy rainfall on regional to global scales are associated with El Niño and La Niña conditions (Gershunov and Cayan, 2003; Grimm and Tedeschi, 2009;

Alexander et al., 2009). Moreover, the links between monthly/daily rainfall extremes and ENSO have also been explored over land and tropical oceans. More recently, it has been shown that there is a fundamental natural mode of variability in tropical rainfall accumulation extremes with the changing phases of ENSO (Sukhatme and Venugopal, 2015). Thus, the long-term natural variability of tropical extremes is itself robust and significant (Chang et al., 2015) and not easily disentangled from the secular changes due to warming.

This volume presents an up-to-date collection of contributions on various aspects related to observed tropical extremes, ranging from regional rainfall extremes and their impacts to global sea level rise. It also aims to present a modelling perspective on the challenges that remain in representing or capturing changes in these extremes, be they from natural variability or from warming. A brief overview of these contributions follows.

The volume begins with an atlas of precipitation extremes throughout the tropics. The focus here is to provide users with a novel "Integrated Data Viewer," which is flexible and allows them to combine multiple data streams (e.g., reanalysis, satellite), create "bundles" (subsets), and get a detailed visual and quantitative characterization of a given extreme event. Examples of its use are included in detail, and synoptic conditions that preceded and may have caused these extreme events are discussed.

Chapter 2 delves into South Asian monsoon extremes, with an emphasis on the Indian region. Here, the narrative is on extremes of two kinds, namely, climate and short-term extremes. The occurrence of droughts (climate extremes) is discussed both in the context of external forcing (ENSO) and internal variability (e.g., intraseasonal oscillations), and open scientific questions are highlighted. Moving on to longer timescales, the discussion introduces the concept of a "mega drought," given the propensity of higher frequency of droughts during one of the phases of this monsoonal multidecadal mode. This contribution concludes with a contextual discussion of increase in short-duration extreme rainfall events, their clustering, and possible causal elements.

Moving to the Southern Hemisphere, the observed austral summer monsoon extremes over South America are presented in Chapter 3. At the outset, features of the large-scale circulation, synoptic and mesoscale systems, and their role in the evolution of the South American monsoon are described. This is followed by a discussion on the role of climate modes on different timescales, in modulating the conditions responsible for extremes (heavy rainfall and droughts). The chapter then considers case studies of regional extremes, including possible mechanisms, and concludes with a discussion on observed trends in indices related to short-duration precipitation extremes.

Staying with the theme of monsoon systems, Chapter 4 presents a statistical analysis of observed West African Monsoon Extremes, a subject that has not received due attention in the scientific literature. In particular, it reports on a significant intensification of the rainfall over central Sahel in the past two decades. Emphasis is then shifted to understanding the physical mechanisms behind extreme events recorded during this period. The concurrence of East African Waves and large-scale moisture anomalies appear to be the most prominent drivers of these intense events. These changes, such as intensification, are subsequently put in the context of the so-called decadal scale recovery of Sahelian rainfall.

Given the importance of the state (present) and fate (projections) of rainfall extremes, it is important that key processes, which govern them, are faithfully represented in climate models. These challenging issues are discussed in Chapter 5, along with a summary evaluation of atmospheric and coupled models. The emphasis is on their ability to capture observed variability and is used to evaluate simulations of current and future changes in extremes of tropical rainfall. An important issue addressed in this chapter is the complicated response of precipitation to warming-related increases in atmospheric moisture content. This includes a dynamical component wherein the rainy regions are seen to shift in space with time and is further complicated by decadal variability in the land ocean partitioning of rainfall. Despite these pitfalls, emerging tropics-wide constraints on increasing precipitation extremes are put forth and discussed. Local changes in these extremes are still a matter of much uncertainty and it is emphasized that their projection remains an outstanding challenge for the climate community.

From regional monsoon hotspots and an overview of the ability of models to capture observed changes in tropical extremes, the volume moves on to the strongest natural mode of the global climate system, namely, ENSO. In particular, the emphasis in Chapter 6 is on "extreme" El Niño events, which appear to have a return period of ~20 years. The underlying dynamics of such a class of events is put forth in the context of recently introduced idea of "ENSO diversity." Furthermore, the fate of such events in a warming environment is discussed in detail.

A clear and present danger of a warming environment is that associated with rising sea level. This issue, which has profound social and economic implications for coastal tropical regions, is discussed in detail in Chapter 7. Beginning with an introduction to the concept of relative sea level (RSL), the chapter combines recent (two decades) satellite observations with longer-term tide gauge data to present a global map of RSL hotspots and their trends. In addition, the challenges involved in separating variations in absolute sea level due to land movement and secular changes that could potentially be linked to warming are also presented.

Staying with the theme of impacts of climate change but exploring their relevance in the context of policy, Chapter 8 presents a case study that examines the complex interactions of climate, environment, and society in Peru during the extreme El Niño event of 1997–98. The analysis presented is interdisciplinary and explores in detail the role of the past and present social and health conditions in compounding the physical climate risk befalling an ENSO hotspot such as Peru. The chapter concludes with several pointers aimed at public policy makers, in relation to climate change adaptation and resilience to hydrometeorological risks associated with large climate anomalies such as El Niño.

The book concludes with a paradigm of tropical convection revisited. It presents a broad-brush overview of the existing literature on modeling changes in tropical climate using idealized configurations. The modeling studies presented in this chapter discuss the strong role that the representation of deep moist convection plays in determining the nature of tropical circulation and its changes.

REFERENCES

Alexander, L.V., Uotila, P., Nicholls, N., 2009. Influence of sea surface temperature variability on global temperature and precipitation extremes. J. Geophys. Res. Atmos. 114, D18116. https://doi.org/10.1029/2009JD012301.

Alexander, L.V., 2016. Global observed long-term changes in temperature and precipitation extremes: a review of progress and limitations in IPCC assessments and beyond. Weather Clim. Extrem. 11, 4–16. https://doi.org/10.1016/j.wace.2015.10.007.

Allan, R.P., Soden, B.J., John, V.O., Ingram, W., Good, P., 2010. Current changes in tropical precipitation. Environ. Res. Lett. 5, 025205. https://doi.org/10.1088/1748-9326/5/2/025205.

Allen, M.R., Ingram, W.J., 2002. Constraints on future changes in climate and the hydrologic cycle. Nature 419, 224–232.

Chang, C.-P., Ghil, M., Latif, M., Wallace, J.M. (Eds.), 2015. Climate Change: Multidecadal and Beyond. World Scientific Publishing, Singapore. 376 p.

Chou, C., Chiang, J.C.H., Lan, C.-W., Chung, C.-H., Liao, Y.-C., Lee, C.-J., 2013. Increase in the range between wet and dry season precipitation. Nat. Geosci. 6, 263–267. https://doi.org/10.1038/NGEO1744.

Dai, A., Wigley, T.M.L., 2000. Global patterns of ENSO-induced precipitation. Geophys. Res. Lett. 27, 1283–1286. https://doi.org/10.1029/1999GL011140.

Easterling, D.R., Meehl, G.A., Parmesan, C., Changnon, S.A., Karl, T.R., Mearns, L.O., 2000. Climate extremes: observations, modeling, and impacts. Science 289, 2068–2074.

Ebi, K.L., Bowen, K., 2016. Extreme events as sources of health vulnerability: Drought as an example. Weather Clim. Extrem. 11, 95–102. https://doi.org/10.1016/j.wace.2015.10.001.

Emanuel, K., 2005. Increasing destructiveness of tropical cyclones over the past 30 years. Nature 436 (7051), 686–688.

Gershunov, A., Cayan, D.R., 2003. Heavy daily precipitation frequency over the contiguous U.S. Sources of climatic variability and seasonal predictability. J. Clim. 16, 2752–2765.

Ghosh, S., Das, D., Kao, S.-C., Ganguly, A.R., 2012. Lack of uniform trends but increasing spatial variability in observed Indian rainfall extremes. Nat. Clim. Chang. 2, 86–91. https://doi.org/10.1038/nclimate1327.

Goswami, B.N., Venugopal, V., Sengupta, D., Madhusoodanan, M., Xavier, P.K., 2006. Increasing trend of extreme rain events over India in a warming environment. Science 314, 1442–1446. https://doi.org/10.1126/science.1132027.

Greve, P., Orlowsky, B., Mueller, B., Sheffield, J., Reichstein, M., Seneviratne, S.I., 2014. Global assessment of trends in wetting and drying over land. Nat. Geosci. 7, 716–721.

Grimm, A.M., Tedeschi, R.G., 2009. ENSO and extreme rainfall events in South America. J. Clim. 22, 1589–1609.

Groisman, P.Y.A., Knight, R.W., Easterling, D.R., Karl, T.R., Hegerl, G.C., Razuvaev, V.N., 2005. Trends in intense precipitation in the climate record. J. Clim. 18, 1326–1350.

Held, I.M., Soden, B.J., 2006. Robust responses of the hydrological cycle to global warming. J. Clim. 19, 5686–5699.

Hoeppe, P., 2016. Trends in weather related disasters – consequences for insurers and society. Weather Clim. Extrem. 11, 70–79. https://doi.org/10.1016/j.wace.2015.10.002.

IPCC, 2013. In: Stocker, et al. (Ed.), Climate Change 2013: The Physical Science Basis. Contribution of Working Group I to the Fifth Assessment Report of the Intergovernmental Panel on Climate Change. Cambridge University Press. https://doi.org/10.1017/CBO9781107415324. 1535 p.

Lau, W.K.-M., Wu, H.-T., Kim, K.-M., 2013. A canonical response of precipitation characteristics to global warming from CMIP5 models. Geophys. Res. Lett. 40, 3163–3169. https://doi.org/10.1002/grl.50420.

McPhaden, M.J., Zebiak, S.E., Glantz, M.H., 2006. ENSO as an integrating concept in Earth Science. Science 314, 1740–1745. https://doi.org/10.1126/science.1132588.

Nicholls, R.J., et al., 2008. Ranking port cities with high exposure and vulnerability to climate extremes: exposure estimates. In: OECD Environment Working Papers, No. 1. OECD Publishing. http://doi.org/10.1787/011766488208.

Roxy, M.K., Ghosh, S., Pathak, A., Athulya, R., Mujumdar, M., Murtugudde, R., Terray, P., Rajeevan, M., 2017. A threefold rise in widespread extreme rain events over central India. Nat. Commun. 8 (1), 708,. https://doi.org/10.1038/s41467-017-00744-9.

Sukhatme, J., Venugopal, V., 2015. Waxing and waning of observed extreme annual tropical rainfall. Q. J. R. Meteorol. Soc. https://doi.org/10.1002/qj.2633.

Trenberth, K.E., 1999. Conceptual framework for changes of extremes of the hydrological cycle with climate change. Clim. Chang. 42, 327–339.

UNISDR Report, 2015. The Human Cost of Weather Related Disasters: 1995–2015. https://www.unisdr.org/we/inform/publications/46796.

Webster, P.J., 2013. Improve weather forecasts for the developing world. Nature 493, 17–19. https://doi.org/10.1038/493017a.

Yang, Q., Bitz, C.M., Doherty, S.J., 2014. Offsetting effects of aerosols on Arctic and global climate in the late 20th century. Atmos. Chem. Phys. 14, 3969–3975.

CHAPTER 1

A Global Atlas of Tropical Precipitation Extremes

Brian C. Matilla, Brian E. Mapes
Rosenstiel School of Marine and Atmospheric Science (RSMAS), University of Miami, Miami, Florida, USA

Contents

1. INTRODUCTION

Extreme accumulations of precipitation, while sometimes beneficial as a water source, can also pose a great hazard to life and property. Precipitation is a richly variable field, offering "extremes" across many octaves of space and timescale. The size and response characteristics of catchment basins define the relevant scales for river flooding. These scales range from urban canyons to half continent–sized basins, which respond to diverse meteorological drivers ranging from single thunderstorms cells to long atmospheric stationary waves. At small scales (involving the happenstance of a few convective cells), there may be little opportunity for generalizable science because the atmosphere is not globally well observed or analyzed on these scales, which are also fundamentally unpredictable in detail beyond "nowcast" range. For large-scale extremes, sophisticated hydrologic modeling may be required because precipitation is just one factor (along with evaporation and runoff, which can be quite nonlocal). For these reasons, our view here focuses on intermediate-scale extreme events: 1–3 days in rain

Tropical Extremes: Natural Variability and Trends
ISBN 978-0-12-809248-4
https://doi.org/10.1016/B978-0-12-809248-4.00001-7

accumulation time, corresponding to phenomena with reasonably well-observed spatial scales of many tens to hundreds of km.

In meteorological parlance, convective scales (related to the depth of the troposphere, ~10 km) and synoptic scales (slow enough for the Earth to turn significantly; spatially >2000 km) bracket a "meso" or middle-scale range with few definitive fundamental constraints, but with long enough space and time spans to be resolved by global satellite and perhaps latest-generation reanalysis data. The large end of mesoscale weather is thus an excellent candidate for data-driven meteorological study.

Our burning question, therefore, is *what can we learn from multiple case studies of precipitation extremes on such large mesoscales? Is there a common set of ingredients we can identify, and can we discern or infer their relative importance?*

Even just within the United States, Maddox et al. (1979) lamented that "elusive characteristics further complicate a difficult forecast problem." Nonetheless, "a number of features were common to many of the events," such as "an advancing middle-level, short-wave trough," and they noted that "many of the intense rainfalls occurred during nighttime hours," which we speculate may involve reduced competition from widespread ordinary convection.

An "ingredients-based methodology" was advanced by Doswell et al. (1996). Accumulated precipitation P is the product of average rainfall rate R and duration dt: $P = R\ dt$. Extremes of P therefore involve large R, large dt, or both. R can be related to the upward flux of vapor through cloud base as $R = Ewq$, where E is a precipitation efficiency defined by this very relation (the ratio between R and upward water vapor flux). The flux involves vertical air velocity w times water vapor mixing ratio q. The updraft w may be related to instability in the case of convectively driven updrafts, or to secondary circulations involved in maintaining long-lived balanced flows, or in many cases both: balanced flows orchestrating convective instability (Raymond et al., 2015) in its most realistic sense with q as a component along with lapse rate. Because R goes as the product of E, w, and q, it can increase quite steeply when all the factors become large. The further product of high rainfall rate with persistence can yield long probability tails of extremes, as in multiplicative cascades (Over and Gupta, 1994).

The "ingredients" formulation may seem trivial, but its vagueness is its strength because the exact *recipe* for extreme precipitation varies from case to case. Some types of events leading to P extremes include organized mesoscale convective systems (Schumacher and Johnson, 2005; Schumacher, 2008; Moore et al., 2012), topographic forcing (Romatschke and Houze, 2011; Houze et al., 2011; Galarneau et al., 2012), tropical cyclones

(Galarneau et al., 2010; Chien and Kuo, 2011), and subtropical jet-front systems (Allen and Mapes, 2017; Yokoyama et al., 2017), with different combinations of ingredients emphasized in different settings. The skill with which these events can be predicted, or even simulated and interpreted after the fact, continues to challenge meteorology (Schumacher, 2017). Case studies remain a relevant approach, so long as the screening process for what gets called a "case" is clear, the mysterious complicated ones are treated equally with the tidy, more easily explained ones.

In this chapter, we describe an online atlas for extreme case study selection based on the time series maximum at each cell within a geographical grid. Data resources for case studies will be steadily improved there, as our own efforts continue. The atlas is at http://weather.rsmas.miami.edu/links/HeavyRains_clickmaps. Our hope is that a community of users might emerge and share results, both to seek commonalities in mechanisms for heavy rains and to feed back improvements to the case study tools and resources there, making the atlas ever more powerful for its subsequent users. This document is an extension to Mapes (2011), describing an earlier instance of the atlas.

Data sources and software used in the case study below are described in Section 2. Section 3 shows a sample of the atlas's maps and statistics of the 3B42 rainfall extremes by resolution. Section 4 shows a case study from the Bay of Bengal. Next steps are discussed at the end.

2. DATA AND METHODS

2.1 Data Sets

The principal global rainfall product for this work is the Tropical Rainfall Measuring Mission 3B42 (TRMM 3B42) precipitation product (Huffman et al., 2007). TRMM 3B42 provides data at 3-hourly intervals on a 0.25×0.25 degree spatial mesh for the latitude belt 50S–50N. We used the merged precipitation variable in TRMM 3B42, which is a calibrated blend of its "high quality" but gappy microwave estimates and inferior but always available infrared-based estimates. Data from version 7 of this product are used, with improved detection and intensity of rainfall events (Liu, 2015).

The native-resolution data were also coarsened to 1, 2, and 4 degrees spatial meshes using climate data operators (cdo) second-order conservative remapping. Time averaging was then performed on each mesh, using boxcar smoothers to make centered averages of 9-point (27-hour) and 27-point (75-hour) duration. The absolute maximum of the time series for each

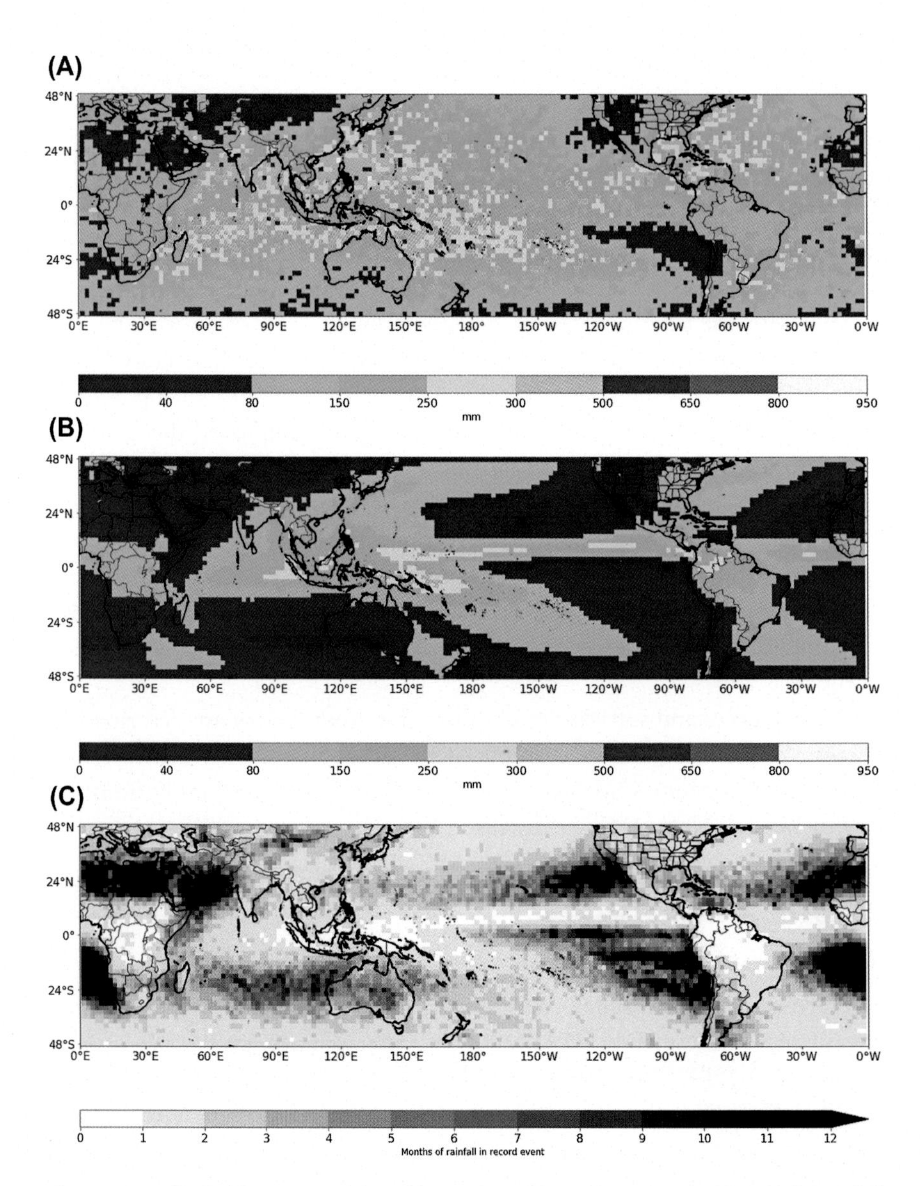

Figure 1.1 Shaded contour plots of precipitation amounts in the Tropical Rainfall Measuring Mission 3B42 product domain in 1998–2016. (A) Record 3-day accumulation (units mm) as it appears in the atlas. (B) Climatological annual rainfall expressed as mm month^{-1}. (C) Ratio A/B.

space and time resolution was logged and displayed in the atlas as images like Fig. 1.1A. By clicking any geographical grid cell, the user is directed to instructions on how to obtain data and displays centered on the chosen event. Because nearby grid cells often have their absolute record rainfalls set by different weather events, this time-only approach can still be used to see an ensemble of cases representing several types of extreme rain-making weather situations for a region.

The meteorological state of the atmosphere for each selected event draws on reanalysis products from various sources, including:

- The Modern-Era Retrospective analysis for Research and Applications (MERRA; Rienecker et al., 2011), version 2 (MERRA-2; Gelaro et al., 2017) provided by the National Aeronautics and Space Administration (NASA), 1–6 hourly on 0.5 × 0.67 degree grids. We use (1) 1-hour time averaged, single-level meteorology (MERRA-2 collection M2T1NXSLV), (2) hourly instantaneous column-integrated atmospheric variables (M2I1NXINT), and (3) 3-hourly 3D assimilated fields (M2I3NPASM).
- ERA-Interim (ERA-I; Dee et al., 2011) from the European Centre for Medium-range Weather Forecasting, 6-hourly on 0.7 × 0.7 degree grids.
- The Climate Forecast System Reanalysis (CFSR; Saha et al., 2010), 6-hourly on 0.5 × 0.5 degree grids.
- The Japanese 55-year Reanalysis (JRA-55; Kobayashi et al., 2015) from the Japanese Meteorological Agency, 6-hourly on 1.25 × 1.25 grids.

These data sets are used to interpret synoptic weather, bracketing analysis uncertainties.

2.2 Other Precipitation Data Sources

To bracket the uncertainties of the TRMM 3B42 precipitation estimate driving our case selection, several other precipitation estimates are also compared, all on a common 1-degree, daily grid:

- Global Land Data Assimilation System (Rodell et al., 2004).
- Climate Prediction Center Unified gauge-based analysis.
- Climate Hazards Group InfraRed Precipitation with Station data (Funk et al., 2015).
- Precipitation Estimation from Remotely Sensed Information using Artificial Neural Networks (PERSIANN; Sooroshian et al., 2014).
- Multi-Source Weighted-Ensemble Precipitation.
- Global Satellite Mapping of Precipitation (GSMAP; Okamoto et al., 2005).
- Global Precipitation Climatology Project.
- CPC MORPHing Technique (CMORPH; Joyce et al., 2004).

Altogether, 14 data sets (including reanalysis model-produced precipitation) provide estimates of daily accumulation around each event. A Python code (at https://github.com/bmatilla/Precip_MultiPanel) allows the user to specify a start and an end date and lat–lon bounding box. OPeNDAP links to the data sets are used to subset daily data sets from the authors' repository to create a figure like Fig. 1.2.

2.3 Capture and Visualization of Data

At the writing of Mapes (2011), only a link to satellite imagery was offered for the extreme events. Here we report a major improvement: the ability to download a small .isl file containing a script that can be opened with the Integrated Data Viewer (IDV). Created by Unidata, a US NSF-supported consortium, the IDV is a powerful open-source, click-to-install, all-platform 3D visualization tool for geophysical data sets, with many special functions for atmospheric research. The IDV saves its state in the form of a "bundle," an XML code that contains links to data sources, arbitrarily numerous displays of selected variables, and the exact 3D viewpoint for the visualization. Vertical cross sections, soundings, and 3D isosurfaces can be used along with 2D plan views to provide insight into the meteorology of a case study. Instructions for obtaining the IDV are offered through the atlas when case studies are selected there. Section 4 shows static images of such an IDV case study.

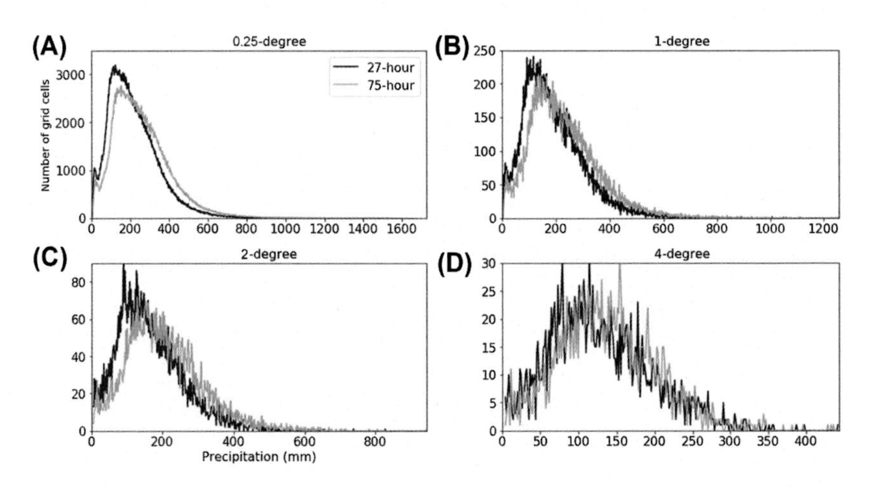

Figure 1.2 Histograms of the spatial structure of extreme (maximum in the time domain) Tropical Rainfall Measuring Mission 3B42 accumulations over 27 hours (blue) and 75 hours (orange) sliding time windows. Units: mm.

3. GLOBAL STATISTICS OF RECORD RAIN AMOUNTS

The atlas' map of 3-day, 2-degree record rainfalls is shown in Fig. 1.1A. For comparison, annual climatology is in Fig. 1.1B. The greatest record amounts are well off the equator, while climatology emphasizes places in the deep tropics with long rainy seasons. Dividing the precipitation records by climatology, Fig. 1.1C emphasizes how many "typical months" of rain fell in the record 3-day period. Many semiarid places received >1 year's typical rainfall in one 3-day event, making those cases consequential locally and thus worthy of study, even if they are not the largest in absolute amount.

Histograms of these maps of record amounts are depicted in Fig. 1.2 for various regridding mesh sizes. Going from panels A–D, the amounts decrease only modestly, with right-hand tail values changing from ~600–500 mm to 400–300 mm, even as the area of grid cells increases by factors of 16, 16 × 4, and 16 × 4 × 4. This modest decline indicates that these heavy rainfall events are much broader than convective scale, with strong clumping or coherence on scales up to 4 degrees. In time, however, events are much less coherent: 75-hour record amounts (orange) are only slightly greater than 27-hour records (blue). In wet regions, the 2 days straddling a 1-day record event add typically <30% to its total, judged by the roughly <30% horizontal offset of upper quantiles of the orange versus blue curves. For lower quantiles (that is, in drier places, such as over land, as seen in Fig. 1.1A), the blue–orange horizontal offset is a greater fraction of the total, meaning that the adjacent 2 days tend to contribute a larger fraction on top of the central wet day's rainfall.

Returning attention to Fig. 1.1A, the densest concentration of the most extreme record rainfalls (orange–red pixels) is located over north tropical Asia, around 10–20 N. With its warm seas, the multitude of mesoscale and synoptic scale features that can develop in such a broad area of activity, and perhaps also slow motions of features in this area, there are surely many interesting events (and lessons to be learned) from this area alone. As an example of extreme extremes, consider the magenta (>650 mm) pixel that touches India's east coast.

4. AN EXAMPLE CASE STUDY: BAY OF BENGAL

The 2-degree cell centered at latitude 19N, 87E had a record 75-hour rainfall exceeding 592 mm, as a reader may verify on the atlas (byte compression limits the accuracy of this metric). The central time of this record 3-day accumulation was at 21:00 UTC on June 29, 2006. This event was evidently part of the development of Deep Depression Bay of Bengal 02, named by the Indian Meteorological Department as the second classified storm of the season (https://en.m.wikipedia.org/wiki/2006_North_Indian_Ocean_cyclone_season). The storm claimed 131 lives via landslides, flooding, and collapsing infrastructure due to the extreme precipitation accumulation.

A comparison of various 5-day precipitation estimates encompassing this event is shown in Fig. 1.3. Land-only data sets (panels A–C) necessarily missed the offshore rain. TRMM 3B42 (Fig. 1.3D) shows the highest amounts, with 1-degree averages exceeding >700 mm—consistent with the fact that this product's extremity was our selection criterion. Some other satellite data sets (GSMAP and CMORPH, panels J and M) also showed high precipitation totals in the right area, whereas others appeared to blur or miss the peak (e.g., PERSIANN in Fig. 1.3E). The mean of all the non-land-confined data sets (Fig. 1.3O) peaks at a mere ~250 mm, whereas the ensemble root mean square (rms) (Fig. 1.3P) is double that, because the different estimates are so skewed. Clearly there are data challenges in this case. Happily, however, MERRA-2 (rainfall produced by a global reanalysis model) has a decent depiction of the scale and location of the 3B42-observed heavy rainfalls, albeit offset somewhat to the northeast (Fig. 1.3H). This makes MERRA-2's meteorological depiction worth considering in more detail.

Meteorology depictions for the case are summarized in Fig. 1.4 by sea level pressure (SLP) from MERRA-2, ERA-I, CFSR, and JRA-55, along with TRMM 3B42 rain rates at the time of most intense rainfall (July 1 at 00:00 UTC). Although all the reanalyses have a low pressure center in the region, only MERRA-2 (red) has a tight center in the right place. The intense rainfall observed by satellite was in westerlies to the south of the cyclone center.

Upward motion is closely coupled to rainfall, and vortex stretching by its associated horizontal convergence can lead to cyclone intensification (e.g., Grimes and Mercer, 2015). For this reason, Fig. 1.5 shows cross sections of MERRA-2 vertical pressure velocity (ω) and potential vorticity (PV) at the time of Fig. 1.4. Consistent with its precipitation offset (Fig. 1.3H), MERRA-2's updraft is offset to the northeast relative to satellite rainfall (3B42, green) but is still distinct to the southwest of the cyclone center. The analyzed updraft peaks at mid-levels, with weak values at low levels.

PV is a key field in cyclone structure and evolution (e.g., Molinari et al., 1998). The PV here is revealed in Fig. 1.5C to be centered at mid-levels, with

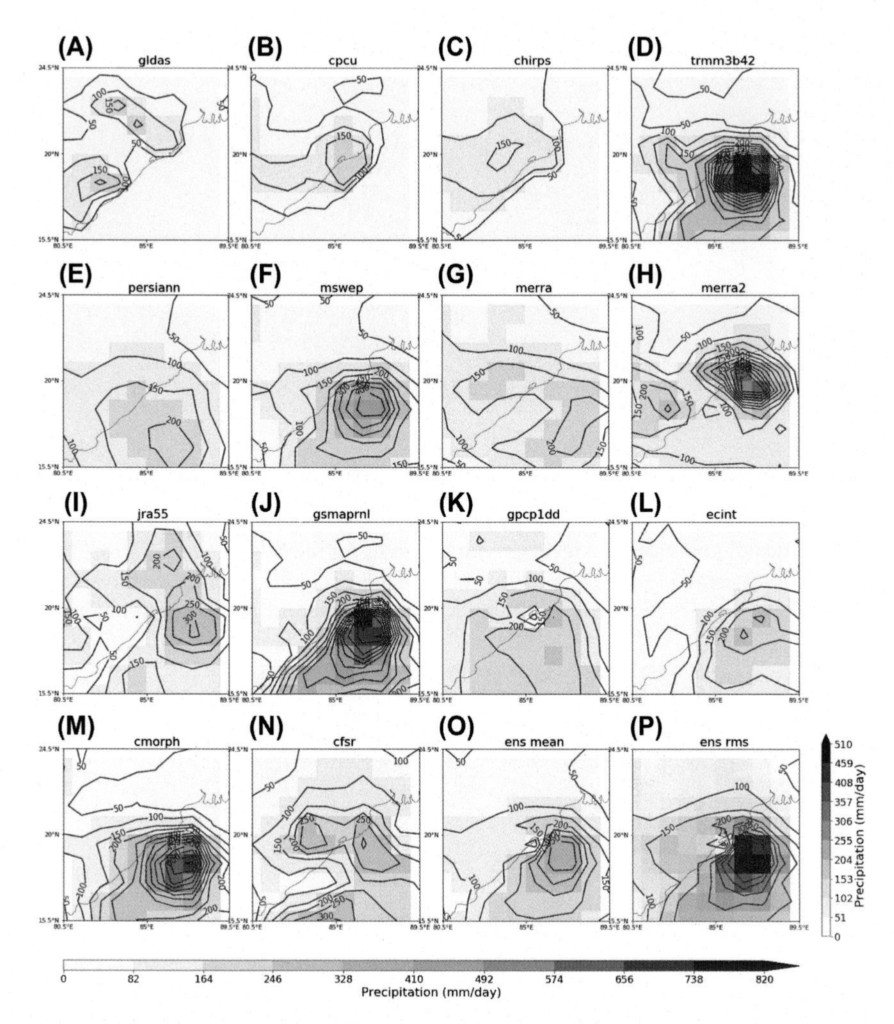

Figure 1.3 Five-day (June 28–July 2, 2006) rainfall accumulation estimates for all observational and reanalysis data sets on a 1-degree grid, for the 75-hour 2-degree record event centered at 19N, 87E. Color intervals are 82 mm for all plots except ensemble RMS (51 mm), whereas contour intervals are 50 mm.

maximum just above the 5.8 km altitude where the blue Z500 contours intersect the cross section. In other words, the cyclone has a cool core at low levels and warm core at upper levels, typical of monsoon cyclones (Boos et al., 2015, 2017). Cool core cyclones in shear can induce upward motions at low levels, which may be a mechanism at play in this rain event. An area of weaker PV is also seen at mid-levels at left in Fig. 1.5C, corresponding to weak, broad troughs in the SLP and Z500 fields west and southwest of the main center.

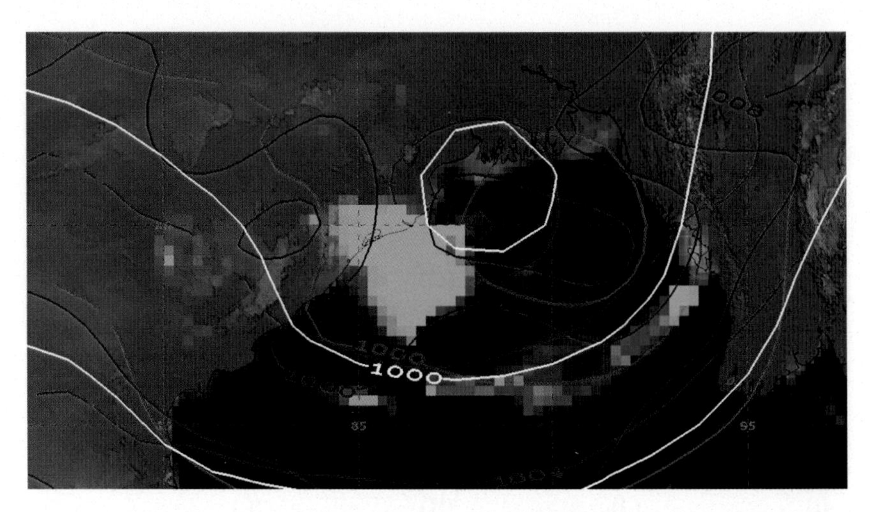

Figure 1.4 Top view on July 1, 2006 at 00:00 UTC of TRMM 3B42 instantaneous rainfall (*green shading*) and mean sea level pressure analysis for MERRA-2 (*red contours*), ERA-I (*pink contours*), JRA-55 (*white contours*), and CFSR (*blue contours*). Contour intervals are 4 hPa. Parallels and meridians are indicated by red dash marks with orange numbering.

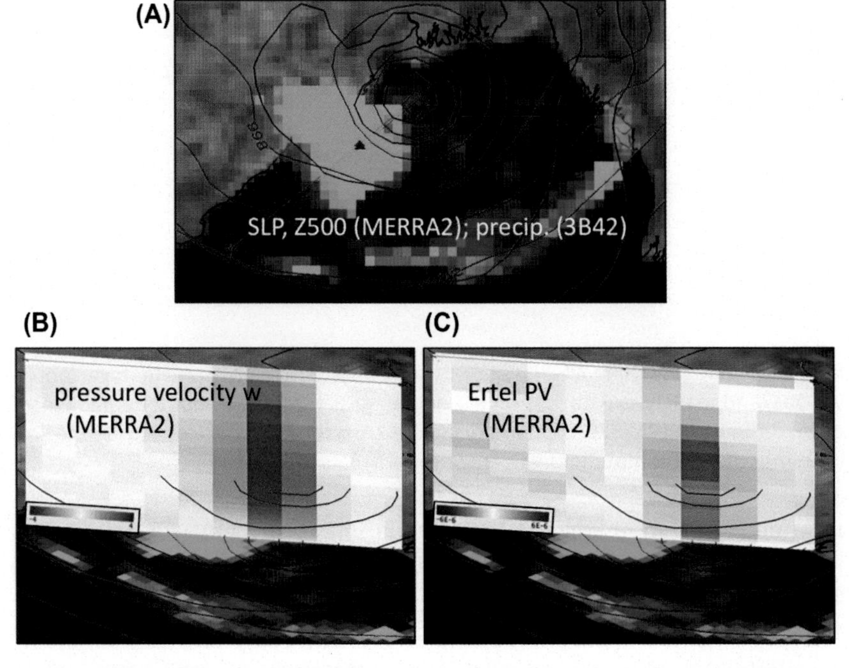

Figure 1.5 Cross sections of MERRA-2's vertical velocity (−4 to 4 Pa s^{-1}) and potential vorticity (−6 to 6 PVU) on July 1, 2006 at 00:00 UTC. The vertical range of the cross section is 0–16 km, and 500 hPa height contours (blue, 10 m interval) are at about 5800 m altitude. The cross-section position is shown in (A), then viewed from the southeast in (B) and (C).

We speculate that interactions of these asymmetries, as well as the mean monsoonal easterly shear, may be inducing the updraft and rainfall in the southwest sector (see Boos et al., 2015, 2017)—at least in the MERRA-2 model's storm, which we have shown is plausibly (albeit imperfectly) related to the 3B42 rain event that motivated this case study. More definitive meteorological conclusions will require closer scrutiny, and perhaps active model experimentation. Short of that, circumstantial evidence and better speculations may still be obtainable from such visual inspection of more cases.

5. FUTURE WORK

There are many more fields and displays one could examine about a case, just from existing global data archives. Broadly, these are (1) thermodynamic (humidity and instability as ingredients of convection), (2) dynamics-related (low-level winds relevant to moisture flux; vorticity and PV balance processes that modulate cool cores and thus lapse-rate instability), and (3) boundary condition factors (topography; sea surface temperature; antecedent soil moisture for runoff modeling of flood hazards). The causes of rare events are always multiple: by definition, these are coincidences of factors with once-in-two-decades likelihood. To explain the combinatorics of that unlikelihood, standardized anomalies of the ingredient fields would be valuable, but require climatological mean and standard deviation data sets to be computed (once).

Active modeling experiments with these ingredients will be necessary to fully understand the "recipe" behind each individual case. We hope to automate the conversion of the various reanalyses into ensemble boundary conditions for explicit convection simulations, whose closed budgets will permit more robust diagnosis of the true rain-making processes, if the event is reasonably well simulated. In addition, experiments with initial and boundary conditions and uncertain processes, centered around such a control run ensemble, could further illuminate the knowability and predictability of such events. There is plenty to do, but a crowd-sourced approach could be powerful, if the technical tools and frameworks of interpretation can be simplified and standardized. We continue to work toward that goal—and we seek allies, dear reader.

6. CONCLUDING REMARKS

Precipitation events are like shadows on the Earth's surface, cast by complex 4D meteorological phenomena. Our atlas offers rich 4D analyses of extreme events via the IDV and also multirainfall-estimate comparisons. We hope these resources may encourage the examination of numerous case studies by fellow scholars and students of the atmosphere, building up a clearer view of the range of processes behind these important events. From a library of cases, grounded in the context of the whole data record, overarching questions of attribution—thermodynamics versus dynamics, initial versus boundary effects—can be addressed. Better understanding of these mechanisms, their predictability, and the limits of both (due to chaos) are our science's ultimate goal, our offering to the social value of protecting lives and vital interests. Collaborations are welcomed.

ACKNOWLEDGMENTS

The authors gratefully acknowledge financial support from NASA NEWS grant NNX15AD11G, NNX14AR75G, and NOAA grant NA13OAR4310156. Jason B. (Brent) Roberts of NASA-MSFC assembled the database of daily rainfall estimates behind Fig. 1.3.

The authors gratefully acknowledge the financial support given by the Earth System Science Organization, Ministry of Earth Sciences, Government of India (Grant no./Project no. MM/SERP/Univ_Miami_USA/2013/INT-1/002) to conduct this research under Monsoon Mission.

REFERENCES

Allen, T.L., Mapes, B.E., 2017. The late spring Caribbean rain-belt: climatology and dynamics. Int. J. Climatol. 37, 4981–4993. https://doi.org/10.1002/joc.5136.

Boos, W.R., Hurley, J.V., Murthy, V.S., 2015. Adiabatic westward drift of Indian monsoon depressions. Q. J. R. Meteorol. Soc. 2015 (141), 1035–1048.

Boos, W.R., Mapes, B.E., Murthy, V.S., 2017. Potential vorticity structure and propagation mechanism of Indian monsoon depressions. In: The Global Monsoon System: Research and Forecast. World Scientific Publishing Company, pp. 187–199.

Chien, F.-C., Kuo, H.-C., 2011. On the extreme rainfall of Typhoon Morakot (2009). J. Geophys. Res. 116, D05104.

Dee, D.P., et al., 2011. The ERA-interim reanalysis: configuration and performance of the data assimilation system. Q. J. R. Meteorol. Soc. 137, 553–597.

Doswell, C.A., Brooks, H.E., Maddox, R.A., 1996. Flash flood forecasting: an ingredients-based methodology. Weather Forecast. 11, 560–581.

Funk, C., et al., 2015. The climate hazards infrared precipitation with stations—a new environmental record for monitoring extremes. Sci. Data 2.

Galarneau, T.J., Bosart, L.F., Schumacher, R.S., 2010. Predecessor rain events ahead of tropical cyclones. Mon. Weather Rev. 138, 3272–3297.

Galarneau, T.J., Hamill, T.M., Dole, R.M., Perlwitz, J., 2012. A multiscale analysis of the extreme weather events over western Russia and northern Pakistan during July 2010. Mon. Weather Rev. 140, 1639–1664.

Gelaro, R., et al., 2017. The modern-era retrospective analysis for research and applications, version 2 (MERRA-2). J. Clim. 30, 5419–5454.

Grimes, A., Mercer, A.E., 2015. Synoptic-scale precursors to tropical cyclone rapid intensification in the atlantic basin. Adv. Meteorol. 2015, 16 Article ID 814043.

Houze, R.A., Rasmussen, K.A., Medina, S., Brodzik, S.R., 2011. Anomalous atmospheric events leading to the summer 2010 floods in Pakistan. Bull. Am. Meteorol. Soc. 92, 291–298.

Huffman, G.J., et al., 2007. The TRMM multisatellite precipitation analysis (TMPA): quasi-global, multiyear, combined-sensor precipitation estimates at fine scales. J. Hydrometeorol. 8, 38–55.

Joyce, R.J., Janowiak, J.E., Arkin, P.A., Xie, P., 2004. CMORPH: a method that produces global precipitation estimates from passive microwave and infrared data at high spatial and temporal resolution. J. Hydrometeorol. 5, 487–503.

Kobayashi, S., et al., 2015. The JRA-55 reanalysis: general specifications and basic characteristics. J. Meteorol. Soc. Jpn. 1, 5–48.

Liu, Z., 2015. Comparison of precipitation estimates between version 7 3-hourly TRMM multi-satellite precipitation analysis (TMPA) near-real-time and research products. Atmos. Res. 153, 119–133.

Maddox, R.A., Chappell, C.F., Hoxit, L.R., 1979. Synoptic and meso-α scale aspects of flash flood events. Bull. Am. Meteorol. Soc. 60, 115–123.

Mapes, B.E., 2011. Heaviest precipitation events, 1998-2007: a near-global survey. In: The Global Monsoon System: Research and Forecast. World Scientific Publishing Company, pp. 15–22.

Molinari, J., Skubis, S., Vollaro, D., Alsheimer, F., Willoughby, H.E., 1998. Potential vorticity analysis of tropical cyclone intensification. J. Atmos. Sci. 55, 2632–2644.

Moore, B.J., Neiman, P.J., Ralph, F.M., Barthold, F.E., 2012. Physical processes associated with heavy flooding rainfall in Nashville, Tennessee, and vicinity during 1–2 May 2010: the role of an atmospheric river and mesoscale convective systems. Mon. Weather Rev. 140, 358–378.

Okamoto, K.I., Ushio, T., Iguchi, T., Takahashi, N., Iwanami, K., 2005. The global satellite mapping of precipitation (GSMaP) project. In: Proc. 25th Int. Symp. On Geoscience and Remote Sensing. IEEE, Seoul, South Korea, pp. 3414–3416.

Over, T.M., Gupta, V.K., 1994. Statistical analysis of mesoscale rainfall: dependence of a random cascade generator on large-scale forcing. J. Appl. Meteorol. 33, 1526–1542.

Raymond, D., Fuchs, Ž., Gjorgjievska, S., Sessions, S., 2015. Balanced dynamics and convection in the tropical troposphere. J. Adv. Model. Earth Syst. 7, 1093–1116. https://doi.org/10.1002/2015MS000467.

Rienecker, M.M., et al., 2011. MERRA: NASA's modern-era retrospective analysis for research and applications. J. Clim. 24, 3624–3648.

Rodell, M., et al., 2004. The global land data assimilation system. Bull. Am. Meteorol. Soc. 85, 381–394.

Romatschke, U., Houze Jr., R.A., 2011. Characteristics of precipitating convective systems in the South Asian monsoon. J. Hydrometeorol. 12, 3–26.

Saha, S., et al., 2010. The NCEP climate forecast system reanalysis. Bull. Am. Meteorol. Soc. 91, 1015–1057.

Schumacher, R.S., Johnson, R.H., 2005. Organization and environmental properties of extreme-rain-producing mesoscale convective systems. Mon. Weather Rev. 133, 961–976.

Schumacher, R.S., 2008. Mesoscale processes contributing to extreme rainfall in a midlatitude warm-season flash flood. Mon. Weather Rev. 136, 3964–3986.

Schumacher, R.S., 2017. Heavy Rainfall and Flash Flooding. Invited Contribution, Oxford Research Encyclopedia of Natural Hazard Science.

Sooroshian, S.P., Nguyen, P., Sellars, S., Braithwaite, D., AghaKouchak, A., Hsu, K., 2014. Satellite-based remote sensing estimation of precipitation for early warning systems. In: Ismail-Zadeh, A., Fucugauchi, J.U., Kijko, A., Takeuchi, K., Zaliapin, I. (Eds.), Extreme Natural Hazards, Disaster Risks and Societal Implications. Cambridge University Press, pp. 99–111.

Yokoyama, C., Takayabu, Y.N., Horinouchi, T., 2017. Precipitation characteristics over East Asia in early summer: effects of the subtropical jet and lower-tropospheric convective instability. J. Clim. 30, 8127–8147. https://doi.org/10.1175/JCLI-D-16-0724.1.

CHAPTER 2

South Asian Monsoon Extremes

B.N. Goswami[1], V. Venugopal[2], R. Chattopadhyay[3]
[1]Cotton University, Guwahati, India; [2]Centre for Atmospheric and Oceanic Sciences & Divecha Centre for Climate Change, Indian Institute of Science, Bangalore, India; [3]Indian Institute of Tropical Meteorology, Pune, India

Contents

1. INTRODUCTION

For one-sixth of world's population residing in the region, the South Asian Monsoon (SAM) represents a lifeline, providing water and food security, and influences the economy of the region significantly. As a result, the monsoon season and monsoon rainfall are deeply etched into the socioeconomic fabric and literature of the region (Fein and Stephens, 1987). The seasonal contrast of rainfall being the hallmark of all monsoons, the SAM is characterized by a very wet summer and a dry winter and therefore, Indian summer monsoon (ISM) is often defined by June–September (JJAS) rainfall over India (Indian summer monsoon rainfall [ISMR]). The ISMR is part of the annual cycle of rainfall over the region, with an abrupt "onset" around the beginning of June and a slow withdrawal in late September or early October. The ISMR is a manifestation of the seasonal northward migration of the tropical rain band in this region known as the Intertropical Convergence Zone (ITCZ; Ramage, 1971) driven primarily by the north–south gradient of the tropospheric temperature (TT; Xavier et al., 2007) facilitated by the land–ocean contrast. Major shifts in the SAM (or ISMR) have shaped the rise and fall of civilizations in this region. The abrupt weakening of the ISM and sustained weak monsoon for about 200 years around

Tropical Extremes: Natural Variability and Trends
ISBN 978-0-12-809248-4
https://doi.org/10.1016/B978-0-12-809248-4.00002-9

4.1 ky (Dixit et al., 2014) may have been responsible for the termination of urban Harappan civilization in the Indus valley (Staubwasser et al., 2003).

On shorter timescales, year-to-year variations of the relatively stable ISMR in the form of floods (strong ISM) and droughts (weak ISM) result in death, destruction, and misery to the people of the region. The great famine in India during 1877–78 was a result of the failure of monsoon in 1876, followed by a severe large-scale drought in 1877. Similar back-to-back droughts during 1965 and 1966 caused serious food grain deficiency in India. The crisis was so bad that the country had to import large amount of food grains from other countries, leading to a balance of payments economic crisis in the country. Against the backdrop of a serious groundwater decrease in most parts of India over the past few decades, India has become more vulnerable to back-to-back monsoon droughts in recent years as was illustrated by the serious water crisis in many parts of India due to failure of monsoon during 2014 and 2015. Thus, these "monsoon *climate* extremes" have highly visible impact on the people of the region and its economy. We shall discuss the drivers of these "monsoon *climate* extremes," their potential for, and possibility of prediction. We shall also speculate on what may be happening to these "monsoon climate" extremes in a warming environment.

The monsoon "climate" can be seen as an aggregate of monsoon "weather" and monsoon "intraseasonal oscillations" (MISOs). The monsoon "weather," namely, the lows, depressions, and mesoscale disturbances, as well as the MISOs, has its own extremes. Although the frequency of occurrence of tropical cyclones (TCs) in the north Indian Ocean (IO) peaks during premonsoon (April–May) and postmonsoon (October–November) months, an occasional TC is known to occur even during the monsoon months, thus making them an integral part of monsoon weather systems. Unlike climate extremes such as large-scale droughts and floods that could affect the whole country, weather extremes are by definition short-lived and localized. However, due to their much larger intensity, weather extremes have substantial local impact, and are often devastating to the local community. The Mumbai flood of 2005 arising from 95 cm rain over the city in 1 day (Kumar et al., 2008), the Uttarakhand flood and landslide of 2013 (Kotal et al., 2014), the Kashmir flood of 2014 (Gogoi and Bhatt, 2014), and the Orissa super-cyclone of 1999 (Mohanty et al., 2004) are a few examples of devastating weather extremes over India in the recent past. In the backdrop of global warming and increased moisture content in the atmosphere, the intensity of "extreme" weather events is expected to increase (Trenberth et al., 2003;

Emanuel, 1987; Webster et al., 2005). Is the observed change as expected in the historical increasing greenhouse gas (GHG) period? At what rate is it expected to increase in the future under different GHG scenarios? Although it may be easy to understand why the intensity of the "extreme" events should increase in a warming environment, do we expect the frequency of occurrence of the "extreme" events also to increase in a warming environment? If so, at what spatial scale do we expect it to increase? We shall address and explore answers to these questions in this chapter. We realize that in a warming climate, the "extreme temperatures" or "extreme heat waves" could also increase. We shall also examine what is happening and what is to be expected to the "temperature extremes" over the Indian monsoon region.

2. MONSOON "CLIMATE" EXTREMES

Two of the most important variations in SAM rainfall are the interannual variability of the seasonal mean rainfall over India (Fig. 2.1A) and the subseasonal variability of daily rainfall (Fig. 2.1B). The departure of the seasonal mean ISMR from long-term mean, normalized by its own standard deviation, is shown in Fig. 2.1A, and the daily rainfall (mm day^{-1}) over central India (CI) during the monsoon season of 1 year is shown in Fig. 2.1B. Flood and drought years are those years that depart significantly from the long-term mean ISMR and have normalized values greater than +1 and less than −1, respectively (Fig. 2.1A). In Fig. 2.1B, positive deviation of daily rainfall (blue) from daily climatology (smooth curve) represents "active" spells, whereas negative deviations (red) represent "break" spells.

The seasonal contrast of rainfall and winds is a hallmark of the SAM and is a result of the seasonal migration of the ITCZ in the region, facilitated by the land–ocean contrast and manifested in the gradient of the TT (Xavier et al., 2007). External factors that could modulate the annual cycle of rainfall in the region (annual cycle of the ITCZ) could, therefore, influence the year-to-year variations of the ISMR. Global-scale low-frequency climate modes such as the El Niño–Southern Oscillation (ENSO) that influence the interannual variation of the ISMR could be considered as "external" drivers of the SAM. Many monsoon floods and droughts are unrelated to external forcing such as ENSO and instead are of "internal" origin. Although the Indian monsoon as represented by ISMR is a reasonably stable system with the interannual standard deviation of the seasonal mean being 10% of its long-term mean, the extremes of Indian monsoon rainfall as defined by deviations of ISMR from the mean (Fig. 2.1A) have a large

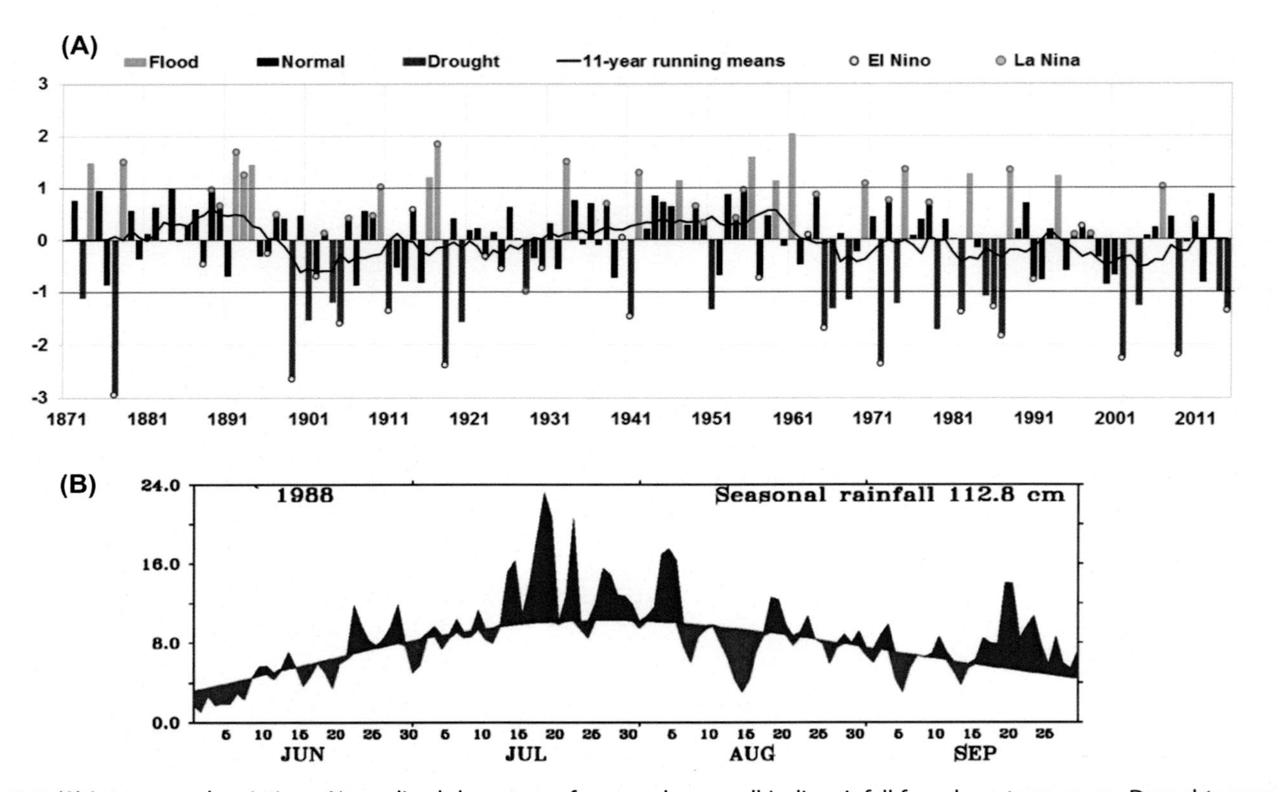

Figure 2.1 (A) Inter-annual variations. Normalised departure of seasonal mean all India rainfall from long term mean. Drought years (normalised departure < -1) are indicated by red while flood years (normalised departure >+1) are indicated by blue. (B) Intra-seasonal variations. Daily rainfall averaged over CI (mm/day) with the smooth curve indicating the long term climatology.

socioeconomic impact on the region. The interannual variability of the Indian monsoon rainfall within this period (1871–2015) appears to be roughly distributed normally with drought extremes (26) slightly outnumbering flood extremes (21). The worst drought recorded till date was in 1877, with ISMR recording nearly -30% departure from long-term mean, whereas the worst flood happened in 1961 with ISMR recording nearly +20% departure from long-term mean. Although these monsoon climate extremes represent strong signals and hence may be predictable, they represent only 30% of the total number of years. For the remaining 70% of the years, the Indian monsoon is "normal" and represents a weak signal with limited predictability. It may be noted that many droughts (e.g., 1872, 1901, 1904, 1919, 1951, 1966, 1968, 1974, 1980, 2004) are not concurrent with an El Niño, and many floods (e.g., 1873, 1894, 1916, 1947, 1956, 1959, 1961, 1983) are not concurrent with a La Niña. Although the worst drought in history (1877) was associated with an El Niño, the worst flood of 1961 was not associated with a La Niña. Thus, monsoon climate extremes could be driven either by "external" factors such as the ENSO or by "internal" dynamics. What actually produces the "internal" interannual variability of the monsoon has been a subject of intense study over the past couple decades as it appears to be a factor for limiting the predictability of the seasonal mean monsoon (Goswami, 1998). The vigorous MISOs have emerged as one of the key factors that produces "internal" interannual variability of the seasonal mean monsoon through linear (residual of seasonal mean of MISO anomalies) and nonlinear effects (Goswami and Xavier, 2005; Goswami et al., 2006). The other factor that could produce "internal" interannual variability is the land surface and soil moisture feedback (Saha et al., 2012). The Indian Ocean Dipole (IOD; Saji et al., 1999) arising from air–sea interaction in the tropical IO also influences the Indian monsoon (Ashok et al., 2001; Saji and Yamagata, 2003), contributing to the "internal" interannual variability of the Indian monsoon. Also, regional air–sea interactions that lead to a quasibiennial tropospheric oscillation in the region (Meehl, 1994; Webster el al., 1998; Chang and Li, 2000; Yu et al., 2003) add to internal interannual variability of SAM. How these processes individually or in combination conspire to produce an "extreme" monsoon climate event (an extreme drought or an extreme flood event) remains an open question. One simple result has emerged from some recent studies (Gadgil et al., 2004) that almost all ISMR extremes (normalized anomaly >1.5 or <-1.5; Fig. 2.1B) seem to occur when both ENSO and IOD are favorable or both are unfavorable.

Is there a trend in the occurrence and/or intensity of the monsoon climate extremes (floods or droughts) within a decade or 15 years in the instrumental record and what do we expect trends of these extremes to be, in different warming scenarios? A count of floods and droughts in a moving window of 15 years from the available records (Fig. 2.1A) would indicate that no significant trend could be seen in these counts while they have a multidecadal variation. We shall discuss this multidecadal variation later in this section. The frequency of occurrence of monsoon climate extremes, therefore, seems to be related to the modulation of the mean monsoon flow and thermodynamic conditions by low-frequency climate variability. As for the future, with the exception of a small number of models, the CMIP5 models project a substantial increase of 8–10% of the seasonal mean by the end of the 21st century, under different GHG scenarios (Chaturvedi et al., 2012). However, it has been contested recently (Sabeerali et al., 2013) that this result may not be reliable because of an intrinsic limitation in the formulation of cloud and rain processes in most models. Also, most models underestimate the observed interannual variability during the historical simulations. As a result, the projections of frequency of occurrence of monsoon climate extremes are also not reliable.

Droughts typically tend to have a stronger impact on society. However, droughts defined based on precipitation alone may not be adequate to properly quantify the intensity of droughts; in particular, temperature variations could increase or decrease demand for moisture and dictate the severity of the drought. Among various drought indices, the Palmer Drought Severity Index (PDSI; Palmer, 1965) is one of the most widely used. PDSI is a climatic water balance index that combines precipitation and evapotranspiration anomalies, with soil water–holding capacity. The PDSI is the most prominent index of meteorological drought used in the United States (Dai, 2011) and has been used to quantify long-term changes in aridity (Dai et al., 1998). Therefore, if we are interested in the trends in drought intensity, it may be more meaningful to look at the trend of drought indices such as the PDSI. Using observed temperature, evaporation minus precipitation, and streamflow data, Dai et al. (2004) estimate PDSI and show that the global aridity over land shows a significant increasing trend from 1920 till 2010, with a stronger increase after 1950s. Using CMIP3 and CMIP5 model projections, this study also suggests severe and widespread droughts in the next 30–90 years over many land areas, resulting from either decreased precipitation and/or increased evaporation. In the context of the Indian monsoon, Niranjan Kumar et al. (2013) examined the variability and trend of droughts using high-resolution temperature and precipitation data during 1901–2010. They argue that PDSI is not suitable for looking at the strength and trends of monsoon droughts because monsoon droughts are multiscale,with some droughts being short-lived while some could be longer than a year. Recently a new drought index, the standardized precipitation evapotranspiration index (SPEI), has been proposed (Begueria et al., 2010;Vicente-Serrano et al., 2010a,b) to quantify drought conditions over any given area. The SPEI considers not only precipitation but also temperature data in its estimation, allowing for a more complete approach. The SPEI can be calculated at different timescales to adapt to the characteristic times of response of the target natural and economic systems to drought. Although the accumulated SPEI over different timescales (6, 12, 18 and 24 months) indicate decadal oscillations (Fig. 1 of Niranjan Kumar et al., 2013), it also shows a decreasing trend since the 1950s. The variations of SPEI seem to be closely related to variations of ISMR, indicating that monsoon rainfall has a strong control on aridity over the region. The monotonic increase in aridity since 1950 is consistent with a decreasing trend of ISMR during the same period and an increasing trend of surface temperature over the region. Although the severity of aridity over the country as whole is increasing, they also find that the percent area affected by moderate drought ($SPEI < -1.0$) at different accumulated periods for the period 1951–2010 shows a statistically significant increasing trend (Fig. 3 of Niranjan Kumar et al., 2013). This result suggests that the country now must prepare to cope with increased severity of droughts, together with their increased spatial extentin the coming years. As stated earlier, there is considerable uncertainty in the projection of ISMR by the climate models under different GHG scenarios (Sabeerali et al., 2013). However, there is no uncertainty that the temperature is going to increase by 2–3°C by the end of the century. Therefore, we could be reasonably certain that the severity of droughts is likely to be much higher in the following decades, compared with the present decade.

The impact of climate extremes on society becomes severe when two of the same type occur back-to-back. Floods and droughts seem to affect food production and GDP of the country differently (Gadgil and Gadgil, 2006), with droughts having a stronger impact. For example, the twin droughts during 1965–66, 1986–87, and 2014–15 had serious impact on the food production and water resources of the country. Extending the argument, an epoch where frequency of occurrence of droughts is higher than that of the floods, would result in severe water stress. It is noted that the ISMR has a multidecadal mode (MDM) of variability (see the 11-year running mean curve in Fig. 2.1A), with approximately three decades of above normal condition followed by three decades of below normal condition; the exception seems to be that the most recent below normal phase seems to have extended for nearly six decades. The frequency of droughts is about twice (half) than that of floods during the negative (positive) phase of this multidecadal oscillation (Table 2.1). The recent extended negative phase of the monsoon MDM (MMDM) is of particular interest, as this negative phase has not only persisted but also the frequency of occurrence of droughts during this phase is more than twice (actually two and half times, see Table 2.1) than that of floods. Such an extended below normal phase with higher than normal frequency of droughts could be termed as a "megadrought." Such "megadroughts" are yet another kind of climate "extremes" with very large potential societal impact. The past "megadroughts" of ISM have been associated with disastrous socioeconomic impacts over the region (Government of Maharashtra, 1973; Sinha et al., 2007).

Reliable prediction of the phases and duration of the MDM, therefore, could be highly valuable for policy makers and water resource managers. What drives these "megadroughts"? Is the recent monsoon "megadrought" part of the monsoon MDM? Paleoclimate reconstruction of SAM rainfall with annual resolution has recently extended the record to more than 500 years back using tree-ring width (Borgaonkar et al., 2010). Furthermore, using $\delta^{18}O$ of cave stalagmites, reconstruction of ISM rainfall has been extended to about 900 years (Sinha et al., 2007) and about 2000 years (Sinha et al., 2011) with approximately yearly resolution. Even though these reconstructions of ISM rainfall may still not be useful for a detailed study of interannual variability, they seem to represent decadal to multidecadal variability as well. A detailed study of several such long reconstructed time series

Table 2.1 Frequency of floods (ISMR > 1) and droughts (ISMR < −1) during different phases of the monsoon multidecadal oscillation

Period (phase of multi-decadal oscillation)	Number floods	No of droughts
1871–1990 (positive)	**5**	3
1891–1930 (negative)	3	**7**
1931–1960 (positive)	**6**	3
1961–2015 (negative)	6	**15**

The bold-faced numbers indicate more frequent category (flood or drought) within each phase of multi-decadal variability.

of ISM rainfall brings out a statistically significant 50–80 year MDM of variability (Goswami et al., 2015). A wavelet spectrum analysis of these time series (Goswami et al., 2015) shows that the mode is nonstationary with one period persisting for a couple of centuries while shifting to another period for another extended span of time. The nonstationarity also contributes to the broadband aperiodic nature of the mode; in other words, extended weak periods of 5–6 decades are not uncommon. Thus, the current decreasing trend of ISM rainfall may still be within its natural variability.

It is thus imperative to understand the following: What drives the multidecadal variability of the ISM rainfall? How is the MMDM related to other known MDMs of variability such as the Atlantic Multidecadal Oscillation (AMO), the Pacific Decadal Oscillation (PDO), and the multidecadal ENSO? Although it has been demonstrated that the AMO represents an "internal" oscillation of the climate system, involving ocean–atmosphere–land interactions and Atlantic meridional overturning circulation (Delworth and Mann, 2000), a clear picture of what drives the MMDM is not currently available. Here, we propose a conceptual mechanistic model of the MMDM as illustrated in Fig. 2.2. There are three interacting elements required to explain the observed features of the 50–80 year MMDM. Prominent multidecadal climate variations elsewhere could drive multidecadal variations of ISM through teleconnections. For example, the AMO is correlated with the multidecadal oscillation of ISM rainfall (Goswami et al., 2006) through a teleconnection via modification of the TT gradient over the Indian monsoon region. It has also been shown that the multidecadal oscillation of ISM

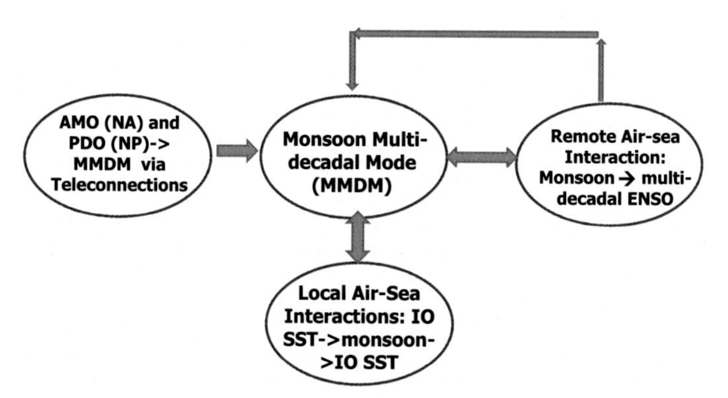

Figure 2.2 A conceptual framework for the MMDM. *AMO*, the Atlantic Multidecadal Oscillation; *ENSO*, El Niño–Southern Oscillation; *IO*, Indian Ocean; *MMDM*, monsoon multidecadal mode; *PDO*, the Pacific Decadal Oscillation; *SST*, sea surface temperature.

is related to a MDM of variability of ENSO through a teleconnection that is similar to the one that operates for ENSO–monsoon relationship on the interannual timescale (Krishnamurthy and Goswami, 2000). Thus, multidecadal climate variability with epicenters elsewhere (AMO, PDO, multidecadal ENSO) could impart a multidecadal component in ISM variability. Also, local ocean–atmosphere–land interaction, involving the tropical IO sea surface temperature (SST) and monsoon winds, could also give rise to a potentially multidecadal oscillation of the ISM. It has been shown recently that the decreasing trend of ISM rainfall over the past 5–6 decades is indeed driven by the increasing trend of equatorial IO SST (Roxy et al., 2015) by making the north–south gradient ofTT to decrease as IO SST increases. The equatorial SST does this by making the TT over the southern oceanic box increase at a much faster rate than that over the northern largely land box. Thus, the equatorial warm SST has the tendency to weaken the ISM by weakening the TT gradient. But what is driving the increasing trend of equatorial IO SST? It has also been shown that the weakening surface winds associated with weakening strength of the ISM changes the heat flux and ocean circulation in such a way so as to increase the SST further (Swapna et al., 2015). Therefore, it is clear that the large-scale ocean–atmosphere interaction is responsible largely, if not fully, for the current decreasing trend of the ISM rainfall establishing that such regional air–sea interactions contribute to the multidecadal variability of the ISM. For this air–sea interaction to lead to a multidecadal oscillation, however, would require that it give rise to a negative feedback (something similar to a delayed oscillator). We propose that it happens through a remote air–sea interaction, involving the ISM and the tropical Pacific. Weakening ISM and the associated increasing maritime precipitation drives a multidecadal La Niña in the tropical Pacific, which in turn strengthens the ISM through its teleconnection. The observed multidecadal variability of the ISM is a result of contributions from the local air–sea interactions, together with contributions from teleconnections with multidecadal ENSO and AMO. The intrinsic timescale associated with the multidecadal variability arising from air–sea interaction, involving IO SST and monsoon, is not yet well established. Analysis of recent decreasing trend of ISM and increasing trend of IO SST indicates that this timescale could be longer than 60 years. The PDO and the multidecadal ENSO are associated with a timescale of 50–70 years. The AMO also has timescale of about 50–70 years. It is likely that the broad band 50- to 80-year periodicity of the MMDM is a result of interactions between the timescales involved with the three different drivers.

In terms of societal impact and impact on food production and water resources, the "megadroughts" are the real monsoon climate extremes. Forewarning of such megadroughts would be of tremendous benefit to the policy-makers. With an amplitude of 2% of the long-term mean as compared with 10% of the mean being the amplitude of the interannual variability, the MMDM represents a reasonably large potentially predictable signal. The problem of predicting the phases of the MMDM is related to the general problem of "decadal prediction" of climate, which itself is still in a nascent stage with most climate models struggling to simulate the big decadal/multidecadal signals such as the AMO and PDO. Current climate models have serious systematic errors in simulating the seasonal mean climate (Goswami and Goswami, 2016; Sperber et al., 2013); moreover, CMIP5 models have difficulty in simulating the recent decreasing trend of the ISM (Saha et al., 2014) making them mostly incapable of simulating the MMDM with enough fidelity. As a result, forewarning monsoon "megadroughts" may not be feasible with current climate models. However, with the constant improvement of the climate models, forewarning of "monsoon megadroughts" may be possible within the next decade.

3. MONSOON "WEATHER" EXTREMES

Weather systems bringing rain during the monsoon season consist of lows, depressions, and an occasional TC. Convective cloud clusters, largely within these synoptic disturbances, bring heavy rain. The synoptic disturbances are modulated by the monsoon intraseasonal oscillations (MISOs, see Goswami, 2012). Lows and depressions, collectively termed as low-pressure systems (LPS), make the largest contribution to the seasonal mean rainfall. These LPS have typical time and length scales of 3–5 days and 1000–2000 km, respectively. The maximum wind speed associated with lows is less than $8.5\,\mathrm{m\,s^{-1}}$, whereas that with the depressions is between 8.5 and $16.5\,\mathrm{m\,s^{-1}}$. Lows and depressions are shear instabilities energized by moist convection (Shukla, 1978; Goswami et al., 1980; Mak, 1975). The meridional shear of low-level winds associated with the ITCZ (of which the "monsoon trough" is a quasistationary structure) provides the genesis ambience for the lows. Although lows could occur almost anywhere in the north IO (Fig. 2.3, also see Fig. 1 of Krishnamurthy and Ajaya Mohan, 2010), the north Bay of Bengal is the preferred region for frequent genesis of LPS due to the warm ocean with a shallow mixed layer overlaid by the "monsoon trough." Intensification of lows through moist convective feedback leads to

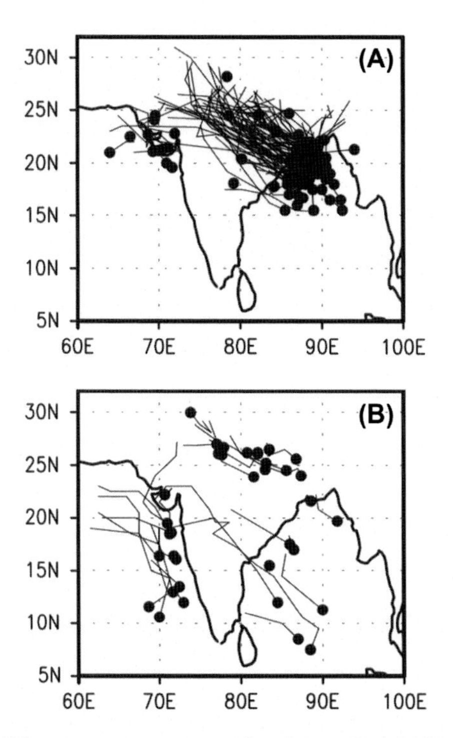

Figure 2.3 Tracks of low-pressure systems for the period 1954–83 during extreme phases of monsoon intraseasonal oscillation (ISO). (A) "Active" ISO phase (MISI > +1) and (B) "break" ISO phase (MISI < 1). *Dark dots* represent the genesis point, and their *lines* show the tracks. *(Reproduced from Goswami, B.N., Ajaya Mohan, R.S., Xavier, P.K., Sengupta, D., 2003. Clustering of low pressure systems during the Indian summer monsoon by intraseasonal oscillations. Geophys. Res. Lett. 30 (8), 1431. https://doi.org/10.1029/2002GL016734.)*

depressions, and further intensification of depressions leads to TCs. After formation, the lows and depressions move west–northwest direction due to beta-effect, nonlinear vorticity advection, and interaction with convective heating (Goswami, 1987). The rarity of TCs during the summer monsoon season (SMS) is due to two factors. Firstly, most depressions form over the north Bay of Bengal and move over to the continent within 2–3 days, thus not providing enough time over the ocean for intensification. With reduced moisture supply over land, the depressions have little chance of intensification to a TC. Furthermore, because of the strong easterly jet in the upper levels, the vertical shear of winds during the monsoon season is large and the resulting ventilation makes it difficult for the depressions to intensify into a TC. Thus, the monsoon depressions are the monsoon weather extremes.

During the SMS, until recently 6–7 monsoon depressions used to occur out of a total of about 13 LPS that occur in the IO, north of 10°N (Goswami et al., 2003; Krishnamurthy and Ajaya Mohan, 2010). However, the number of monsoon depressions influencing the Indian monsoon system has reduced significantly, with the years 2009, 2010, and 2011 seeing only around 1, 0, and 1 per season, respectively. Examining the frequency of occurrence of lows and depressions and storms since 1989, Dash et al. (2004) find that the total number of LPS remains roughly constant. Both lows and depressions and storms show significant multidecadal variability with an unmistakable decreasing trend of depressions and storms since 1975, with compensating increasing trend of lows (Fig. 1 in Dash et al., 2004). As the monsoon depressions contribute significantly to the seasonal mean rainfall, there is a suggestion that the recent decreasing trend of seasonal mean rainfall over CI could be due to this decreasing trend of occurrence of monsoon depressions (Pattanaik, 2007). With the total frequency of LPS remaining nearly constant, some studies (Prajeesh et al., 2013; Krishnamurti et al., 2013) tried to identify the dynamic and thermodynamic factors that prevented the lows from intensifying into depressions and storms. On the one hand, Prajeesh et al. (2013) suggest that the observed weakening of the frequency of monsoon depressions, south of 20°N in the Bay of Bengal since 1950s, is likely due to a declining trend in the mid-tropospheric relative humidity over the Indian region. On the other hand, Krishnamurti et al. (2013) indicate that the increased level of pollution in recent years leads to the enhancement of cloud condensation nuclei, resulting in disruption of the organization of convection and preventing formation of monsoon depressions. Recently, Cohen and Boos (2014) used two different vortex-tracking algorithms, along with criteria similar to the ones used by the India Meteorological Department (IMD, 2011), to identify monsoon depressions, failed to find a decreasing trend of monsoon depressions in the two reanalysis data sets. This raises an important question on the robustness of the reported trends in monsoon depressions in recent years. Although errors could exist in the IMD data set, alternate data sets also have issues that may bias their long-term trends, making it difficult to definitively answer the question of decreasing trend of monsoon depressions in the recent years.

3.1 Monsoon "Rainfall" Extremes

Lows and depressions are very important in producing widespread rain on synoptic scales (~1000 km) across the country, with maximum precipitation occurring southwest of the vortex center (Godbole, 1977; Sikka, 1977;

Ding and Sikka, 2006). The composite maximum precipitation in a monsoon depression is about 100 mm day^{-1}, decreasing to less than half the maximum within 100 km away from the location of maximum precipitation (Godbole, 1977). The composite daily rainfall contribution of 34 monsoon depressions from the Tropical Rainfall Measuring Mission (TRMM) precipitation data along the track of the depressions and west of Western Ghats is about 20 mm day^{-1}, whereas maximum daily contribution is about 40 mm day^{-1} (Hunt et al., 2016). However, many parts of India experience heavy to very heavy rainfall (>100 mm day^{-1}) on some days during the monsoon season. The Mumbai rainfall event in July 2005 where 940 mm rain fell in 1 day is an extreme example of such daily "extreme" rainfall events. Such rainfall "extremes" are highly localized and mostly restricted to less than 500 sq.km. However, the resulting flash floods, landslides, and other related disasters can inflict heavy loss of life and property. The intensity of the disaster depends on the environmental conditions such as the topography, soil type, vegetation cover, and prevailing land-use practices. General degradation of the environment due to human activity has made most places vulnerable to such extreme precipitation-related disasters. One such example is the Malin landslide that occurred on July 30, 2014, killing 151 people. (Malin is a village near the city of Pune, and the landslide was caused by 10.8 cm rain on July 29, 2014.) Environmental degradation due to deforestation and uncontrolled "development" appears to have contributed to the intensity of the disaster from a relatively modest extreme rainfall event (ERE).

In terms of hydrological impact of daily rainfall, anything greater than 10 cm day^{-1} could be considered "extreme." Hydrological impact also depends on the local "mean" rainfall. For example, a 10 cm day^{-1} rain event in Rajasthan where the mean annual rainfall is ~40 cm will have a much bigger impact compared with a place near Western Ghats or in the northeast India, where the annual mean rainfall is of the order of 200 cm. Mathematically speaking, however, an "extreme" must be defined in terms of the probability distribution function of daily rainfall at any place. For example, events contributing to exceedance of 99th percentile could be termed as "extreme." These daily heavy ("extreme") rainfall events occur when a tropical mesoscale convective complex (MCC, Houze, 2004) gets organized usually within the environment of a synoptic disturbance such as lows and depressions. Francis and Gadgil (2006), examining rainfall extremes over the western coast of India, find that about 60% of these events are associated with synoptic disturbances. MCCs could, however, form even in the absence of a synoptic disturbance. Tropical MCCs have a horizontal scale of ~100 km, and hence precipitation averaged over 100 km × 100 km (or 1° × 1°) is a good horizontal unit for examining tropical daily rainfall "extremes." With increasing temperature in the Indian region (Kothawale and Rupa Kumar, 2005; Kothawale et al., 2010), increased moisture content in the atmosphere is potentially likely to drive stronger convective events and stronger rainfall "extremes" (Trenberth, 1999, 2011; Trenberth et al., 2003). To this end, we examine the following questions: Is the intensity of rainfall extremes increasing over India? What about their frequency of occurrence?

A large fraction of the EREs may arise from processes such as severe thunderstorms, which at any given time may be almost randomly distributed in space; this could be due to inherent thermodynamic nonlinearity and small-scale changes in the initial and boundary conditions (e.g., orography). Even though the total number of extreme events over a homogeneous large-scale environment may have an increasing trend, a particular station may miss it because of sampling or local inhomogeneity. Therefore, it is meaningful to examine trends of such rainfall extremes aggregated over a reasonably large area (approximately the size of a synoptic disturbance, ~10° × 10° box), where the rainfall variability is also roughly homogeneous. The climatological variance of daily rainfall during the SMS (June 1–September 30) has large spatial variability in the southern peninsula, northeast, and the western part of the country. However, it is rather homogeneous over central India (CI; 74.5°E–86.5°E, 16.5°N–26.5°N). Therefore, CI is the region where we choose to examine the trend of rainfall extremes aggregated over the region. For this purpose, we use the daily gridded rainfall data at 1° × 1° resolution from IMD, based on 1803 stations (Rajeevan et al., 2006) for the period 1951–2003. The maximum 1-day rainfall during the summer monsoon of 1951–2003 in any box over CI is ~58 cm. The climatological seasonal mean over CI during this period is 7.35 mm day^{-1}, whereas the standard deviation of the daily anomalies is 14.85 mm day^{-1}. As the climatological mean and the daily variance during the monsoon season are homogeneous (at least weak second order) over the CI, it is reasonable to consider a fixed threshold for defining "extreme" rainfall over this region. To see whether the probability density function (PDF) of rainfall over CI has changed in recent years, the PDF during the summer monsoon season of 1951–70 and 1981–2000 was estimated separately. The tails of the PDF (or, equivalently, the histogram as shown in Fig. 2.4) suggest a larger number of extreme events (>100 mm rain) in recent years (1981–2000) compared to

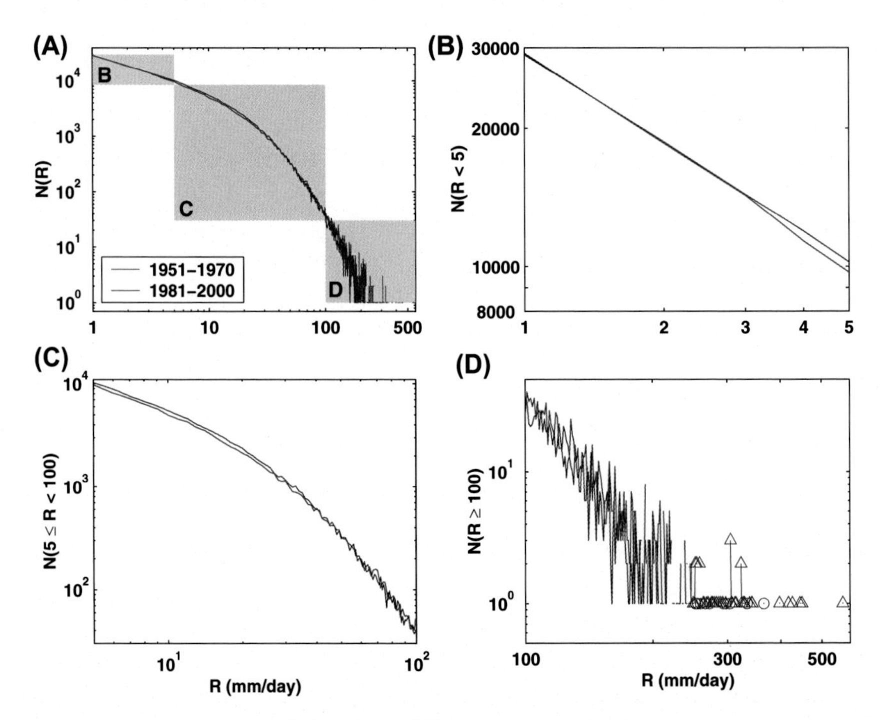

Figure 2.4 (A) Histogram of of daily rainfall over central India during summer monsoon for two periods, 1950–70 and 1980–2000. The grey-shaded regions in the top left panel correspond to different ranges of rain rate, and are magnified in the top right (B) and bottom panels (C,D). *(Reproduced from supplementary material of Goswami, B.N., Venugopal, V., Sengupta, D., Madhusoodanan, M. S., Xavier, P.K., 2006. Increasing trend of extreme rain events over India in a warming environment. Science 314, 1442–1445.)*

the previous era (1951–70). On the other hand, the number of weak and moderate events (>5 mm but <100 mm) has decreased during the recent years compared with the previous period. In fact, the frequency of heavy events (rain >100 mm; blue in Fig. 2.5) and very heavy events (rain >150 mm; Fig. 2.5 lowest panel) over CI shows clear and significant increasing trends, whereas that of weak and moderate events show a clear and significant decreasing trend (green in Fig. 2.5). There is a 10% increase per decade in the level of heavy rainfall activity since the early 1950s, whereas the number of very heavy events has more than doubled (Fig. 2.5); this increase translates to an increase in the disaster potential. Trends of both heavy and very heavy events are significant at 95% significance level. On the other hand, the decreasing trend of weak and moderate events is significant at 90% significance level. These findings are in tune with model projections (Semenov and Bengtsson, 2002) and some observations (Brunetti et al., 2001) that

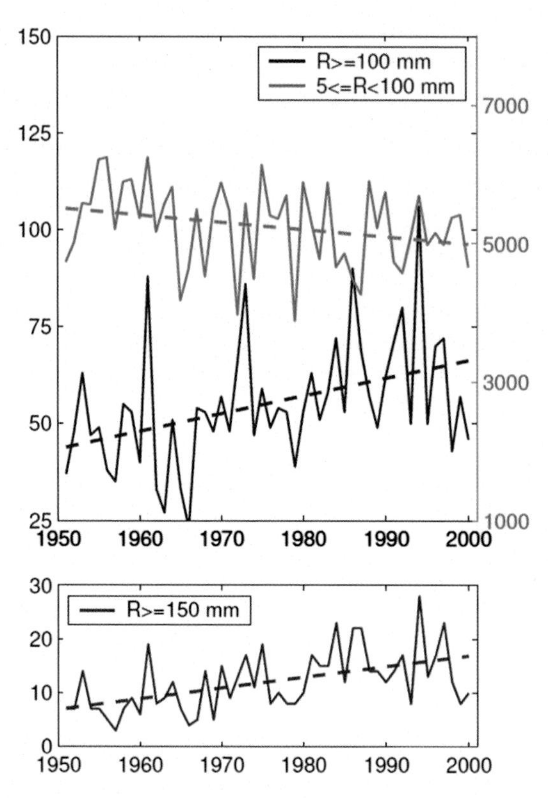

Figure 2.5 Temporal variation of the number of heavy (blue), very heavy (red), and moderate (green) daily rain events between 1950 and 2000 during the summer monsoon season over central India. *(Reproduced from (Figure content same, but in colour) Goswami, B.N., Venugopal, V., Sengupta, D., Madhusoodanan, M. S., Xavier, P.K., 2006. Increasing trend of extreme rain events over India in a warming environment. Science 314, 1442–1445.)*

indicate an increase in heavy rain events and decrease in weak events under global warming scenarios. To see whether the intensity of the "extreme" rainfall events is also increased, the average intensity of the heaviest four events over CI, during each monsoon season, is shown in Fig. 2.6. There is approximately a 10% per decade increase in the intensity of the heaviest rain events during the 50-year period, significant at 99% significance level.

With the availability of a longer daily gridded rainfall data, Rajeevan et al. (2008) examined the trend of daily "extreme" rainfall events over CI during the period 1901–2004 and found that the frequency of EREs over CI shows significant interannual and interdecadal variations in addition to a statistically significant long-term trend of 6% per decade with larger trend

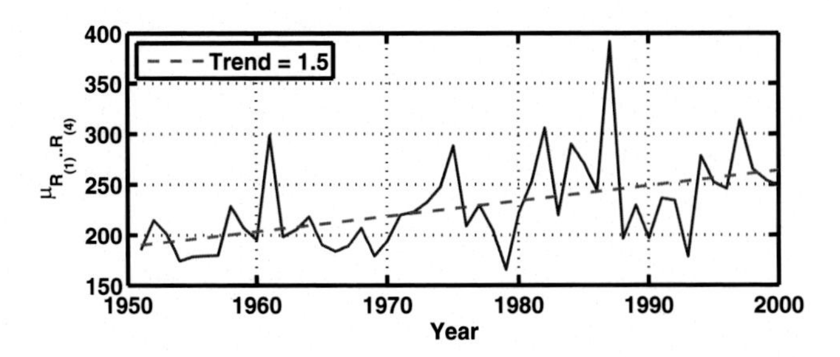

Figure 2.6 Temporal variation of the mean of the highest four daily values over central India per season. *(Reproduced from Goswami, B.N., Venugopal, V., Sengupta, D., Madhusoodanan, M. S., Xavier, P.K., 2006. Increasing trend of extreme rain events over India in a warming environment. Science 314, 1442–1445.)*

during the period 1951–2004 as described above and first reported in Goswami et al. (2006). There is a tantalizing association between the increasing trend and multidecadal variability of SST over the tropical IO and that of the frequency of occurrence of "extreme" rainfall events (Goswami et al., 2006; Rajeevan et al., 2008); this suggests a possible strong role of the IO SST in the variability and trend of "extreme" rainfall events over the central Indian region.

EREs are a result of convective instability of the atmosphere. An increasing trend of the convective available potential energy (CAPE) and a decreasing trend of convective inhibition energy (CINE) over the CI region during JJAS, based on calculation of daily CAPE and CINE from ERA40 (Neena et al., 2009), indicate that the convective instability is increasing and thus is consistent with the increasing trend of "extreme" rainfall events in the region.

On interannual timescale, the northeast regions of India (NEI) tends to go out of phase with CI (Shukla, 1987), indicating opposing dynamic and thermodynamic conditions in the two regions. As CAPE depends on the dynamic and thermodynamic conditions of the region, and as "extreme" rain events depend on CAPE, it is possible that the trend of "extreme" rainfall events in NEI may be different from that found in CI. NEI is different from CI in that the mean rainfall (and variability) is almost twice as high as in CI; moreover, it is geographically inhomogeneous with the Brahmaputra valley surrounded by moderately high orography. Collecting rainfall observations from a set of representative stations, Goswami et al. (2010) find that there is a decrease in frequency of occurrence of "extreme" rainfall events in the region between 1975 and 2006.

Using satellite and reanalysis data, they were also able to decipher how a multiscale interaction of circulation with orography is responsible for the "extreme" rain events in this region. Their finding, which is in contrast to the trends reported in CI, is consistent with the finding that the mean CAPE and CINE in the region are decreasing and increasing, respectively (Goswami et al., 2010). Thus, the large-scale dynamic and thermodynamic conditions, which could be modulated by multidecadal modes of variability, are critical to the frequency of occurrence and intensity of "extreme" rainfall events in a region.

On a related note, the idea of aggregating "extreme" events on $1° \times 1°$ boxes on a larger region like CI has been criticized by some (Krishnamurthy et al., 2009; Ghosh et al., 2011), arguing that "extreme" rainfall events are caused by "storms" of much larger size than $1° \times 1°$ boxes and if we aggregate one storm may be counted more than once thereby biasing the count and the trend. As mentioned earlier, the "extreme" rainfall events do not occur everywhere on the area covered by a synoptic disturbance and are associated with MCCs of size ~500 km^2. Therefore, the $1° \times 1°$ boxes are a good unit to average rainfall and if there are two counts of "extreme" events in a storm environment, there are actually two "extreme" events and not one. The claim of double counting, therefore, is misplaced. Hence, the idea of aggregating the "extreme" rain events is physically based and the only meaningful way to look at trends of EREs.

3.2 Clustering of "Extreme" Rainfall Events in Time

With the robust increasing trend of EREs over CI found in these studies (Goswami et al., 2006; Rajeevan et al., 2008), it is imperative that a strategy for adaptation and impact mitigation for related disasters be put in place. Such a policy or strategy becomes even more urgent as most climate models agree (even if they differ on the trends in seasonal mean; Chaturvedi et al., 2012; Kharin et al., 2013) that the EREs over the SAM region are expected to show an increase in the next several decades. However, for effective disaster prevention measures to be implemented, CI is too big an area and the monsoon season (JJAS) is too big a time window. It will be highly desirable if either or both of these windows could be narrowed. The monsoon intraseasonal oscillations (MISOs) with a dominant period of ~40 days, and relatively narrow meridional scale (Goswami, 2012), provide a potential platform for spatiotemporal clustering of the EREs. It has been demonstrated that synoptic events such as the LPS during the monsoon season are clustered by the MISOs (Goswami et al., 2003; Krishnamurthy and Ajaya

Mohan, 2010). Increased horizontal shear of low-level winds along the "monsoon trough" and higher level of moisture during active spells leads to occurrence of four times larger number of LPS during active spells, compared to that during break spells. As a significant fraction of EREs occur in association with synoptic events, it is logical to assume that EREs may also be clustered by the active phase of the monsoon. With this background, here, we demonstrate that the EREs are indeed clustered during active conditions and the previously documented increasing trend of EREs (Goswami et al., 2006) essentially occurs due to EREs during the active spells, providing a scientific basis for better disaster preparedness. With the advance of extended range forecasting using coupled ocean–atmosphere models, the active and break spells can now be predicted with a useful skill of 15–20 days in advance (Abhilash et al., 2014). Clustering of EREs during active spells could, therefore, provide a relatively narrow window of about 15 days for disaster managers to concentrate efforts, for which a forewarning could be given more than 15 days in advance.

The total number of EREs, as defined by daily rainfall (in a grid box) greater than or equal to 120 mm day^{-1}, during active and break spells of the SMS (JJAS) from 1901 till 2013 clearly shows that EREs are indeed strongly clustered by active spells (Fig. 2.7A). It is found that nearly six times larger number of events (long-term average ~ 30) occur during active phases compared to those which occur during break phases (approximately 5, Fig. 2.7B). This appears to be consistent with the observations that the LPS are similarly clustered by the MISOs (Goswami et al., 2003). Although there is considerable year-to-year variation in the number of EREs during active phases, a statistically significant (*P*-value of .05) increasing trend of the EREs during active phases is apparent with an increase of about two events per decade (Fig. 2.7A). However, the small increase in the number of EREs during break phases (Fig. 2.7B) and during the 112-year period is not statistically significant. Thus, the reported increasing trend of EREs (Goswami et al., 2006; Rajeevan et al., 2008) comes primarily from the EREs during the active spells of the Indian monsoon. The year-to-year variability of EREs seems to have a low-frequency interannual variability (Fig. 2.7), whereas the amplitude of interannual variability shows a multidecadal variability of EREs similar to that found by Rajeevan et al. (2008). As indicated in that study, the multidecadal variation of the EREs may be related to that of SST over the equatorial IO. The clustering of the EREs by active spells and the increasing trend of EREs occurring only during active spells are, however, robust results as they are not dependent on the threshold used for

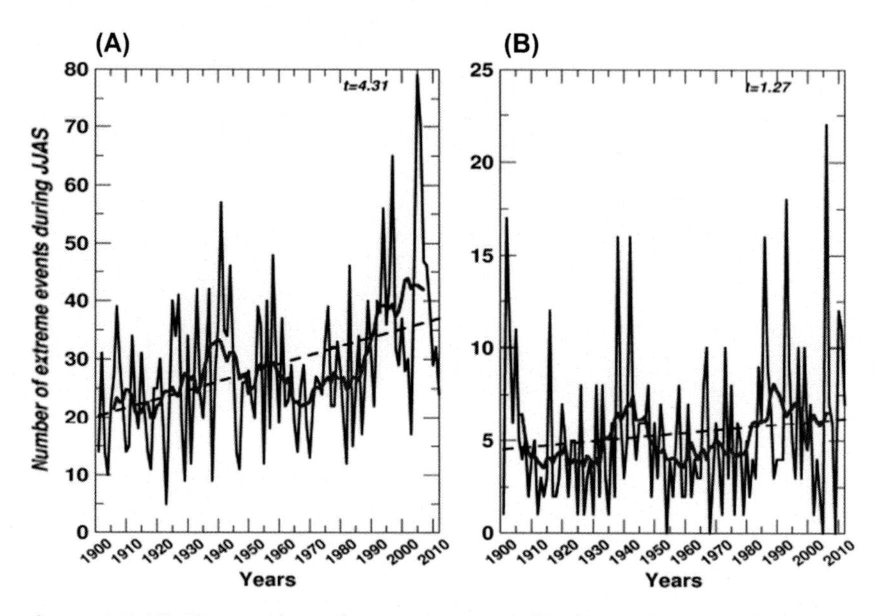

Figure 2.7 (A) The number of extreme events (*black curve*) over Indian region (10°N–30°N; 70°E–90°E) during the June–September (JJAS) monsoon season for the active phases. The threshold for extreme event is taken as 120 mm day^{-1}. Also shown are the linear trends (*dashed line*) and the 11-year running mean (*blue curve*). The t-value is quoted at the top right corner. (B), same as (A) but for break phases.

defining EREs. A statistically significant increasing trend is clearly evident even if the threshold for defining EREs increased to 150 mm day^{-1} or reduced to 100 mm day^{-1} (Fig. 2.8). It is clear from the three figures that the increasing trend of EREs is stronger for very heavy events compared with that for heavy events (Fig. 2.8). The clustering also remains strong with the average number of EREs during active spells being more than five times compared with the number during break spells.

During an active spell, the favorable condition for EREs is enhanced manifold compared with that during break spells, primarily due to the availability of substantially more moisture content in those spells. Therefore, if the total number of active spell days during the season increases, we may also expect a larger number of EREs during the season. The total number of "active days" and "break days" within each season was counted following the same definition of active and break conditions, as described in *Methods*, and is shown in Fig. 2.9. Although there is considerable year-to-year variation for both "active days" and "break days," the average number of either spell is roughly same. Interestingly, however, "active days" show a significant

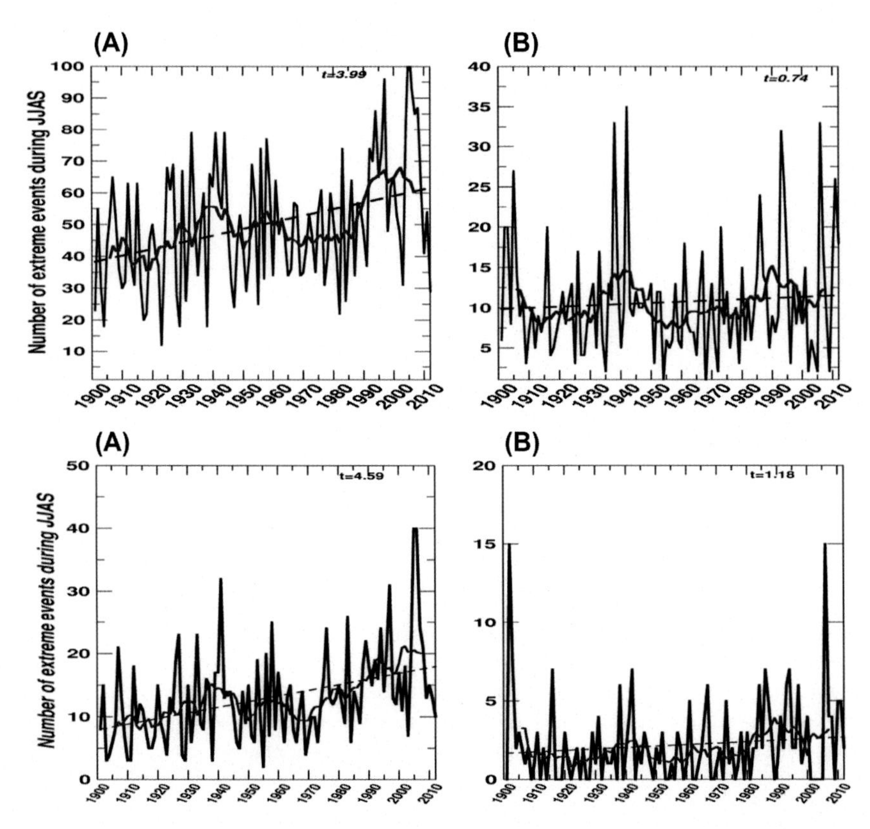

Figure 2.8 Same as Fig. 2.7 but for extreme rain events defined with threshold 100 mm day^{-1} (top) and 150 mm day^{-1} (bottom). (A) Active phase. (B) Break phase.

increasing trend, whereas the "break days" show no significant trend. This means that the overall conducive environment for EREs to occur during the active spells of the monsoon season is slowly expanding over the years, and thus also contributing to the observed increasing trend of EREs. We consider this as an important new finding providing an insight why the EREs are increasing in the monsoon trough region of CI. Our finding also provides a basis for the finding of Ajayamohan et al. (2010) that the LPS have an increasing trend over the period. Global warming may have the potential to increase the moisture content everywhere thereby increasing the potential for an ERE to occur. However, this potential could be realized only if the large-scale environment is conducive. The active spells of the northward propagating MISOs provide this fertile environment over the monsoon trough region of CI, and the expanding wing of the active spells helps realize the potential of larger number of LPS and EREs.

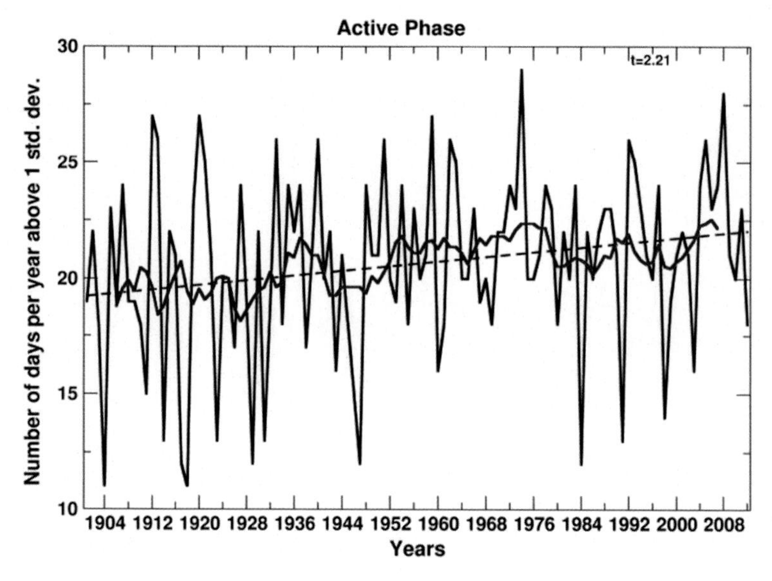

Figure 2.9 Number of days above one standard deviation for the 20- to 90-day filtered time series area averaged over Indian region (10°N–30°N; 70°E–90°E) during the June–September season. The *dashed line* shows linear regressed trend, and the blue curve shows 11-year running average.

A much higher level of convectively unstable environment during the active spells compared to that during break spells also indicates that the intensity of the EREs is likely to be significantly higher during active spells compared with that during break spells. Therefore, the intensity of EREs is also likely to be clustered by the active spells. The average intensity of the top four most intense events over the region during the season (Fig. 2.10) shows that the intensity of the active spell EREs is more than two times larger than that during break spells in recent decades. Also the EREs during active spells show a strong increasing trend (*P*-value of .01), whereas that during break spells have a very weak increasing trend. It is also interesting to note that the year-to-year variation of maximum intensity of EREs during the whole season correlates strongly with that during the active spells. This suggests that the highest intensity of EREs during the season is essentially governed by that during the active spells. The significant increasing trend of EREs during the wet active spells contributes to the observed increasing trend of the intensity of the wet spells themselves in recent years (Singh et al., 2014). With more than two times weaker and far fewer EREs during the break spells, the disaster potential during break phases is minimal

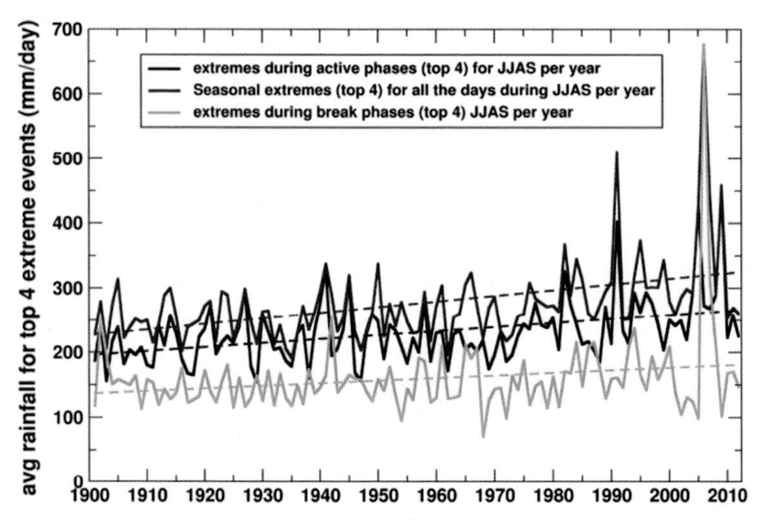

Figure 2.10 The average rainfall (intensity) for top four extreme events during June–September (JJAS) active phase (*black curve*), during JJAS break phase (*green curve*), and taking all the days within a JJAS season (*red curve*). The *dashed curves* show the corresponding trends for each case.

and hence it would be prudent to concentrate disaster prevention measures primarily during the active spells.

Next, we examine the region sensitiveness of our findings on ERE. As MISOs have a large spatial scale, the active and break spells have been identified with the help of a daily rainfall time series averaged over a reasonably large area (10°N–30°N, 70°E–100°E); also, EREs are aggregated over the same region. To examine whether the choice of this area, which is slightly larger than CI as defined in Goswami et al. (2006) (CI, 74.5°E–86.5°E and 16.5°N–26.5°N), had any impact on our conclusions on the clustering and the trend of the EREs, the calculations were repeated exactly over their CI region. We find that, both the clustering and the increasing trend of EREs during the active spells are independent of a strict choice of the region (Figure not shown).

As the active and break spells are a manifestation of the northward propagating MISOs, and as these MISOs have a distinct spatial structure with a longer zonal scale and a relatively short meridional scale (Goswami, 2012), the EREs are expected to be clustered in the spatially active regions. This spatial clustering is seen in Fig. 2.11, where we show the frequency of occurrence of EREs at each 1° × 1° boxes during active (Fig. 2.11A) and break (Fig. 2.11B) spells covering the entire period (1901–2012) when the spells were located

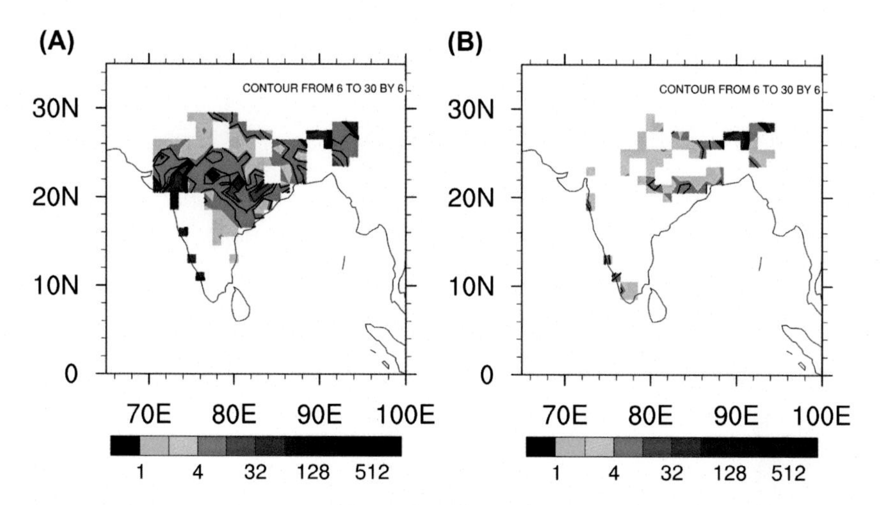

Figure 2.11 The spatial pattern of the number of extreme event during (A) active and (B) break phases of the monsoon seasons of 1901–2012.

over CI (15°N–25°N, 70°E–95°E). Although some EREs do occur over the foothills of the Himalayas during both active and break spells, the EREs are primarily clustered over the core monsoon zone of CI during active spells. This figure also shows that, within the core monsoon region, there are a few pockets of higher probability of occurrence of EREs providing a basis for prioritizing these areas for disaster mitigation activity in the first place.

To get further insight into the spatial clustering of the EREs during active and break spells, we calculate the frequency of occurrence of 1, 2, 3, up to 20 EREs in an active (or a break) day over the domain (Fig. 2.12A and B). Although 43% of active days may be without an ERE anywhere over continental India, nearly 80% of break days do not experience an ERE. Thus, there is a three times higher probability of an ERE occurring during active spells compared with that during break spells. The probability of occurrence of higher number of EREs in a day decreases exponentially, decreasing much faster during break than during active spells. When more than one ERE occurs in a day, how close or separated are they in space? To examine this question, the frequency distribution of occurrence of two EREs in a day with Euclidian distance between them ranging from 0 unit, 1 unit, 2 units, etc., is calculated during active and break spells over the entire period (Fig. 2.12C and D). Because the data are on a 1° × 1° grid, one unit is 1°. A distance of 0 unit means that there was only one ERE over the entire domain. Similarly, a distance of 1 unit means that two EREs are contiguous,

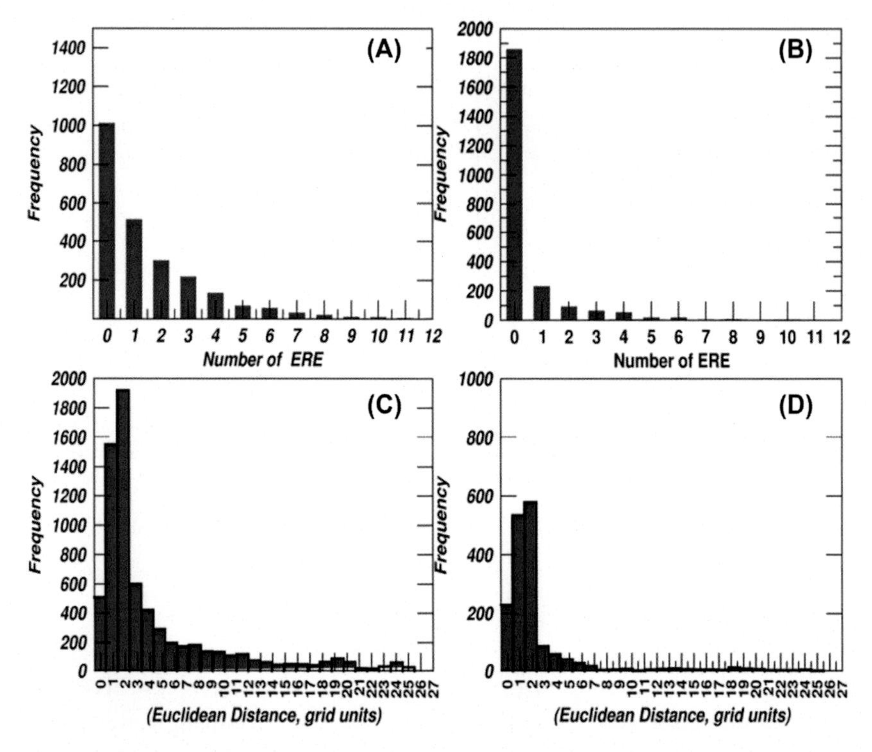

Figure 2.12 Histogram of the frequency of extreme rainfall events (EREs) during (A) active, (B) break phases. Bottom panels show the histogram of the spatial separation (Euclidean distance) between two ERE grid points (in units of grid scale) for (C) active and for (D) break phases. The frequency count of zero (0) Euclidean distance in (C) or (D) implies a single grid extreme event in the domain. The calculation is made for the monsoon seasons (June–September) of 1901–2012. The spatial domain is 5–30°N; 70–90°E.

i.e., occurring side by side. The distribution is highly skewed (Fig. 2.12C and D) for both active and break spells with the mode at EREs with separation of one unit. The contiguous EREs are likely to be part of mesoscale convective complex (MCC) or synoptic events such as LPS. However, they account for 25% of total EREs during the active spells and 20% of total EREs during break spells. Thus, a significant fraction of the total EREs (more than 75%) is not associated with either MCC or LPS and occurs anywhere in the generally favorable large-scale environment of the MISO, namely the active spell. The predictability of these submesoscale events being limited (~24 hours), it is difficult to get a long-lead advisory for disaster preparation from numerical weather prediction models. Thus, disaster preparation in such cases may have to be based on probabilistic information. The fact that the EREs are clustered during the active spells provides the scientific basis

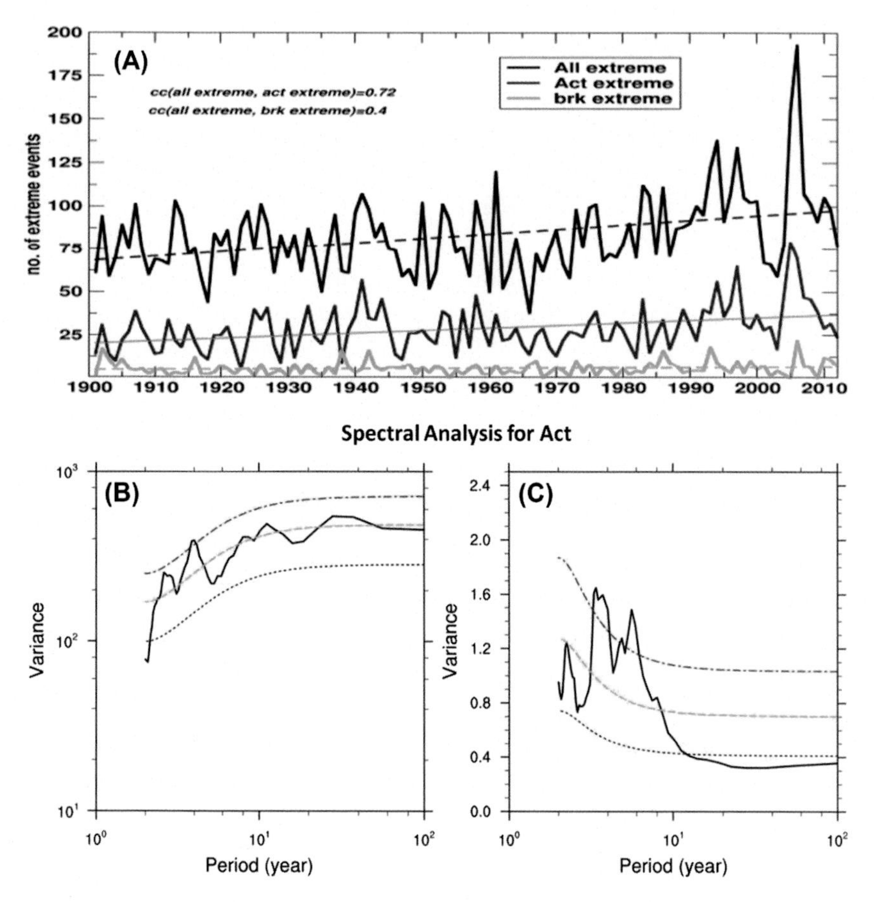

Figure 2.13 (A) Comparison between the trends in extreme rainfall events (EREs) during active, break, and all days (inclusive of active, break, and normal days). (B) The spectral power plot for the ERE counts for active phases during the June–September (JJAS) season. (C) same as (B) but for the JJAS sea surface temperature anomalies averaged over the Nino3.4 region using ERSSTv3b data.

for long-lead preparation for disasters associated with EREs as these phases of MISO could be predicted with 15–20 day lead time.

The large year-to-year variability of the count of EREs over the CI (Fig. 2.13A) is rather intriguing. What is responsible for this significant interannual variability of the EREs? The power spectra of the ERE count (Fig. 2.13B) and that of Nino3.4 SST (Fig. 2.13C) indicate a possibility that the count of EREs may be modulated by ENSO. While similarity in the spectral peaks is not a definitive indication of association, it is plausible that modulation of the large scale monsoon climate by the ENSO teleconnection could modulate the ERE count. With increasing frequency and intensity of EREs over continental India, there is an urgent need for effective disaster mitigation strategy against associated flash floods and landslides. Unlike land falling TCs in the Indian region, where disaster prevention has been successfully implemented recently (Harrman, 2015), disaster prevention against EREs has remained a big challenge. Compared with the skillful long-lead forecasts (3–5 days ahead) available for the TCs, the predictability of the EREs is limited (~1 day) and has been largely responsible for the inability to make adequate preparation for effective disaster prevention. Therefore, disaster prevention strategy in the case of EREs must be based on long-lead probabilistic forecasts of occurrence of the EREs.

4. SUMMARY

In this chapter, we focus on monsoon extremes of various kinds—climate extremes, such as floods and droughts, and weather extremes, such as those triggered by lows and depressions.

On the climate extremes front, we note that neither all of the droughts are triggered under El Nino-like conditions nor are all floods concurrent with La Niña. In other words, these climate extremes are driven by external factors such as El Nino or by internal dynamics ("natural variability"). Among the factors that can potentially play a role in driving the internal dynamics of the monsoon are MISOs (by way of modulating the seasonal mean monsoon), IOD, as well as land surface, and soil moisture feedbacks. To what extent many of these factors contribute to the observed departures from "normal" (in particular, an extreme drought or flood), and whether they act independently or in unison, remains an open and important question.

We also explore the question of whether there is a trend in the observed occurrence and/or intensity of these floods and droughts, and what one would expect these trends, if at all, to be under warming. The most striking evidence from instrumental records is that the frequency of occurrence of monsoon climate extremes seems to be closely tied to the modulation of the mean monsoon flow by an underlying (multi)decadal variability. Specifically, the seasonal mean monsoon has approximately three decades of above normal, followed by three decades of below normal condition. The observed frequency of droughts is about twice (half) that of floods during the negative (positive) phase of the monsoonal multidecadal oscillation. The most recent negative phase of this multidecadal oscillation, which appears to have persisted since 1960, is of particular interest, as this has also seen the ratio of frequency of droughts to floods is more than during earlier negative phases. This persistence is given the term "megadrought," and is a climate "extreme" in its own right. Given the potential societal and economic impact of a megadrought, its prediction would be of immense value. Based on paleoclimate reconstruction (of nearly 1000 years), we show that there is a significant 50–80 year MDM of variability; this mode appears to be nonstationary with one period persisting for a couple of centuries, whereas shifting to another period for an extended span of time. The nonstationarity also contributes to the broad band aperiodic nature of the mode; in other words, extended weak periods of 5–6 decades are not uncommon. Thus, the current decreasing trend of ISM rainfall may still be within its natural variability. Thus, an answer to the question, what drives the multidecadal variability of the monsoon, would also offer a handle on understanding these "megadroughts," which might be considered as real monsoon climate extremes. Finally, on the model front, given the inability of even the best coupled models to reproduce the observed interannual variability during historical simulations, the projections of any change in the frequency of occurrence of monsoon climate extremes need to be viewed with due caution. In addition, the problem of decadal prediction is still in infancy with most climate models having difficulty to simulate large multidecadal signals such as the PDO/AMO. Thus forewarning megadroughts may not be feasible at present; however, with the constant (accelerated) improvement of climate models, the prediction of these multidecade long negative phases of monsoon, which seem to be associated with an increased incidence of droughts, might very well be possible sooner than later.

Moving on to monsoon weather extremes, we show that the frequency and intensity of heavy rainfall (>100 mm day^{-1}) and very heavy rainfall (150 mm day^{-1}) has seen a significant change over the past five decades over CI. This increase coupled with a decrease in moderate intensity rainfall appears to offer a partial answer to the stability of the monsoon rainfall over the past 50 years. Longer-term studies have pointed to the presence of interdecadal variability of these monsoon weather extremes with an accelerated increase in the past 50 years. Given that EREs are a by-product of convective instability, we also report on the role of CAPE (increasing trend) and CINE (decreasing trend) in driving the observed trend in extreme rain.

Following this, we explore space–time clustering of these EREs. Our finding is that not only the frequency of occurrence but also the intensity of the EREs are strongly clustered by the active spells of the monsoon; this implies a high probability of occurrence of EREs during these spells and provides a potential scientific basis for planning effective disaster prevention. The temporal clustering by the active spells also indicates a spatial clustering due to the spatial structure and northward propagating character of the MISOs (Goswami, 2012). When it is "active" spell in the southern peninsular part of India (5°N–16°N), the core monsoon zone (16°N–28°N) is going through a transition and 10 days later the core monsoon zone comes under "active" spell, whereas the southern India would prepare to go toward a "break" phase. Therefore, if the northward propagation of the MISO could be predicted, the disaster prevention activity could be concentrated in the "active" zone. Together with this finding, the fact that the coupled climate models can now make useful forecasts of these spells 15–20 days in advance (Abhilash et al., 2014, 2015) indicates that probabilistic forecasts of occurrence of EREs could be made with sufficiently long-lead time (~15 days). This is highly significant as disaster prevention efforts so far have been ineffective because of the lack of such long-lead forecasts of EREs.

On a broader issue, we also highlight the fact that the large-scale environment (circulation and thermodynamics) provides a strong control on the occurrence of the EREs. Although in a warming environment the EREs are expected to increase globally (Trenberth et al., 2003; Allan and Soden, 2008), there are regions where the EREs have either no trend or even a decreasing trend (Easterling et al., 2000; Goswami et al., 2010). These regional differences in the trends of EREs are essentially related to availability or lack of such a favorable large-scale environment.

5. MATERIALS AND METHODS

The rainfall data used in this study are the daily gridded data at 1° × 1° spatial resolution prepared by India Meteorological Department (IMD; Rajeevan et al., 2006). The data are created by using the station gauge rainfall data with samples collected from all over India. The gridded version was prepared based on interpolation of these station data at daily frequency rate. The data were originally prepared for the period 1901–2004, which is later extended until recent period. The length of the data set used here is from 1901 to 2012. The extreme events during the JJAS monsoon season were defined by setting up a threshold value. We have used the threshold value of 120 mm day^{-1} for any grid point on a given day to be counted as an extreme event. Also we have varied this threshold to 150 and 100 mm day^{-1}. The threshold criteria are adapted based on Goswami et al. (2006). Similarly, rainfall events at any grid are classified as a "low to moderate" event if the measured rainfall is in the range 5–100 mm day^{-1}.

MISOs are northward propagating convective rain bands with large-scale organization of convection. The active and break spells over the Indian region are manifestations of the MISOs. The active spells are characterized by large-scale temporally and spatially persisting rainy days over Indian region, whereas the break spells are characterized by large-scale subdued rainfall activities over the Indian region. To define the active and break spells associated with MISO, we use the 20- to 90-day Lanczos filtering method (Duchon, 1979). The total numbers of weights used were 201. Before the filtering is applied to the rainfall data, some standard preprocessing is done. First, the anomalous data are created by removing a long-term climatology (1901–2004) from the total rainfall for each day and at each grid point. Then, a multiyear data set is created and the filtering is applied. Finally, *standardized anomaly* was made by dividing the anomalies by daily long-term (climatology) standard deviation at each grid point. Then the data are area averaged over a large region (10°N–30°N; 70°E–90°E). Based on this area averaged, standardized filtered rainfall anomaly data, we define active (break) spells as days with rainfall above (below) 1(−1) standard deviation.

REFERENCES

Abhilash, S., et al., 2014. Prediction and monitoring of monsoon intraseasonal oscillations over Indian monsoon region in an ensemble prediction system using CFSv2. Clim. Dyn. 42 (9–10), 2801–2815.

Abhilash, S., et al., 2015. Improved spread-error relationship and probabilistic prediction from CFS based grand ensemble prediction system. J. Appl. Meteorol. Climatol. 1569–1578.

Ajayamohan, R.S., Merryfield, W.J., Kharin, V.V., 2010. Increasing trend of synoptic activity and its relationship with extreme rain events over Central India. J. Clim. 23, 1004–1013. https://doi.org/10.1175/2009JCLI2918.1.

Allan, R.P., Soden, B.J., 2008. Atmospheric warming and the amplification of precipitation extremes. Science 321 (5895), 1481–1484.

Ashok, K., Guan, Z., Yamagata, T., 2001. Impact of the Indian ocean dipole on the relationship between the Indian monsoon rainfall and ENSO. Geophys. Res. Lett. 28, 4499–4502.

Begueria, S., Vicente Serrano, S.M., Martínez, M.A., 2010. A multi scalar global drought data set: the SPEI base: a new gridded product for the analysis of drought variability and impacts. Bull. Am. Meteorol. Soc. 10, 1351–1356.

Borgaonkar, H.P., Sikdera, A.B., Rama, S., Panta, G.B., 2010. El Niño and related monsoon drought signals in 523-year-long ring width records of teak (Tectona grandis L.F.) trees from south India. Palaeogeogr. Palaeoclimatol. Palaeoecol. 285, 74e84.

Brunetti, M., Colacino, M., Maugeri, M., Nanni, T., 2001. Trends in the daily intensity of precipitation in Italy from 1951 to 1996. Int. J. Climatol. 21, 269–284.

Chang, C.P., Li, T., 2000. A theory of the tropospheric biennial oscillation. J. Atmos. Sci. 57, 2209–2224.

Chaturvedi, R.K., Joshi, J., Jayaraman, M., Bala, G., Ravindranath, N.H., 2012. Multi-model climate change projections for India under representative concentration pathways. Curr. Sci. 103 (7), 791–802.

Cohen, N.Y., Boos, W.R., 2014. Has the number of Indian summer monsoon depressions decreased over the last 30 years? Geophys. Res. Lett. 41, 7846–7853. https://doi.org/10.1002/2014GL061895.

Dai, A., 2011. Drought under global warming: a review. Wiley Interdiscip. Rev. Clim. Change 2 (1), 45–65.

Dai, A., Trenberth, K.E., Karl, T.R., 1998. Global Variations in droughts and wet spells: 1900–1995. Geophys. Res. Lett. 25, 3367–3370.

Dai, A., Trenberth, K.E., Qian, T., 2004. A global data set of Palmer Drought Severity Index for 1870–2002: relationship with soil moisture and effects of surface warming. J. Hydrometeorol. 5, 1117–1130.

Dash, S.K., Rajendra Kumar, J., Shekhar, M.S., 2004. On the decreasing frequency of monsoon depressions over the Indian region. Curr. Sci. 86 (10), 1404–1411.

Delworth, T., Mann, M., 2000. Observed and simulated multi-decadal variability in the northern hemisphere. Clim. Dyn. 16, 661–676.

Ding, Y., Sikka, D.R., 2006. Synoptic systems and weather, Chapter 4. In: Wang, B. (Ed.), The Asian Monsoon. Praxis Publishing Pvt. Ltd., Chichester, UK. 786 pp.

Dixit, Y., Hodell, D.A., Petrie, C.A., 2014. Abrupt weakening of the summer monsoon in northwest India ~4100 yr ago. Geology 42 (2), 339–342. https://doi.org/10.1130/G35236.1.

Duchon, C.E., 1979. Lanczos filtering in one and two dimensions. J. Appl. Meteorol. 18 (8), 1016–1022.

Emanuel, K.A., 1987. Dependence of Hurricane intensity on climate. Nature 326, 483–485.

Easterling, D.R., et al., 2000. Observed variability and trends in extreme climate events: a brief review. Bull. Am. Meteorol. Soc. 81 (3), 417–425.

Fein, J.S., Stephens, P.L., 1987. Monsoons. John Wiley and Sons, New York.

Francis, P.A., Gadgil, S., 2006. Intense rainfall events over the west coast of India. Meteorol. Atmos. Phys. 94, 27–42. https://doi.org/10.1007/s00703-005-0167-2.

Gadgil, S., Vinayachandran, P.N., Francis, P.A., Gadgil, S., 2004. Extremes of the Indian summer monsoon rainfall, ENSO and equatorial Indian Ocean oscillation. Geophys. Res. Lett. 31, L12213. https://doi.org/10.1029/2004GL019733.

Gadgil, S., Gadgil, S., 2006. The Indian monsoon, GDP and agriculture. Econ. Polit. Wkly. 4887–4895.

Ghosh, S., Das, D., Kao, S.-C., Ganguly, A.R., 2011. Lack of uniform trends but increasing spatial variability in observed Indian rainfall extremes. Nat. Clim. Change 2. https://doi.org/10.1038/NCLIMATE1327.

Godbole, R.V., 1977. The composite structure of the monsoon depression. Tellus 29 (1), 25–40.

Gogoi, E., Bhatt, M., 2014. Kashmir flood disaster – How the next one could be avoided. Wall Str. J. (India). Available at: http://blogs.wsj.com/indiarealtime/2014/09/12/kashmir-flood-disaster-how-the-next-one-could-be-avoided/.

Goswami, B.N., Keshavamurty, R.N., Satyan, V., 1980. Role of barotropic-baroclinic instability on the growth of monsoon depressions and mid-tropospheric cyclones. Proc. Indian Acad. Sci. (Earth Planet. Sci.) 89, 79–97.

Goswami, B.N., 1987. A mechanism for the West Northwest movement of the monsoon depressions. Nature 326, 370–376.

Goswami, B.N., 1998. Inter-annual variation of Indian summer monsoon in a GCM: external conditions versus internal feedbacks. J. Clim. 11, 501–522.

Goswami, B.N., Ajaya Mohan, R.S., Xavier, P.K., Sengupta, D., 2003. Clustering of low pressure systems during the Indian summer monsoon by intraseasonal oscillations. Geophys. Res. Lett. 30 (8), 1431. https://doi.org/10.1029/2002GL016734.

Goswami, B.N., Xavier, P.K., 2005. Dynamics of 'internal' inter-annual variability of indian summer monsoon in a GCM. J. Geophys. Res. 110, D24104. https://doi.org/10.1029/2005JD006042.

Goswami, B.N., Madhusoodanan, M.S., Neema, C.P., Sengupta, D., 2006. A physical mechanism for North Atlantic SST influence on the Indian summer monsoon. Geophys. Res. Lett. 33, L02706. https://doi.org/10.1029/2005GL024803.

Goswami, B.N., 2012. South Asian monsoon, Chapter 2. In: Lau, W.K.M., Waliser, D.E. (Eds.), Intraseasonal Variability in the Atmosphere-Ocean Climate System, second ed. Springer Praxix, Heidelberg. https://doi.org/10.1007/978-3-642-1394-7.

Goswami, B.N., Kriplani, R.H., Borgaonkar, H.P., Preethi, B., 2015. Multi-decadal variability of Indian summer monsoon rainfall using proxy data. In: Chang, C.P., Ghil, M., Latif, M., Wallace, M. (Eds.), Climate Change: Multi-Decadal and beyond, Chapt. 21. World Scientific, New Jersey, London, Singapore, Beijing, Chennai, pp. 327–346.

Goswami, B.B., Goswami, B.N., November 2016. A road map for improving dry-bias in simulating the South Asian monsoon precipitation by climate models. Clim. Dyn. https://doi.org/10.1007/s00382-016-3439-2.

Goswami, B.B., Mukhopadhyay, P., Mahanta, R., Goswami, B.N., 2010. Multiscale interaction with topography and extreme rainfall events in the north-East Indian region. J. Geophys. Res. 115, D12114. https://doi.org/10.1029/2009JD012275.

Government of Maharashtra, 1973. Report of the Fact Finding Committee for Survey of Scarcity Areas Maharashtra State (Mumbai, India), p. 310.

Harrman, L., 2015. Cyclone Phailin in India: Early Warning and Timely Actions Saved Lives. UNEP Sioux Falls. Available at: http://na.unep.net/geas/getUNEPPageWithArticleIDScript.php?article_id=106.

Houze Jr., R.A., 2004. Mesoscale convective systems. Rev. Geophys. 42, RG4003. https://doi.org/10.1029/2004RG000150.

Hunt, K.M.R., Turner, A.G., Inness, P.M., Parker, D.E., Levine, R.C., 2016. On the structure and dynamics of indian monsoon depressions. Mon. Weather Rev. 144, 3391–3416. https://doi.org/10.1175/MWR-D-15-0138.1.

India Meteorological Department, 2011. Tracks of cyclones and depressions over North Indian Ocean (from 1891 onwards). In: Tech. Note Version 2.0, Cyclone Warning and Research Centre India Meteorological Department Regional Meteorological Centre, Chennai, India.

Kharin, V.V., Zwiers, F.W., Zhang, X., Wehner, M., 2013. Changes in temperature and precipitation extremes in the CMIP5 ensemble. Clim. Change 119 (2), 345–357.

Kothawale, D.R., Rupa Kumar, K., 2005. On the recent changes in surface temperature trends over India. Geophys. Res. Lett. 32, L18714. https://doi.org/10.1029/2005GL023528.

Kothawale, D.R., Revadekar, J.V., Rupa Kumar, K., 2010. Recent trends in pre-monsoon daily temperature extremes over India. J. Earth Syst. Sci. 119 (1), 51–65.

Krishnamurti, T.N., Martin, A., Krishnamurti, R., Simon, A., Thomas, A., Kumar, V., 2013. Impacts of enhanced CCN on the organization of convection and recent reduced counts of monsoon depressions. Clim. Dyn. 2013 (41), 117–134. https://doi.org/10.1007/s00382-012-1638-z.

Krishnamurthy, V., Goswami, B.N., 2000. Indian monsoon-ENSO relationship on inter decadal time scales. J. Clim. 13, 579–595.

Krishnamurthy, V., Ajaya Mohan, R.S., 2010. Composite structure of monsoon low pressure systems and its relation to Indian rainfall. J. Clim. 23. https://doi.org/10.1175/2010JCLI2953.1.

Krishnamurthy, C.K.B., Lall, U., Kwon, H.,-H., 2009. Changing frequency and intensity of rainfall extremes over India from 1951 to 2003. J. Clim. 22. https://doi.org/10.1175/2009JCLI2896.1.

Kotal, S.D., Roy, S.S., Bhowmik, S.K.R., 2014. Catastrophic heavy rainfall episode over Uttarakhand during 16-18 June 2013-observational aspects. Curr. Sci. 107 (2), 234–245.

Kumar, A., Dudhia, J., Rotunno, R., Niyogi, D., Mohanty, U.C., 2008. Analysis of the 26 July 2005 heavy rain event over Mumbai, India using the Weather Research and Forecasting (WRF) model. Q. J. R. Meteorol. Soc. 134 (636), 1897–1910.

Mak, M.K., 1975. The monsoonal mid-troposhperic cyclogenesis. J. Atmos. Sci. 32, 2246–2253.
Meehl, G.A., 1994. Coupled land-ocean-atmosphere processes and south Asian monsoon variability. Science 266, 263–267.
Mohanty, U.C., Mandal, M., Raman, S., 2004. Simulation of Orissa super cyclone (1999) using PSU/NCAR mesoscale model. Nat. Hazards 31, 373–390.
Neena Mani, J., Suhas, E., Goswami, B.N., 2009. Can global warming make Indian monsoon weather less predictable? Geophys. Res. Lett. 36, L08811. https://doi.org/10.1029/2009GL037989.
Niranjan Kumar, K., Rajeevan, M., Pai, D.S., Srivastava, A.K., Preethi, B., 2013. On the observed variability of monsoon droughts over India. Weather Clim. Extreme. 1, 42–50.
Palmer, W.C., 1965. Meteorological Drought. US Department of Commerce, Weather Bureau, Washington, DC, USA.
Prajeesh, A.G., Ashok, K., Bhaskar Rao, D.V., 2013. Falling monsoon depression frequency: a Gray-Sikka conditions perspective. Sci. Rep. 3, 2989. https://doi.org/10.1038/srep02989.
Pattanaik, D.R., 2007. Analysis of rainfall over different homogeneous regions of India in relation to variability in westward movement frequency of monsoon depressions. Nat. Hazards 40, 635–646. https://doi.org/10.1007/s11069-006-9014-0.
Rajeevan, M., Bhate, J., Kale, K.D., Lal, B., 2006. High resolution daily gridded rainfall data for the Indian region: analysis of break and active monsoon spells. Curr. Sci. 91, 296–306.
Rajeevan, M., Bhate, J., Jaswal, A.K., 2008. Analysis of variability and trends of extreme rainfall events over India using 104 years of gridded daily rainfall data. Geophys. Res. Lett. 35, L18707. https://doi.org/10.1029/2008GL035143.
Ramage, C.S., 1971. Monsoon Meteorology (Int.Geophys. Ser. vol. 15). Academic Press, San Diego, California. 296 p.
Roxy, M., Ritika, K., Terray, P., Murtugudde, R., Ashok, K., Goswami, B.N., 2015. Drying of Indian subcontinent by rapid Indian Ocean warming and a weakening land-sea thermal gradient. Nat. Commun. https://doi.org/10.1038/ncomms842.
Sabeerali, C.T., Rao, S.A., Dhakate, A.R., Salunke, K., Goswami, B.N., 2013. Why ensemble mean projection of south Asian monsoon rainfall by CMIP5 models is not reliable? Clim. Dyn. https://doi.org/10.1007/s00382-014-2269-3.
Saha, S.K., Halder, S., Suryachandra Rao, A., Goswami, B.N., 2012. Modulation of ISOs by land-atmosphere feedback and contribution to the interannual variability of Indian summer monsoon. J. Geophys. Res. 117, D13101. https://doi.org/10.1029/2011JD017291.
Saha, A., Ghosh, S., Sahana, A.S., Rao, E.P., 2014. Failure of CMIP5 climate models in simulating post-1950 decreasing trend of Indian monsoon. Geophys. Res. Lett. 41, 7323–7330. https://doi.org/10.1002/2014GL061573.
Saji, N.H., Goswami, B.N., Vinayachandran, P.N., Yamagata, T., 1999. A dipole mode in the tropical Indian Ocean. Nature 401, 360–363.
Saji, N.H., Yamagata, T., 2003. Possible impacts of Indian Ocean Dipole mode events on global climate. Clim. Res. 25, 151–169.
Semenov, V.A., Bengtsson, L., 2002. Secular trends in daily precipitation characteristics: greenhouse gas simulation with a coupled AOGCM. Clim. Dyn. 2002 (19), 123–140. https://doi.org/10.1007/s00382-001-0218-4.
Singh, D., Tsiang, M., Rajaratnam, B., Diffenbaugh, N.S., 2014. Observed changes in extreme wet and dry spells during the South Asian summer monsoon season. Nat. Clim. Change 4 (6), 456–461.
Sikka, D.R., 1977. Some aspects of the life history, structure and movements of monsoon depressions. Pure Appl. Geophys. 115, 1501–1529.

Shukla, J., 1978. CISK, barotropic and baroclinic instability and the growth of monsoon depressions. J. Atmos. Sci. 35, 495–500.

Sinha, A., Berkelhammer, M., Stott, L., Mudelsee, M., Cheng, H., Biswas, J., 2011. The leading mode of Indian summer monsoon precipitation variability during the last millennium. Geophys. Res. Lett. 38, L15703. https://doi.org/10.1029/201 1GL047713.

Sinha, A., Cannariato, K.G., Stott, L.D., Cheng, H., Edwards, R.L., Yadava, M.G., Ramesh, R., Singh, I.B., 2007. A 900-year (600 to 1500 A.D.) record of the Indian summer monsoon precipitation from the core monsoon zone of India. Geophys. Res. Lett. 34, L16707. https://doi.org/10.1029/2007GL030431.

Shukla, J., 1987. Interannual variability of monsoon. In: Fein, J.S., Stephens, P.L. (Eds.), Monsoons. John Wiley and Sons, New York, pp. 399–464.

Sperber, K.R., Annamalai, H., Kang, I.S., et al., 2013. The Asian summer monsoon: an intercomparison of CMIP5 vs. CMIP3 simulations of the late 20th century. Clim. Dyn. 41, 2711–2744. https://doi.org/10.1007/s00382-012-1607-6.

Staubwasser, M., Sirocko, F., Grootes, P.M., Segl, M., 2003. Climate Change at the 4.2 ka BP Termination of the Indus Valley Civilization and Holocene South Asian Monsoon Variability: Geophysical Research Letters, vol. 30, p. 1425. https://doi.org/10.1029/2002GL016822.

Swapna, P., Roxy, M.K., Kulkarni, P.A.G., Ashok, K., Krishnan, R., Moorthi, S., Kumar, A., Goswami, B.N., 2015. IITM earth system model: transformation of a seasonal prediction model to a long term climate model. Bull. Am. Meteorol. Soc. 96, 1351–1367. https://doi.org/10.1175/BAMS-D-13-00276.1.

Trenberth, K.E., 1999. Conceptual framework for changes of extremes of the hydrological cycle with climate change. Clim. Change 42, 327–339.

Trenberth, K.E., 2011. Changes in precipitation with climate change. Clim. Res. 47, 123–138. https://doi.org/10.3354/cr00953.

Trenberth, K.E., Dai, A., Rasmussen, R.M., Parsons, D.B., 2003. Changing character of precipitation. Bull. Am. Meteorol. Soc. 84, 1205.

Vicente-Serrano, S.M., Beguería, S., López-Moreno, J.I., 2010a. A multi-scalar drought index sensitive to global warming: the standardized precipitation evapotranspiration index. J. Clim. 23, 1696–1718.

Vicente-Serrano, S.M., et al., 2010b. A new global 0.5 gridded data set (1901–2006) of a multi scalar drought index: comparison with current drought index datasets based on the Palmer Drought Severity Index. J. Hydrometeorol. 11, 1033–1043.

Webster, P.J., Holland, G.J., Curry, J.A., Chang, H.-R., 2005. Changes in tropical cyclone number, duration, and intensity in a warming environment. Science 309, 1844–1846.

Webster, P.J., Magana, V.O., Palmer, T.N., Shukla, J., Tomas, R.T., Yanai, M., Yasunari, T., 1998. Monsoons: processes, predictability and the prospects of prediction. J. Geophys. Res. 103 (C7), 14451–14510. https://doi.org/10.1029/97JC02719.

Xavier, P.,K., Marzin, C., Goswami, B.N., 2007. An objective definition of the Indian summer monsoon season and a new perspective on the ENSO–monsoon relationship. Q. J. R. Meteorol. Soc. 133, 749–764.

Yu, J.-Y., Weng, S.-P., Ferrara, J.D., 2003. Ocean roles in the TBO transitions of the Indian-Australian monsoon system. J. Clim. 16, 3072–3080.

CHAPTER 3

South American Monsoon and Its Extremes

Alice M. Grimm
Federal University of Parana, Curitiba, Brazil

Contents

Tropical Extremes: Natural Variability and Trends
ISBN 978-0-12-809248-4
https://doi.org/10.1016/B978-0-12-809248-4.00003-0

1. INTRODUCTION

Besides being an important component of the global monsoon system, the South American monsoon (SAM) during the warm season is also responsible for large fraction of the annual precipitation over most of South America (SA), including subtropical regions (Fig. 3.1). In the core monsoon region, in central Brazil, it rains above 10 times more in austral summer than in winter (hereafter the seasons refer to the Southern Hemisphere). Accordingly, most of the precipitation extremes with great social and economic consequences occur during the monsoon season. It is worth pointing out that the worst energy crisis in Brazil (in 2001) and the intense drought affecting southeast Brazil in 2014 were both associated with deficient summer precipitation. On the other hand, one of the worst natural disasters in the continent, associated with heavy rainfall and widespread landslides, happened in the highlands of the Rio de Janeiro state, in southeastern Brazil, on January 2011, when more than 900 people perished and other 35,000 were displaced (Marengo and Alves, 2012).

Local and remote influences contribute to produce rich variability of the SAM, from both temporal and spatial points of view, with modes ranging from synoptic and intraseasonal to the interdecadal temporal scales. Climate variability affects and modulates the synoptic and mesoscale features responsible for extreme events (Grimm and Tedeschi, 2009; Grimm, 2011; Grimm et al., 2015). Therefore, it is important to know this spatial and temporal variability and the mechanisms involved, and it is also important to know how it affects the frequency of the extreme events. Beyond natural climate variability, anthropogenic climate change could produce an increase in monsoon precipitation variability, with a possible increase in the frequency and severity of both droughts and floods (Kitoh et al., 2013), and the impacts of natural variability itself could change (Grimm, 2011).

The objective of this chapter is to provide an overview of (1) the general features of the SAM, (2) its variability in different timescales, (3) the way this climate variability affects the frequency of extreme events, (4) the mechanisms and influences associated with extreme events in different regions of SA, (5) case examples of extreme rainfall or drought events and their impacts in different regions, and (6) observed trends regarding extremes.

2. SOUTH AMERICAN MONSOON: GENERAL FEATURES AND EVOLUTION

It is convenient to summarize the general features of the SA monsoon because their variability produces the extremes that are the focus of this chapter. Further details can be found in several reviews (Nogues-Paegle et al., 2002;

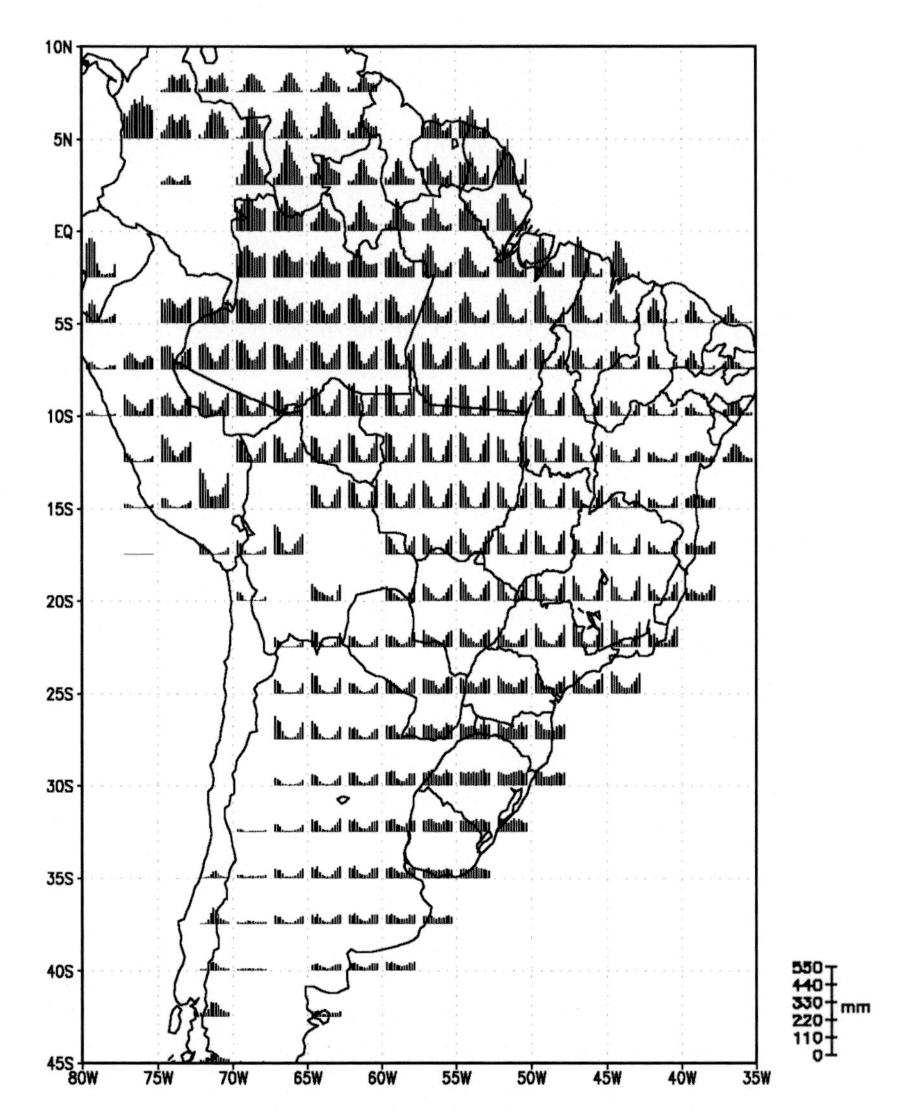

Figure 3.1 Annual cycles of precipitation in South America from at least 25 years of data in the period 1950–2005, for areas of 2.5° × 2.5° latitude–longitude. *(From Grimm, A.M., 2011. Interannual climate variability in South America: impacts on seasonal precipitation, extreme events, and possible effects of climate change. Stoch. Environ. Res. Risk Assess. 25, 537–554. https://doi.org/10.1007/s00477-010-0420-1.)*

Grimm et al., 2005;Vera et al., 2006; Grimm and Silva Dias, 2011; Liebmann and Mechoso, 2011; Marengo et al., 2012; Silva and Kousky, 2012; Grimm et al., 2015).

When the maximum solar radiation reaches the subtropics in the austral summer, a thermal low-pressure system strengthens over the Chaco region, in central SA. Although this low-pressure system exists over northern Argentina and western Paraguay throughout the year, it is strongest during the summer. The interhemispheric pressure gradient between the South American low and the northwestern Sahara high strengthens, and the tropical northeasterly trade winds increase in intensity (Fig. 3.2). Anomalous flow crosses the equator and penetrates the continent, carrying moisture. It becomes northwesterly, is channeled southward by the Andes Mountains, and turns clockwise around the Chaco low. The interaction of the continental low with the South Atlantic high and the northeasterly trade winds produces low-level winds and moisture convergence, resulting in enhanced precipitation over Amazon, central and southeast Brazil, and over other tropical/subtropical regions east of the Andes (Figs. 3.1 and 3.2). The South Atlantic Convergence Zone (SACZ), a southeastward extension of cloudiness and precipitation from central Brazil toward the Atlantic Ocean (Fig. 3.2, lower panel right) enters its most active stage in summer (December–January–February, DJF). The upper-level anticyclonic center moves southward from the Amazon in spring (September–October–November, SON), setting up the "Bolivian High" in summer (Fig. 3.2). East of this high, over the Atlantic Ocean and close to the coast of northeast (NE) Brazil, an upper-level trough develops (Virji, 1981). The circulation features tend to reverse between the lower and the upper troposphere (Fig. 3.2).

Hirata and Grimm (2015) showed that the climatological circulation near the SACZ favors wave accumulation (increased wave energy density) in the region because of the climatological zonal stretching deformation (zonal variation of the climatological zonal wind, $\partial\overline{U}/\partial x$) in summer at 200 hPa. This is related to the zonal wavenumber of synoptic Rossby waves propagating eastward and their wave energy density (Webster and Chang, 1988). Negative zonal stretching deformation, as happens over the SACZ, increases the zonal wavenumber, leading to a reduction of the longitudinal wave speed and increases wave energy density. The local increase in wave energy density (wave accumulation) results in intense convective activity that leads to the diagonal band of enhanced convection.

Embedded within the northwesterly winds along the Andes Mountains is the South American low-level jet (SALLJ), strongest near Bolivia, carrying moisture from the Amazon to the subtropics, producing enhanced rainfall in its exit region (Marengo et al., 2004).

The monsoon onset and its progression depend on the definition of the monsoon onset. For instance, Kousky (1988) defined monsoon onset at a particular location as occurring when outgoing longwave radiation (OLR)

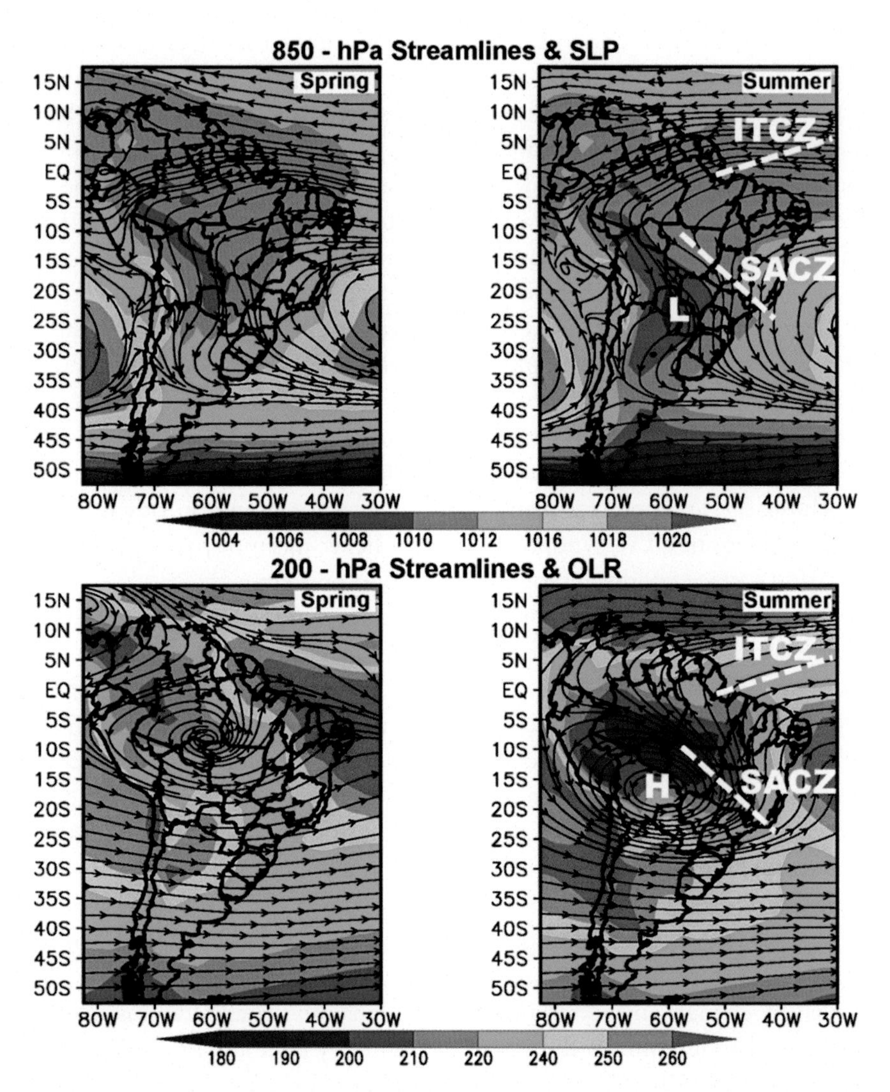

Figure 3.2 Upper panels: streamlines at 850 hPa and sea level pressure (SLP). Lower panels: streamlines at 200 hPa and outgoing longwave radiation (OLR). Left column: spring; right column: summer. *Dashed lines* indicate the South Atlantic Convergence Zone and Intertropical Convergence Zone, and L and H indicate the Chaco Low and the Bolivian High, respectively. *(Adapted from Grimm, A.M., Laureanti, N.C., Rodakoviski, R.B., Gama, C.B., 2016. Interdecadal variability and extreme precipitation events in South America during the monsoon season. Clim. Res. 68, 277–294. https://doi.org/10.3354/cr01375.)*

is less than 220W m^{-2}, provided that it was above the threshold in 10 of 12 preceding pentads and remained below the threshold in 10 of 12 following pentads. Liebmann and Marengo (2001) defined onset as occurring when the accumulation of precipitation exceeds that expected from the annual mean daily average. The first definition results in a northwest to southeast progression of the onset and a southeast to northwest progression of the withdrawal. However, as northwest Amazon is perpetually wet, this threshold does not determine an onset date in this region, where the rainfall peaks in March–May. With this criterion, the wet season onset occurs in mid-October over central Brazil and in mid-November over southeast Brazil. The end occurs in mid-February in southeast Brazil and in early April over central Brazil. The second definition results in the wet season progressing northward from an area just north of the Paraguay border (Liebmann and Mechoso, 2011). There are many other indexes for monsoon onset, based on precipitation, OLR, synoptic-scale flow fields, thermodynamic parameters, land surface conditions, or a combination of them.

The onset of the SAM during the austral spring is characterized by a rapid shift of the intense convection region in Central America and northwestern SA to central-west SA (October), and then southeast Brazil (November). The low-level temperature reaches its maximum in central Brazil on September (Marengo et al., 2012). Land surface warming destabilizes the lower-tropospheric lapse rate from winter to spring. A significant increase of the convection, however, occurs from October to December when more moisture is transported into the region. The land surface warming increases the land–ocean temperature gradient and drives the seasonal changes of circulation. Although the wet season in tropical SA can be initiated rapidly by synoptic systems, the onset of convection is primarily controlled by changes in the regional thermodynamic structure and is mainly related to a moistening of the planetary boundary layer and lowering of the temperature at its top (Fu et al., 1999; Marengo et al., 2001).

Once favorable large-scale thermodynamic conditions are established, the transition from dry to wet season in central Brazil can be rapid and connected to synoptic or intraseasonal variations. For instance, cold fronts may help trigger monsoon onset by enhancing forced ascent in a thermodynamically favorable atmosphere (Li and Fu, 2004; Nieto-Ferreira et al., 2011).

During the premonsoon season, turbulent sensible heating dominates the warming of the subtropics and is confined to the lower atmosphere. This heating is maximum before mid-November, but when deep convection reaches the southeast Brazil highlands latent heat release becomes the dominant heating component, being maximum in the middle and upper troposphere (Zhou and Lau, 1998). These two heat sources are essential in shaping the climatological SAM.

After March, the SAM weakens as the area of deep convection retreats northwestward, and drier conditions return to subtropical SA. Over north and northeast SA precipitation peaks later. During the demise phase of the monsoon, the Atlantic Intertropical Convergence Zone (ITCZ) remains weak. In NE Brazil the rainy season takes place during April through June, when the ITCZ is in its southernmost position.

3. SYNOPTIC AND MESOSCALE VARIABILITY OF THE SAM

As described above, the climatological SAM is driven by seasonally varying large-scale distributions of sensible and latent heating, but its dynamics is influenced by the Andes Mountains to the west and other orographic features such as the Brazilian Plateau in the core monsoon region and the mountains to the east, especially in southeast Brazil, which play important role in anchoring the SACZ (Grimm et al., 2007).

However, there are numerous synoptic and mesoscale features embedded within these large-scale circulation patterns. These features are responsible for the day-to-day weather and high impact rainfall events. Some of these features are here briefly described, and factors that can affect their frequency and intensity are discussed in the following sections. More details can be found in Grimm and Silva Dias (2011) and references therein.

The daily precipitation variability over tropical SA during austral summer results mainly from the combined action of equatorial trades, easterly tropical disturbances, and the equatorward incursions of midlatitude synoptic wave systems.

Over subtropical SA and western Amazon basin, the day-to-day variability of rainfall is largely explained by northward incursions of midlatitude synoptic systems to the east of the Andes. Although particularly large and frequent in winter and transition seasons, they are also present during summer, with about the same periodicity (~7 days), and often reach sufficiently low latitudes to affect the SAM (e.g., Garreaud and Wallace, 1998; Seluchi and Marengo, 2000). The deep northward intrusion of midlatitude systems is favored by the Andes topography, which tends to direct them northward, fostering the advance of cold air incursions (cold surges) well into subtropical (and sometimes tropical) latitudes. During winter the major impact is on temperature while in summer it is on precipitation. This synoptic structure is usually a northwest-southeast–oriented band of enhanced convection ahead of the leading edge of the cooler air, followed by an area of suppressed convection, which propagates equatorward (~10 ms^{-1}), maintaining its identity for about 5 days (e.g., Garreaud and Wallace, 1998). It is the dominant mode of the day-to-day variability of deep convection, contributing ~25% of summer precipitation in the central Amazonia and ~50% over subtropical SA. These bands also influence convection in the SACZ, contributing to the role of transient disturbances in the maintenance of the SACZ. The incursions of extratropical fronts and their interaction with tropical convection can produce different types of spatial organization of tropical convection (Siqueira and Machado, 2004).

The mesoscale organization of deep convection during the monsoon season occurs frequently in certain regions. It is modulated by the diurnal cycle and by transient synoptic systems, as well as influenced by mesoscale effects such as jets (Salio et al., 2007; Durkee et al., 2009) and other topographically forced circulation and surface–atmosphere interactions. The SALLJ is the jet with most extensive influence in the SAM, playing an important role in the transport of moisture from the tropics to the subtropics and producing enhanced rainfall in its exit region. The variability of the SALLJ may be partly caused by changes in the zonal circulation (Byerle and Paegle, 2002) or by anomalous circulation wave trains over SA. The SALLJ events are associated with synoptic variability and can be separated into two groups with different synoptic evolutions: (1) events in which the low-level jet (LLJ) extends farther south, at least to 25°S; (2) events in which the jet leading edge is north of this threshold (Nicolini et al., 2002). Those in the first category are stronger and associated with high moisture convergence and precipitation in southeastern South America (SESA) and low precipitation in the SACZ, whereas those in the second one are weaker and associated with enhanced precipitation in the SACZ and suppressed precipitation in SESA. This behavior is associated with a dipole-like variability between SESA and SACZ. The second category is more frequent than the first one during summer. This dipole-like behavior is also described in connection with climate variability in the next sections.

Mesoscale convective complexes (MCCs) occur frequently during the warm season (October–May) in SESA (an average of 37 MCCs each season for 1998–2007), especially east of the Andes between 20°S and 30°S, over a region comprising southern Brazil, northern Argentina, and Paraguay (Durkee et al., 2009). They move preferentially southeastward in association with the northerly LLJ and enhanced moisture flux convergence

(Machado et al., 1998; Nicolini et al., 2002). These MCCs have, on average, maximum area of 256,500 km^2 and lifetime of 14 hours. Their intensification is related to the position of the upper-level subtropical jet, and its interaction with the low-level warm and moist northerly wind. MCCs are more abundant during SALLJ events. They preferentially initiate in late afternoon and mature during nighttime, which may be partially explained by the diurnal variability of the SALLJ, with late afternoon–evening maximum and by the nocturnal convergence over the Paraná River basin valley.

There are major regional differences in the structure, intensity, and diurnal cycle of rainfall systems during the SAM. Although the La Plata basin, in SESA, is particularly dominated by large and intense mesoscale convective systems (MCSs), the rainfall in the Amazon Basin comes partly from smaller MCSs and partly from frequent showers and thunderstorms. In this region, most convective systems are smaller (average area less than 1×10^5 km^2) and have shorter lifetime (3–6 hours) than MCCs (Carvalho et al., 2002; Nieto Ferreira et al., 2003).

Another example of mesoscale organization of regional convection tied to the diurnal cycle is the afternoon genesis of squall lines in the northeastern coast of SA and subsequent inland propagation in the Amazon Basin (Cohen et al., 1995; Garreaud and Wallace, 1997).

4. CLIMATE VARIABILITY IN THE MONSOON SEASON AND ITS INFLUENCE ON EXTREMES

Some important synoptic and mesoscale features responsible for heavy (or deficient) precipitation are significantly affected in intensity and frequency by climate variability. The main modes of climate variability in different timescales and their impacts on SAM, as well as some other mechanisms of the SAM variability, are briefly described. Some examples of their influence on the frequency of extreme events are presented.

4.1 Intraseasonal Variability

The maximum contribution of the intraseasonal variability (periods in the 10–100 day band) to the total variance of precipitation in SA is concentrated in central-east Brazil, including the SACZ, while the lowest values are in the western part of the continent (Grimm et al., 2005). The main rotated modes from an empirical orthogonal function analysis of intraseasonal precipitation variability confirm that this variability is the strongest in this region (Fig. 3.3, upper panels). The first rotated mode represents oscillations

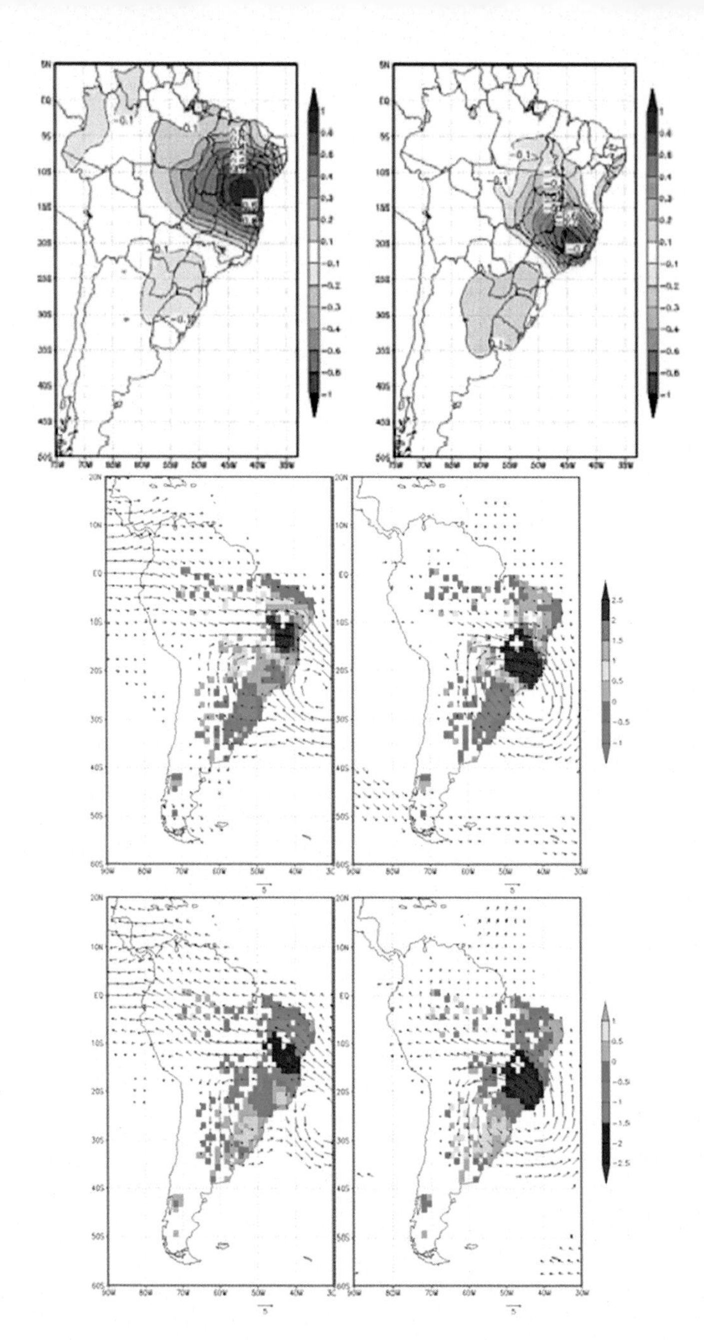

Figure 3.3 Upper panels: (left) first and (right) second rotated empirical orthogonal functions of the intraseasonal precipitation variability in the 30–70 day band. Middle panels: composite of rainfall anomalies and vertically integrated moisture flux for wet phases of these modes. Bottom panels: the same as the middle panels, but for dry phases of these modes. *(From Grimm, A.M., Vera, C.S., Mechoso, C.R., 2005. The South American monsoon system. In: Chang, C.-P., Wang, B., Lau, N.-C.G. (Eds.), The Global Monsoon System: Research and Forecast, pp. 219–238. WMO/TD 1266-TMRP 70. www.wmo.int/pages/prog/arep/tmrp/documents/global_monsoon_system_IWM3.pdf. After Ferraz, S.E.T., 2004. Intraseasonal variations of the summer precipitation over South America. PhD Thesis. Department of Atmospheric Sciences, University of São Paulo.)*

in central-east Brazil, whereas the second one is concentrated in the SACZ. Both modes feature anomalies in the subtropical plains that are opposite in sign to those in the main center, although much weaker in magnitude, featuring a significant "dipole"-like relationship between precipitation anomalies in central-east Brazil/SACZ and the subtropics to the south, although with different magnitude in both centers.

The local circulation anomalies associated with the first two modes are consistent with the "dipole"-like pattern in precipitation. In one extreme phase of the pattern, a cyclonic anomaly around 20°S, 50°W (25°S, 45°W) directs the northwesterly moisture flux into central-east Brazil (SACZ) and decreases the southward transport (Fig. 3.3, middle panels). In the opposite phase, an anticyclonic anomaly enhances the moisture flux toward the subtropical plains (Fig. 3.3, bottom panels). Southwest of this strong circulation anomaly, there is a weaker anomaly of opposite sign. Coherently with this pattern, low-level zonal westerly (easterly) winds over tropical Brazil during summer are associated with an active (inactive) SACZ and net moisture divergence (convergence) over SESA, implying a weak (strong) SALLJ. Other studies have also reported that similar intraseasonal variations in summer low-level wind regimes over central Brazil are linked to breaks and active phases of the SAM (e.g., Herdies et al., 2002; Jones and Carvalho, 2002; Gan et al., 2004). A similar structure is present even on interannual and interdecadal timescales, as shown in the following sections (Robertson and Mechoso, 2000; Grimm, 2003, 2004; Grimm and Zilli, 2009; Grimm and Saboia, 2015; Grimm et al., 2016).

Intraseasonal summer precipitation variability over SA shows different timescales. The main modes shown in Fig. 3.3 appear among the five most important modes in several intraseasonal frequency bands (periods of 10–20, 20–30, and 30–70 day). This means that precipitation over SA results from a complex interaction of different timescales (e.g., Nogues-Paegle et al., 2000). The origin of this variability seems to be associated with wave trains propagating southeastward from west or central Pacific, rounding the southern tip of SA and turning toward the northeast, as part of larger scale systems, originated or modified by associated convection in west and/or central Pacific (Grimm and Silva Dias, 1995; Nogues-Paegle et al., 2000; Grimm, 2018). According to Grimm (2018), based on studies started by Grimm and Silva Dias (1995), the anomalies in central-east Brazil associated with the first modes originate from tropics–tropics teleconnections produced by the equatorial anomalous convection in central Pacific, whereas the anomalies over the SACZ are produced by the subtropical convection anomalies in central-south Pacific. These kinds of anomalies are present not only in the Madden–Julian Oscillation (MJO) but also in other timescales, which explains the variability in these regions also in other intraseasonal periods. The equatorial (subtropical) origin of the circulation anomalies associated with the first (second) mode appears in the composites of Fig. 3.3.

The MJO is the strongest, best known, and most predictable source of intraseasonal variability and has strong impacts on SA in several of its phases (Alvarez et al., 2016; Grimm, 2018). The most extensive significant impacts happen in phases 1 and 5 (defined by Wheeler and Hendon, 2004), with opposite signs.

In addition to the remote influences, the evolution of the summer monsoon can be influenced by regional factors. Grimm et al. (2007) showed that the seasonal evolution of the monsoon in part of SA, including the SACZ, can be influenced by the soil moisture anomalies at the beginning of the season, which are able to produce temperature anomalies and circulation anomalies shaped and enhanced by orography in central-east Brazil, and influenced by sea surface temperature (SST) anomalies off the southeast coast of Brazil.

4.1.1 Influence of the Intraseasonal Madden–Julian Oscillation–Related Variability on Extreme Events

The MJO significantly influences the frequency of extreme events that can lead to disasters because it changes the probability of occurrence of these events (Jones et al., 2004; Alvarez et al., 2016; Grimm, 2018). The strongest and most extensive influence occurs in phase 1 (as defined by Wheeler and Hendon, 2004) near the core monsoon region in central Brazil and near the SACZ, where the probability of occurrence of extreme precipitation events more than doubles during this phase (Grimm, 2018). The role of synoptic and intraseasonal anomalies in the life cycle of summer rainfall extremes in some regions of SA, including SACZ and the middle-lower Parana/La Plata basin, is detailed by Hirata and Grimm (2015).

Hirata and Grimm (2015) showed that intraseasonal variability affecting the subtropical jet modulates the summertime negative zonal stretching deformation around the climatological position of the SACZ and plays an important role in the development of rainfall extremes. Both intraseasonal and synoptic frequency bands are important for the development of rainfall extremes associated with the SACZ. During the pentad prior to the event, intraseasonal negative (positive) 200 hPa geopotential height anomalies are significant in the equatorward (poleward) flank of the South American subtropical jet. These anomalies tend to reduce the zonal wind, further reducing the climatologically negative $\partial\overline{U}/\partial x$ in the region (increasing its absolute value), thus enhancing wave accumulation and convective activity. The rainfall extremes occur when negative geopotential anomalies in both frequency bands become significant and coincide on the northern flank of the jet, south of the SACZ. The extreme events in the middle Paraná/La Plata basin are also influenced by intraseasonal variability, although in this case the synoptic variability is stronger.

4.2 Interannual Variability

Most of the interannual variability of monsoon precipitation comes from ENSO (e.g., Grimm and Zilli, 2009; Grimm, 2011; references therein). On an annual average, during ENSO warm (cold) phase, precipitation is below (above) average in northeastern SA and is enhanced (reduced) in SESA (Fig. 2 of Grimm, 2011), due to a combination of Walker and Hadley circulation anomalies and anomalous Rossby wave activity produced by the anomalous tropical heat sources associated with ENSO events (Grimm, 2003, 2004, 2011; Grimm and Ambrizzi, 2009; references therein). However, these long-established mechanisms do not explain the entire variation observed throughout the monsoon season. Grimm (2003, 2004) found in central-east SA consistent dry (wet) anomalies in the onset of the monsoon (November (0), (0) indicates the year of ENSO onset) during El Niño (EN) (La Niña [LN]), whereas in the peak monsoon (January (+), (+) indicates the following year), the anomalies tend to be opposite (Fig. 3.4). In SESA, wet anomalies are spread and intense throughout the region in November (0) but are restricted to a smaller region in the western part in January (+) and they even change sign in a great part of the region (Fig. 3.4). In the demise of the monsoon, March (+), the anomalies return approximately to the previous spring pattern (not shown).

These different ENSO-related anomalies in different stages of the monsoon are associated with fairly similar EN-related SST anomalies because they do not change much from November (0) to January (+). This indicates that regional surface–atmosphere interactions compete with remote influences during at least part of the monsoon season (Grimm et al., 2007). In spring of an EN year (0), remote influences dominate over the continent because this is the most favorable season for teleconnections between SESA and the tropical/subtropical Pacific Ocean (Cazes-Boezio et al., 2003), whereas in summer, land surface processes prevail. This tendency for reversal of anomalies from spring to summer during the evolution of the monsoon

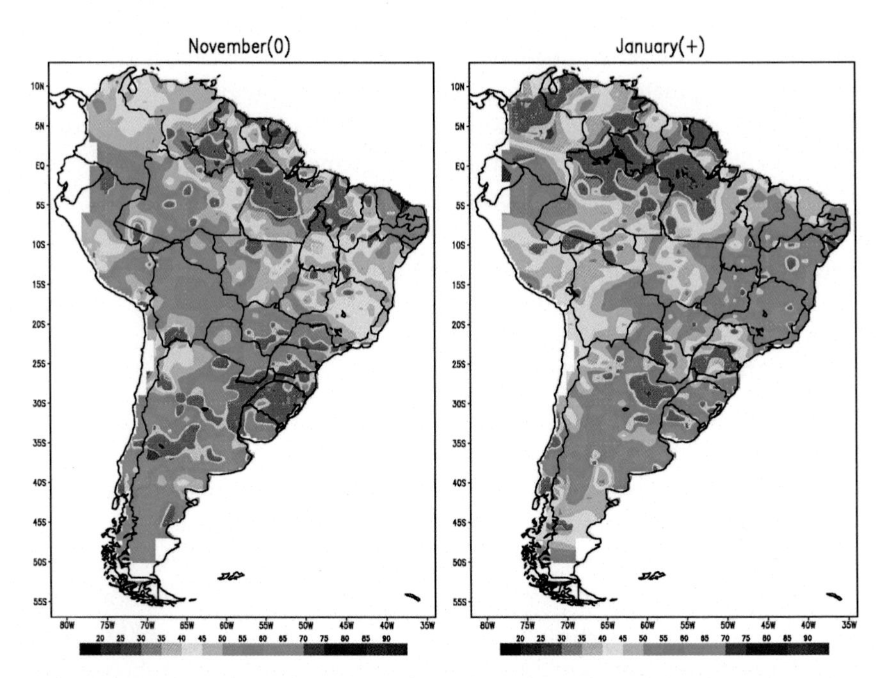

Figure 3.4 Expected precipitation percentiles for El Niño events in (left) November (0), and (right) January (+). *Black and gray areas* indicate anomalies significant to a level better than 90 (85) %. (0) indicates the year of ENSO onset, and (+) indicates the following year. *(From Grimm, A.M., 2003. The El Niño impact on the summer monsoon in Brazil: regional processes versus remote influences. J. Clim. 16, 263–280.)*

can also occur in non-ENSO years, provided that spring anomalies are large (Grimm et al., 2007). It even reflects on the main modes of interannual variability of the continental precipitation in spring and summer (Fig. 3.5). The first mode of spring (Fig. 3.5, left) is positively correlated with the first and second modes of summer (Fig. 3.5, central and right), reflecting the reversal of anomalies. Although the first spring mode and second summer mode are associated with ENSO, the first summer mode is not (Grimm and Zilli, 2009; Grimm, 2011).

The dipole-like pattern of the first modes in spring and summer (Fig. 3.5, left and central) and the tendency of reversing anomalies from one season to the other are associated with an anomalous vortex over southeast Brazil that favors moisture transport from the Atlantic, northern SA and Amazon Basin either into SESA, if it is anticyclonic (as in Fig. 3.5, left), or into central-east Brazil, if it is cyclonic (as in Fig. 3.5, central) (Grimm and Zilli, 2009). This vortex is also present in the second summer mode, related to ENSO, but in this case there is also ENSO-related circulation anomaly in

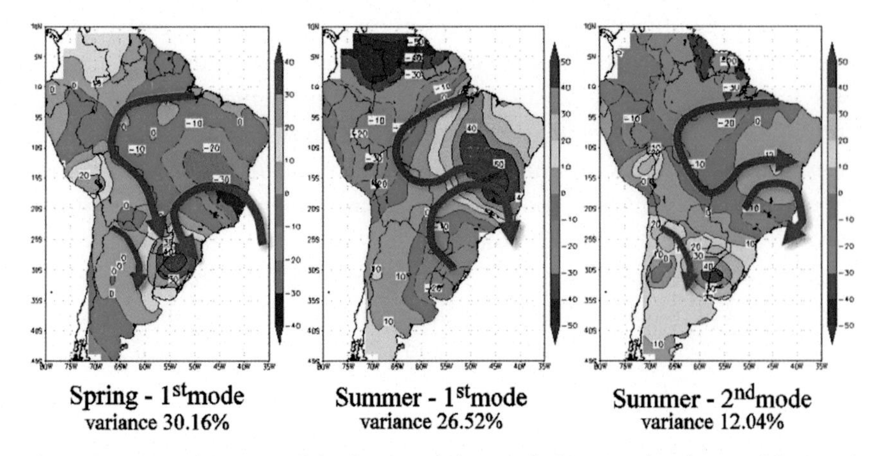

Figure 3.5 Spatial pattern of the first variability mode (empirical orthogonal function) of spring precipitation (left panel), and the first and second modes of summer (central and right panels). The *red lines* indicate the schematic moisture flux or low-level circulation anomalies. *(Adapted from Grimm, A.M., 2011. Interannual climate variability in South America: impacts on seasonal precipitation, extreme events, and possible effects of climate change. Stoch. Environ. Res. Risk Assess. 25, 537–554. https://doi.org/10.1007/s00477-010-0420-1.)*

southwestern SA that still favors anomalous convection in part of SESA, even if this vortex is cyclonic (as in Fig. 3.5, right).

According to the hypothesis proposed by Grimm et al. (2007) to explain the reversal of precipitation anomalies between spring and summer, the precipitation anomalies remotely forced in the spring produce soil moisture and near-surface temperature anomalies that alter the surface pressure and wind divergence over southeast Brazil, with the help of orographic effect of the mountains in this region. This reverses the anomalous vortex over southeast Brazil and the rainfall anomalies in summer. The SST anomalies off the southeast coast of Brazil in spring can enhance this effect. The influence of local processes is favored by the weakening of the ENSO teleconnection to subtropical SA through extratropical latitudes in peak summer (Grimm, 2003; Cazes-Boezio et al., 2003). The mountains in this region also contribute to anchor the SACZ because modeling simulations without the observed orography produce an SACZ displaced to the south (Grimm et al., 2007).

It is well known that ENSO episodes can display different characteristics in relation to the spatial distribution of the SST anomalies. The eastern pacific ENSO (with stronger SST anomalies in the Niño 3 region) and the central pacific ENSO (with stronger SST anomalies in the Niño 4 region) can produce significantly different impacts on precipitation in certain periods of the ENSO cycle and over certain regions of SA. During the monsoon season, the differences in the onset are not significant, but for the intensity of the anomalies (that are stronger in eastern ENSO). These differences increase in peak summer, when the reversal of anomalies is more evident in eastern ENSO and are even greater during the demise of the SAM (Tedeschi et al., 2015, 2016).

Besides the influence of the ENSO-related SST anomalies, there are other connections between SAM precipitation and SST anomalies, although it is not always easy to separate cause and effect in these statistical relationships. For instance, over the tropical Atlantic the precipitation regime is thermally direct, with SST modulating rainfall, whereas over the southwestern Atlantic it is thermally indirect, with SSTs modulated by atmospheric conditions, which determine cloudiness and therefore solar radiation reaching the surface. Therefore, in the tropical Atlantic, SST anomalies have a great influence on rainfall over northeastern SA, especially NE Brazil (e.g., Hastenrath and Heller, 1977; Moura and Shukla, 1981; Nobre and Shukla, 1996), especially when they have opposite signs north and south of the equator, enhancing an anomalous latitudinal SST gradient that produces latitudinal shifts of the ITCZ.

On the other hand, in southwestern Atlantic enhanced (suppressed) precipitation in the SACZ is related with colder (warmer) SST (Robertson and Mechoso, 2000; Doyle and Barros, 2002; Nobre et al., 2012). Grimm (2003) showed that January rainfall in central-east Brazil is positively correlated with November SST in the oceanic SACZ, off the southeast coast of Brazil, and negatively correlated with January SST in the same region. Anomalies of precipitation and circulation in the region, such as those associated with EN events in November, favor increased shortwave radiation and set up warm SST anomalies. On the other hand, enhanced convection and rainfall in January lead to negative SST anomalies. In this case, the atmosphere controls the ocean, but the warmer SST in November may help trigger the regional circulation anomalies that lead to enhanced precipitation in January. Although in the SACZ–SST relationship the SST anomalies seem to be a result of the convection anomalies in the SACZ, there are possible feedback mechanisms between SST and the atmosphere (Robertson et al., 2003; Chaves and Nobre, 2004; Grimm et al., 2007; De Almeida et al., 2007).

Besides the local influence in northeast SA and the SACZ region, the Atlantic SST anomalies are also important to the variability in other regions. For instance, the 2005 drought of southern Amazonia has been attributed to the tropical Atlantic variability (Zeng et al., 2008). Atlantic variability is also considered to play a role in interannual variations of extreme precipitation over southeastern Brazil (Muza et al., 2009). The possible influence of Southern Annular Mode (e.g., Vasconcellos and Cavalcanti, 2010) and the Indian Ocean Dipole (e.g., Chan et al., 2008) has also been suggested.

4.2.1 Influence of the Interannual ENSO-Related Variability on Extreme Events

There are significant ENSO signals in the frequency of extreme events over extensive regions of SA during EN and LN episodes (Grimm and Tedeschi, 2009; Tedeschi et al., 2015). Outstanding examples of this influence are given here for the beginning and peak of the summer monsoon season, focusing on the most populated regions, such as southeast and south Brazil. Although changes in intensity of extreme events show less significance and spatial coherence than in frequency, they are consistently combined with changes in frequency in several regions, especially in SESA (Grimm and Tedeschi, 2009).

At the beginning of the monsoon season EN reduces the frequency of extreme events in central-east Brazil, including SACZ (Fig. 3.6, left, for November (0), after Grimm and Tedeschi, 2009), whereas LN increases it (not shown). Yet this frequency is increased (reduced) in SESA during EN (LN) episodes. As a quantitative example, there are, on average, five times more extreme events during November in EN than in LN in the western part of south Brazil. The MCCs, that are already frequent in SESA in spring, have their frequency and intensity enhanced during EN episodes (Velasco and Fritsch, 1987).

The reason for this opposite behavior resides in the anomalous circulation patterns produced by EN (or LN) in spring. During EN, subsidence prevails over central Brazil, due to a perturbed Walker circulation, while a pair of cyclonic/anticyclonic nearly equivalent barotropic anomalies are produced by Rossby wave propagation over western/eastern subtropical SA (As shown for the first mode in spring in Fig. 3.5 left, but somewhat shifted to the west at the higher levels). At upper levels, these anomalies enhance the subtropical jet and the cyclonic advection over SESA, whereas at lower levels they produce moisture divergence from central Brazil and enhance the SALLJ with its northerly advection of moisture into SESA (Grimm, 2003). Opposite anomalies prevail during LN episodes (Grimm, 2004).

On the other hand, in peak summer (January) EN increases the frequency of extreme rainfall events in central-east Brazil (Fig. 3.6, right), and

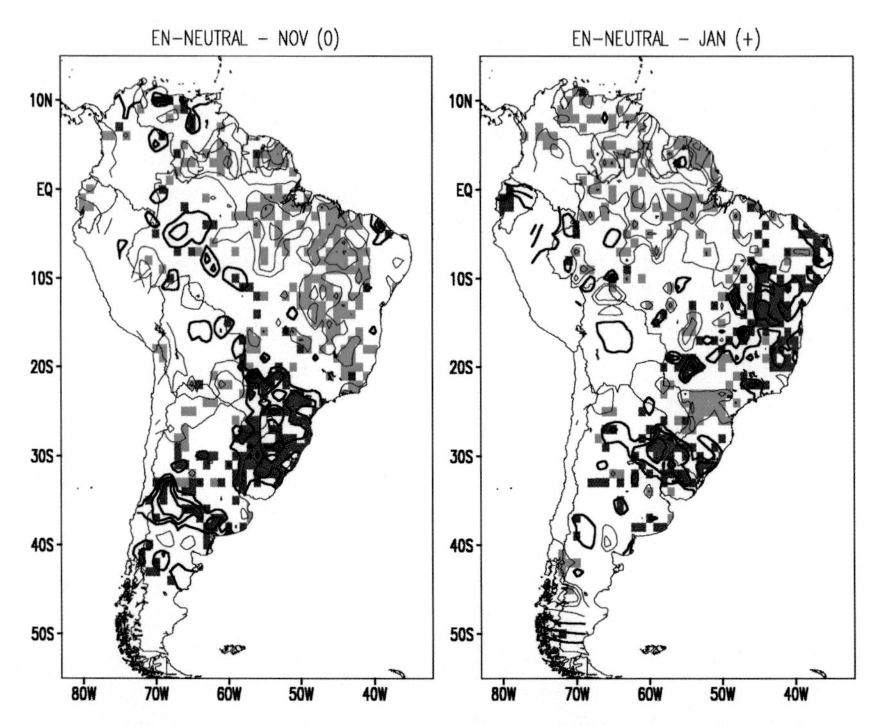

Figure 3.6 Differences between numbers of extreme rainfall events in El Niño years and neutral years in November (0) (left panel) and January (+) (right panel). Extreme events are defined as 3-day running mean precipitation above the 90th percentile. Contour interval is 1 event. Positive (negative) differences significant over the 90% confidence level are represented by red (blue) color. (0) and (+), same as in Fig. 3.4. *(Adapted from Grimm, A.M., Tedeschi, R.G., 2009. ENSO and extreme rainfall events in South America. J. Clim. 22, 1589–1609.)*

extreme precipitation is favored in the SACZ (Carvalho et al., 2004; Grimm and Tedeschi, 2009). Yet in SESA the number of extreme events is reduced in the northernmost part of southern Brazil but remains enhanced in northern Argentina. Impacts during LN are approximately opposite. The reversal of ENSO impact on central-east Brazil from November to January is connected to the inversion of the regional circulation anomaly over southeast Brazil, as explained in the previous section.

ENSO-related significant changes in the frequency of extreme rainfall events are more extensive than changes in monthly rainfall because ENSO influence seems to be stronger on the extreme ranges of daily precipitation (Grimm and Tedeschi, 2009), as can be seen from the comparison of Figs. 3.4 and 3.6. This is an important aspect because the most dramatic consequences of climate variability are due to extreme events. The ENSO influence also extends to extreme drought events or monsoon failures, as can be seen during January of LN events in central-east Brazil (Grimm, 2004).

Different types of ENSO, with major SST anomalies in the central or eastern Pacific, can influence differently the frequency of extreme events (Tedeschi et al., 2015). When the monthly or seasonal atmospheric anomalies associated with a category of ENSO episode are similar to the anomalies associated with extreme precipitation in a certain region, then a significant enhancement of the frequency of extreme events is observed in that region during this type of episode.

4.3 Interdecadal Variability

Interdecadal oscillations can produce significant impacts on the frequency of extreme precipitation events, besides producing the longest lasting (and therefore worst) droughts. The SAM interdecadal variability is significantly affected by the main global SST interdecadal oscillations, the Interdecadal Pacific Oscillation (IPO) and the Atlantic Multidecadal Oscillation (AMO), besides being influenced by other more regional climatic oscillations, oceanic and atmospheric, represented by several climatic indices. A comprehensive continental-scale analysis was carried out by Grimm and Saboia (2015), disclosing interdecadal variability modes for spring and summer and their connections to well-known climatic indices and SST anomalies. The first mode for both spring and summer is dipole-like, displaying opposite anomalies in central-east and southeast SA, similar to the first interannual mode. They also tend to reverse polarity from spring to summer. Yet the mode that affects the core monsoon region in central Brazil and central-northwestern Argentina shows persistence of anomalies from one season to the other, contrary to the first mode. Grimm et al. (2016) repeated the analysis with a longer period of data, obtaining modes consistent with those previously obtained, confirming their robustness. They also clarified the impact of interdecadal oscillations on the frequency of extreme precipitation events over SA in the monsoon season and determined the influence of these oscillations on the daily precipitation frequency distributions.

Some of these results are illustrated in Fig. 3.7, for the first interdecadal variability modes for spring and summer. Their spatial distribution (first row) shows that the precipitation anomalies are similar in both modes, with small shifts, whereas their temporal evolution (second row) and their associated SST anomalies (third row) show nearly similar anomalous patterns but with opposite signs. This means that for similar large-scale SST anomalies, these modes tend to invert precipitation anomalies from spring to summer.

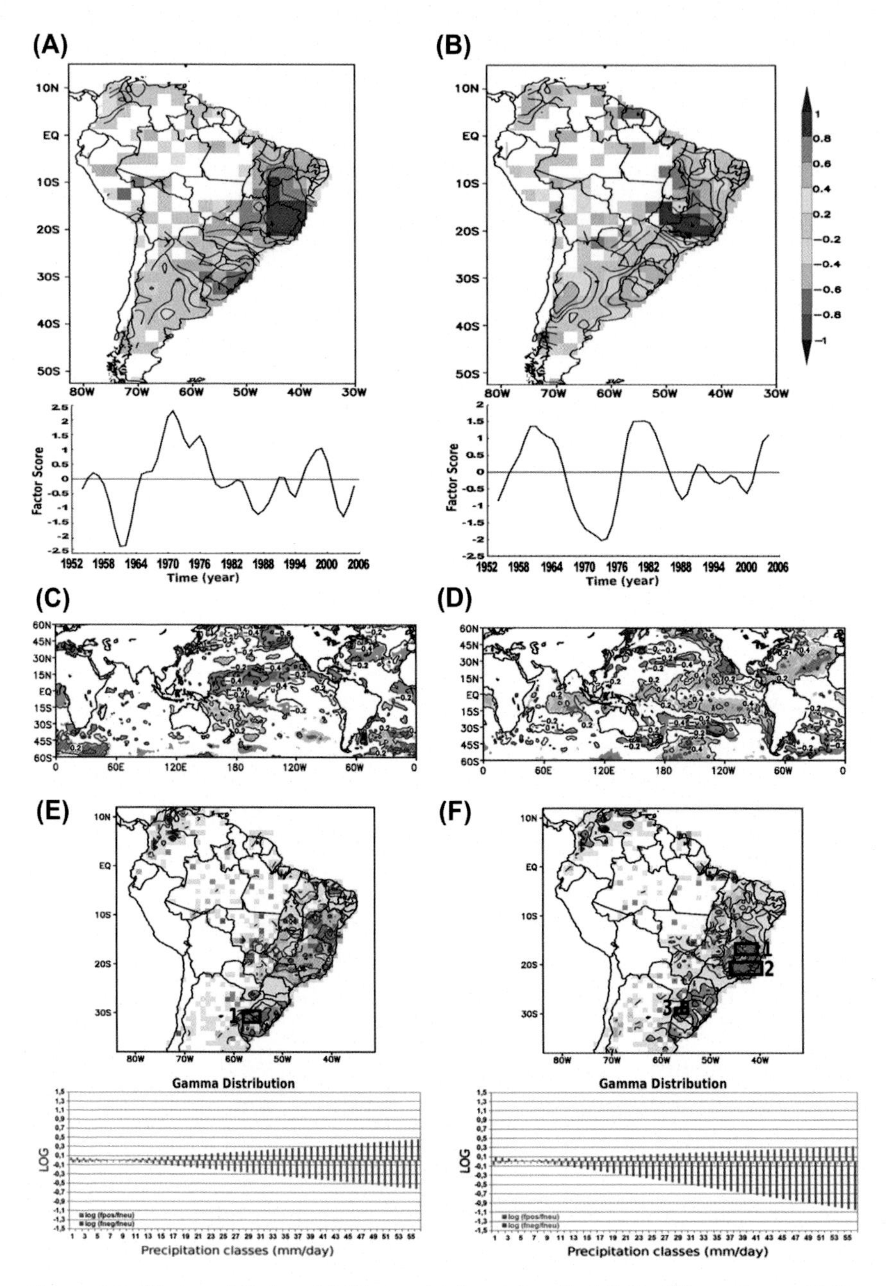

Figure 3.7 (First row) First variability modes for spring (A, 17.5% of the variance, 1950–2009) and summer (B, 15.4% of the variance, 1950–2009); (second and third rows) differences between anomaly composites in their positive and negative phases for sea surface temperature (C and D) and monthly number of extreme rainfall events (E and F); (fourth row) logarithm of the ratio fpos/fneu, in *red bars*, and fneg/fneu, in *blue bars*, for each daily precipitation interval in the box 1 of each map in panels E and F (fpos = frequency in positive phase; fneg = frequency in negative phase; fneu = frequency in neutral phase). *(From Grimm, A.M., Laureanti, N.C., Rodakoviski, R.B., Gama, C.B., 2016. Interdecadal variability and extreme precipitation events in South America during the monsoon season. Clim. Res. 68, 277–294. https://doi.org/10.3354/cr01375.)*

The greatest remote influence for this mode is from the IPO, although there are also connections with other indices. Other interdecadal modes are connected predominantly with AMO, the tropical North and South Atlantic, the North Atlantic Oscillation (NAO), and the Southern Annular Mode (Grimm and Saboia, 2015).

4.3.1 Influence of Interdecadal Variability on Extreme Events

The anomalies of the extreme event frequency associated with the first interdecadal modes of spring and summer precipitation described above display spatial patterns very similar to those of the corresponding modes, which is a good indication of the great contribution of interdecadal variability to extreme rainfall (Fig. 3.7, fourth row). In the regions most affected by the interdecadal variability, the frequency distributions (pdf) of daily rainfall were analyzed for positive and negative phases of these modes, and statistical tests showed that daily precipitation from opposite phases of the interdecadal oscillations pertains to different frequency distributions, proving that they are significantly altered by the interdecadal variability. Furthermore, the influence of interdecadal variability modes on the probability distributions of daily precipitation is relatively greater on the extreme ranges of daily rainfall than on the ranges of moderate and light rainfall (Fig. 3.7, bottom most row).

5. EXTREME EVENTS DURING THE MONSOON IN DIFFERENT REGIONS OF SOUTH AMERICA AND CASE EXAMPLES

5.1 Amazon

Amazon and NE Brazil extreme events have been connected to SST fluctuations over the surrounding ocean basins, Pacific and Atlantic, because the tropical atmosphere is heavily influenced by oceanic and land surface conditions. It is necessary to take into account that these two basins are correlated simultaneously or with a lag, with the Pacific influencing the Atlantic SST. The Pacific influence is mainly due to the ENSO episodes, through perturbations to the east–west Walker circulations (e.g., Grimm and Ambrizzi, 2009). On the other hand, northern and southern tropical Atlantic (NTA and STA) SSTs influence the north–south migration of the ITCZ following warmer SST, and north–south divergent circulation (Hadley circulation).

Over the Amazon, the ENSO influence is generally stronger in the wet season and especially over the eastern Amazon (more rainfall during LN, less during EN), whereas the Atlantic influence is the strongest from NTA

Table 3.1 Major floods of the Negro River in Manaus (http://www2.ana.gov.br/Paginas/imprensa/noticia.aspx?id_noticia=12489)

	Highest water levels (m)	Day/Month	Year
1	29.97	29/05	2012
2	29.77	01/07	2009
3	29.69	09/06	1953
4	29.61	14/06	1976
5	29.43	07/06	2014
6	29.42	03/07	1989
7	29.35	18/06	1922
8	29.33	14/06	2013
9	29.30	24/06	1999
10	29.17	14/06	1909

and over the southern part of the Amazon basin during the Amazon's dry season. On removing the ENSO lagged influence on the North Atlantic SST, it still shows a significant simultaneous correlation with Amazon rainfall (Yoon and Zeng, 2010). Generally, when there is a major ENSO event, the extremes in the Amazon are dominated by the Pacific Ocean influence. When ENSO is weak, but NTA displays SST anomalies, the Atlantic influence predominates. In general, severe droughts are associated with both warm eastern Equatorial Pacific and warm NTA SSTs. Floods were generally associated with a warmer STA without LN (such as in 1953/54 and 2008/09) or with LN episodes, as in 8 of the 12 Negro River (Amazon tributary) floods recorded in Manaus since 1903 (Table 3.1), and droughts were registered during EN events (as in 1964, 1983, 1998, 2016).

5.1.1 Record Drought in Western Amazon in 1926 and Other El Niño-Related Droughts in Amazonia

The 1926 drought year was the most extreme dry period in the century-long record of discharge from the western subbasin of the Amazon, registered at Manaus (3.11°S;60.0°W), on the Rio Negro, and also at Manacapuru (near Manaus) (Williams et al., 2005). There was a pronounced 1925–26 EN episode (one of the six strongest between 1871 and 1949, according to https://www.esrl.noaa.gov/psd/enso/mei.ext/#data), and therefore, dryness in the Amazon basin is expected because of anomalous subsidence in over the region (Section 4.2). Consistently, the regional rainfall anomaly for 1926 was negative over the western Amazon subbasin, but it was positive further east, over NE Brazil, where it is also frequently negative during EN episodes. The reason for this difference is that the precipitation anomalies of the northeast are also associated with anomalous SST in the tropical Atlantic. Although warmer SST in NTA, and therefore unfavorable condition for precipitation in NE Brazil, is common during many EN events, it did not exist during the EN of 1926; on the contrary, the STA was anomalously warm in 1925–26. This condition does not have much impact on western Amazon but favors more rainfall over NE Brazil.

Other extreme EN-related droughts occurred in Amazon in 1982–83 and 1997–98, and recently in 2015–16, which extended over central-west Brazil; this was also associated with record outbreaks of forest fires in Brazil during 2015–16. In addition, the Brazilian grain harvest in 2015–16 saw a reduction of 10.3% (or 21.4 million tons) compared with 2014–15. Besides EN, the NTA warm anomalies also contributed to the droughts of 2005 and 2010 in Amazon.

5.1.2 Floods in Western Amazon in 2012–14 and Drought in Northeast Brazil

In 2011–12 there was an LN episode (Fig. 3.8, upper panel, blue line) which, due to perturbations of the Walker circulation produced by anomalously cold waters in eastern Pacific, favored ascending motion and enhanced convection especially over northwestern Amazon, producing excess rainfall and record floods (Table 3.1). The anomalous convection in this region produced increased subsidence over NE Brazil. Moreover, the predominant negative SST anomalies in the STA (Fig. 3.8, bottom panel, blue line) favored the intensification and northward shift of the South Atlantic high pressure. These effects resulted in moisture transport and convergence into the western Amazon, producing wet condition during the monsoon season, while in northeast dry conditions predominated in the fall rainy season of 2012 (Marengo et al., 2013). An opposite situation occurred during EN 1925/26, when a severe drought affected Amazonia, and NE Brazil was wet (as described in Section 5.1.1), although frequently an EN would produce dry conditions there (Williams et al., 2005).

Although the 2012 anomalous patterns were not observed in other previous LN episodes with enhanced precipitation in Amazon (such as 1989, 1999), LN is the main cause of abundant rainfall in Amazon since 12 floods in the Rio Negro (one of the main tributaries of the Amazon River) recorded at Manaus (capital of the Brazilian State of Amazon) since 1903, 8 occurred during LN events (Table 3.1). The average anomalous patterns for LN events (not shown) show upward flow in tropical SA, east of the Andes, from Amazonia to NE Brazil, with enhanced upper-level divergence from the beginning of summer. However, during 2012 the upward flow and

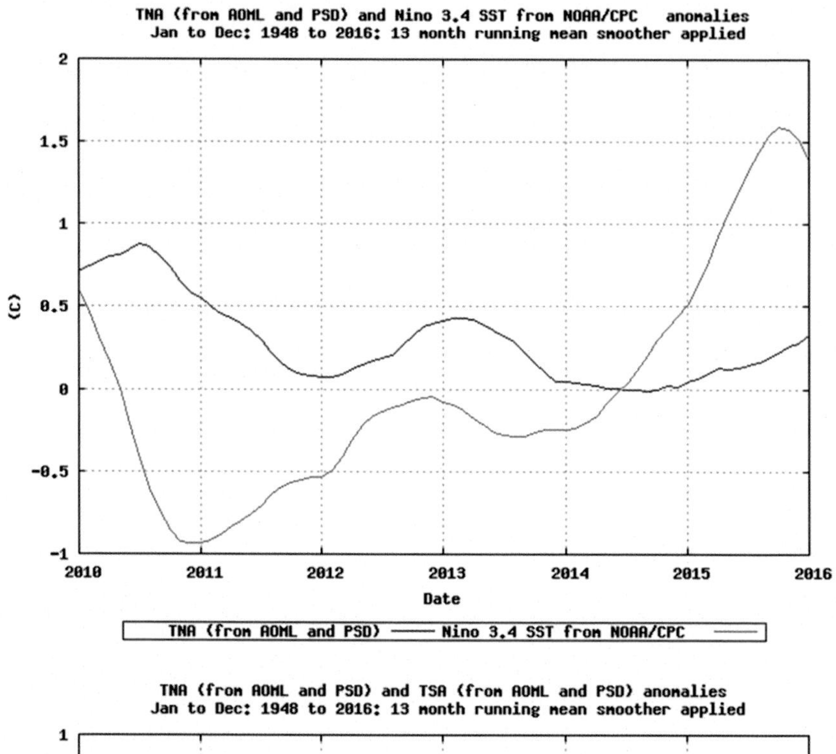

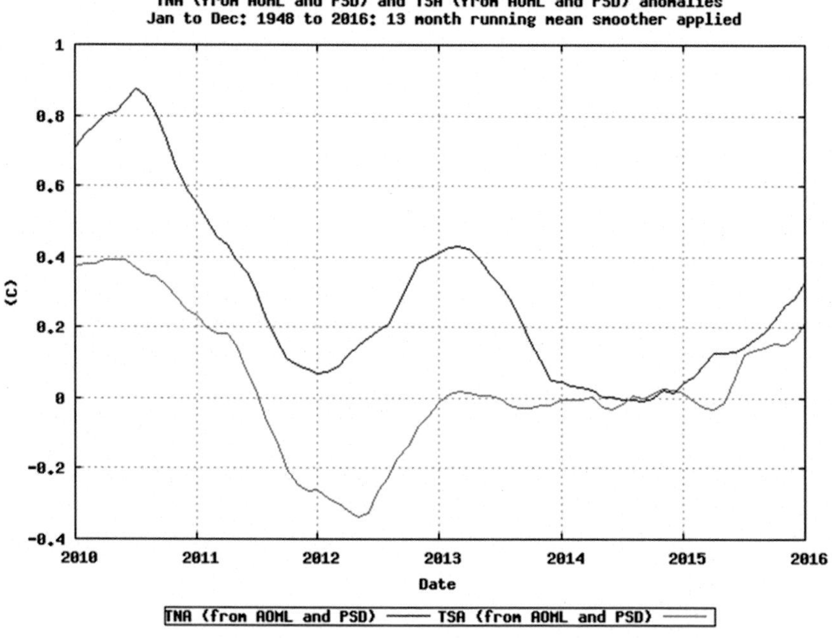

Figure 3.8 (Bottom panel) NTA and STA SST anomalies; (Upper panel) NTA and Niño 3.4 SST anomalies. The series are smoothed with a 13-month running mean. *NTA*, northern tropical Atlantic; *SST*, sea surface temperature; *STA*, southern tropical Atlantic. *(From: ESRL/PSD/NOAA.)*

upper-level divergence was much stronger during summer and fall over western Amazonia, whereas downward motion and subsidence prevailed over NE Brazil in summer and fall 2012.

From 2012 to 2014, wet conditions continued to be predominant over northwest Amazon (Table 3.1 shows that 3 of the 10 highest levels of the Negro River in Manaus since 1902 happened in 2012, 2013, and 2014); on the other hand, dry conditions prevailed over NE Brazil because the main conditions favorable for this combination were still present; specifically, negative Niño 3.4 anomalies favoring Amazon enhanced precipitation, positive NTA, and negative or nearly zero STA anomalies (Fig. 3.8, bottom panel). In 2011, when LN was at its peak and STA was also positive, rainfall was above normal in both regions (Amazon and NE Brazil).

5.2 Northeast Brazil

As in the Amazon, NE Brazil precipitation anomalies and its extreme events have been connected to SST fluctuations over the Pacific and Atlantic oceans, which can be interrelated (e.g., Hastenrath and Heller, 1977; Covey and Hastenrath, 1978; Moura and Shukla, 1981; Nobre and Shukla, 1996). The Pacific influence is mainly due to the ENSO episodes, through perturbations of the east–west Walker circulations, whereas the NTA and STA SSTs influence the north–south migration of the ITCZ following warmer SST. The position of the ITCZ is responsible for the rainy season in NE Brazil.

Over the NE Brazil, particularly its northern part, LN (EN) episodes tend to produce more (less) than normal precipitation. Besides, excess (deficient) rainfall is associated with anomalously cold (warm) SST in the NTA and anomalously warm (cold) SST in the STA. The same kind of influence from the Pacific and Atlantic oceans occurs at interdecadal timescales (Grimm and Saboia, 2015). The southern part of NE Brazil shows different influences on variability: the NTA has less influence, and the correlation between precipitation and STA is negative. A significant influence over this part also comes from the NAO (Grimm and Saboia, 2015).

Besides migration and other social-economic impacts, droughts in NE Brazil caused many deaths till the 1970s. For instance, between 1877 and 1913, estimated 2 million people died because of droughts, and in the prolonged drought 1979–84, estimated 3.5 million people died because of malnutrition. Although deaths are not registered anymore, there are still serious economic and social impacts caused by droughts in NE Brazil.

5.2.1 The Record Drought in Northeast Brazil in 2012–16

As described in Section 5.1.2, from 2012 till 2014 there was enhanced convection over northwest Amazon, producing subsidence over NE Brazil and consequently dry conditions. Besides, other favorable conditions for drought were present: positive NTA and negative or nearly zero STA anomalies (Fig. 3.8, bottom panel). In 2015 and 2016, when Pacific conditions turned to EN and NTA continued warmer than normal (Fig. 3.8, upper panel), the drought continued over NE Brazil, extending also to the Amazon. Therefore, the 2012–16 drought in NE Brazil is already considered the most severe in the last 100 years, since the 1915 drought. Agriculture was severely affected, with losses above US$ 23 billion just between 2013 and 2015. Cities are at risk of water supply collapse, and the metropolitan region of Fortaleza, capital of Ceará, is threatened with rationing. The Castanhão dam, from which the water supplying the nearly 4 million inhabitants of the capital of Ceará comes from, was at just over 5% of its capacity at the end of 2016.

5.2.2 Floods in Northeast Brazil in 2009

Floods in northern NE Brazil are favored during LN events, warm STA, or cold NTA, but MJO can give a decisive contribution, especially during phases 2 and 3 for March–April–May (Alvarez et al., 2016). In April–May 2009 intense floods occurred in NE Brazil (as well as in Amazon), and as a consequence, resulting in 49 deaths and 408,000 people rendered homeless in this region, besides other damages to infrastructure. At the same time, there was a strong MJO episode, with active convection over NE Brazil, besides a higher SST in STA than in NTA. Although the Niño 3.4 SST anomaly was negative, favoring enhanced rainfall, the conditions for a LN were not fulfilled.

5.3 Central-East South America

The Amazon basin displays the highest precipitation rates corresponding to the 95th percentile in austral summer monsoon, followed by the SACZ region and the middle-lower Parana/La Plata basin (parts of southern Brazil, northern Argentina, and southern Paraguay) (Hirata and Grimm, 2015, their Fig. 1). We will focus, as in Hirata and Grimm (2015), on these latter regions because they are the most populous, and extreme events in these regions have greater social and economic impact. The extreme events in these regions are produced mainly by frontal systems and the SACZ (Seluchi and Chou, 2009; Lima et al., 2010; Cavalcanti, 2012). The main feature of rainfall variability in these regions is a dipole pattern between central-east SA (including the SACZ) and southeast SA, which appears in all timescales, from intraseasonal to interdecadal (Nogues-Paegle and Mo, 1997; Nogues-Paegle et al., 2000;

Cunningham and Cavalcanti, 2006; Grimm and Zilli, 2009; Grimm, 2011; Grimm and Saboia, 2015), as shown in Section 4. This dipole pattern is partly related to synoptic waves, and extreme events are associated with the phasing between these waves and the intraseasonal variability, especially in the SACZ region (Hirata and Grimm, 2015). MJO convective activity in the western Pacific (phases 6 and 7) led 31 out of 81 extremes over the SACZ region analyzed by Hirata and Grimm (2015) by nearly 10 days. This means that the extremes would preferentially occur on phases 8 and 1.

In the SACZ, heavy rainfall is also more frequent under EN conditions (Grimm and Tedeschi, 2009). EN and LN have very significant impact on the frequency of extreme events in SACZ and the subtropical plains (middle-lower Parana/La Plata basin). During EN, there is a tendency for less (more) than normal frequency of extreme rainfall events in central-east Brazil (including SACZ) in the onset (peak) of the monsoon season, whereas in SESA opposite tendencies prevail (Fig. 6, Grimm and Tedeschi, 2009; Carvalho et al., 2004). LN exerts fairly opposite type of influence (Grimm and Tedeschi, 2009).

The synoptic evolution of extreme rainfall events in the SACZ during ENSO neutral austral summers analyzed by Hirata and Grimm (2015) shows that 4 days before extreme rainfall, positive sea level pressure (SLP) anomalies are present off SESA, around 30°S, and low-level cyclonic wind and convergence anomalies appear in the SACZ region, around 45°W, 20°S (Fig. 3.9A). At the same time, 500 hPa omega anomalies indicate the beginning of

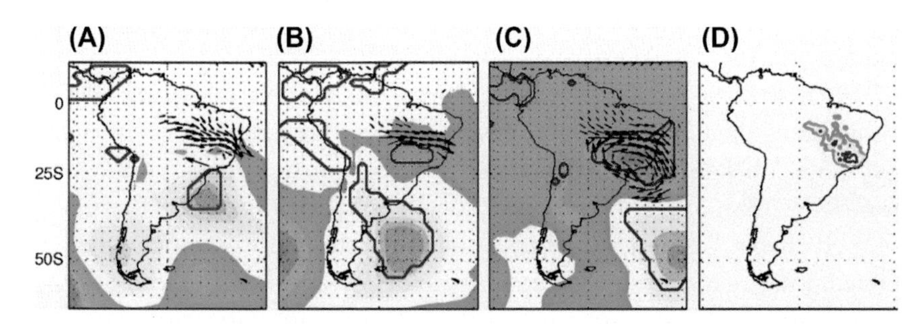

Figure 3.9 Composites of synoptic evolution of extreme rainfall events in the SACZ region for days (A) −4, (B) −2, (C) 0 during neutral ENSO summers. *Shading* represents SLP anomalies, *arrows* represent 850 hPa anomalous wind vectors (only vectors significant at 95% are plotted), and *red contours* represent SLP significance (95%). Panel (D) displays the area-averaged 95th percentile rainfall rate (green), the 95th percentile +5 mm day^{-1} (yellow), and the 95th percentile +10 mm day^{-1} (red) on composite day 0. *ENSO*, El Niño–Southern Oscillation; *SACZ*, South Atlantic Convergence Zone; *SLP*, sea level pressure. *(From Hirata, F.E., Grimm, A.M., 2015. The role of synoptic and intraseasonal anomalies in the life cycle of summer rainfall extremes over South America. Clim. Dyn. https://doi.org/10.1007/s00382-015-2751-6.)*

ascending motion. The negative SLP anomalies and ascending motion are initially stronger in the north of the SACZ region, but 2 days before the extreme events the SLP anomalies extend westward and southward and deepens around 45°W, 20°S (Fig. 3.9B), and vertical motion also spreads in the region (not shown). Positive SLP anomalies weaken around 25°S but increase over southwestern Atlantic, off southern SA and move eastward (Fig. 3.9B and C).

At higher levels, an extratropical wave train extends from the Pacific into subtropical SA, with a nearly barotropic structure in the extratropics, suggesting enhanced extratropical wave activity and eastward propagation in the South Pacific (Fig. 3.10). The approach of the extratropical wave, with the enhancement of the positive SLP anomalies over southern SA, seems to deepen the negative SLP anomalies between 15°S and 25°S (Fig. 3.10B) and extends the vertical ascending motion southward, over the SACZ region, with subsidence to the southwest (not shown). When the negative 200 hPa height anomalies reach a region of climatological negative zonal stretching deformation just south of the strong convection center between 20°S and 30°S their propagation speed is reduced (Fig. 3.10B and C). Therefore the anomalous cyclonic circulation around 22°S, 45°W (Fig. 3.9C) and the ascending motion (not shown) are strengthened over SACZ, whereas the extratropical anomalous high moves eastward with the extratropical wave (Fig. 3.10C). During this synoptic evolution, the 1000–500 hPa layer anomalous thickness (proportional to the layer temperature) is significantly negative south of the 850 hPa cyclonic flow, with no anomalies to the north (not shown), so that the increasing thickness gradient enhances the potential for the development of atmospheric perturbations.

When the wave train is separated into its synoptic and intraseasonal components (Fig. 3.10D–F), it is possible to see that extremes occur when they both are significant and coincide on the northern flank of the subtropical jet (Fig. 3.10F). The intraseasonal anomalies on the northern and southern flanks of the jet tend to reduce the wind between them, further reducing the negative zonal stretching deformation in the region, and the wave accumulation and convective activity (Hirata and Grimm, 2015).

5.3.1 The Floods in Southeast Brazil in 1979

These floods were the worst ones that ever occurred in southeast Brazil, particularly in the states of Minas Gerais and Espírito Santo, due to heavy rainfall between January and February 1979. They resulted in 48,000 becoming homeless, 74 deaths, and ~4500 damaged houses. There was also much anomalous rainfall in central Brazil, following the SACZ cloud band.

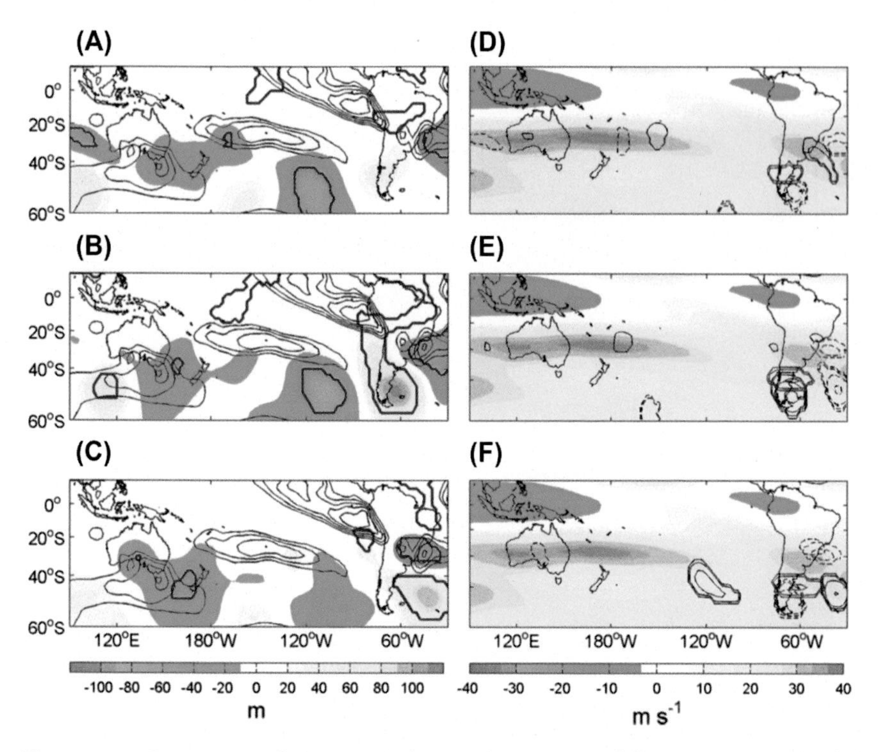

Figure 3.10 Composites of synoptic evolution of extreme rainfall events in the SACZ for days −4 (A and D), −2 (B and E), and 0 (C and F), during neutral ENSO summers. On panels A–C, *shading* represents 200 hPa geopotential height anomalies, with *red contours* indicating significant anomalies at 95%. *Blue contours* represent summertime (December–January–February) 200 hPa negative zonal stretching deformation ($\partial\overline{U}/\partial x < 0$) at $0.2 \times 10^{-7}\,s^{-1}$ intervals. On panels D–F, *shading* represents summertime 200 hPa $\overline{U}$, and the *contours* represent filtered 3- to 10-(blue) and 20- to 90-day (red) 200 hPa geopotential height anomalies at 10 m intervals, with *solid (dashed) contours* indicating positive (negative) anomalies. *ENSO*, El Niño–Southern Oscillation; *SACZ*, South Atlantic Convergence Zone. *(From Hirata, F.E., Grimm, A.M., 2015. The role of synoptic and intraseasonal anomalies in the life cycle of summer rainfall extremes over South America. Clim. Dyn. https://doi.org/10.1007/s00382-015-2751-6.)*

In certain areas, the January precipitation exceeded 600 mm, with anomalies higher than 300 mm (Fig. 3.11A). The heavy precipitation affected a more extensive region than that in January 2011, when flash floods and landslides happened in the highlands of the Rio de Janeiro state, killing more than 900 people and displacing nearly 35,000 people.

During the rainiest period, in January 1979, there was a strong MJO episode in phase 8, which is also visible in the OLR anomalous distribution during the period January 12–February 06 (Fig. 3.11B). The anomalous fields

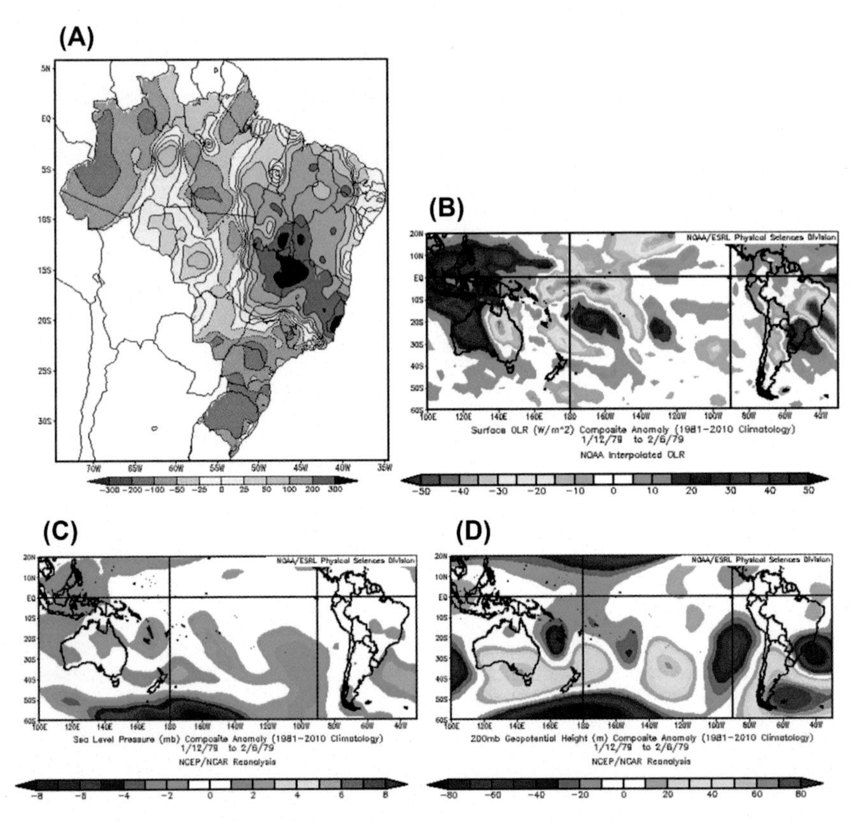

Figure 3.11 (A) Anomalous precipitation in January 1979; (B) anomalous outgoing longwave radiation; (C) anomalous sea level pressure; and (D) anomalous 200 hPa geopotential height during the rainiest period (January 12–February 06, 1979). *((A) From CPTEC/INPE; (D) From ESRL/PSD/NOAA.)*

for the period January 12–February 6 have similar patterns to those composited by Hirata and Grimm (2015). For instance, there are negative SLP anomalies in the SACZ region and positive ones over southwestern Atlantic (see Figs. 3.9C and 3.11C). The wave train in 200 hPa geopotential height over the southwestern Pacific/SA is also similar (cf. Figs. 3.10C and 3.11D).

5.3.2 The Floods in Southeast Brazil in 2013

In December 2013, southeast Brazil endured weeks of torrential, record-setting downpours, comparable only to the extremes in 1979. For example, one of the affected cities received over 850 mm in 27 days. Dozens of cities declared state of emergency, and by the end of the month, devastating flash floods and landslides had killed 45 people and left 70,000 homeless. The cost of infrastructure repairs in the State of Espírito Santo alone was estimated as US $230 million.

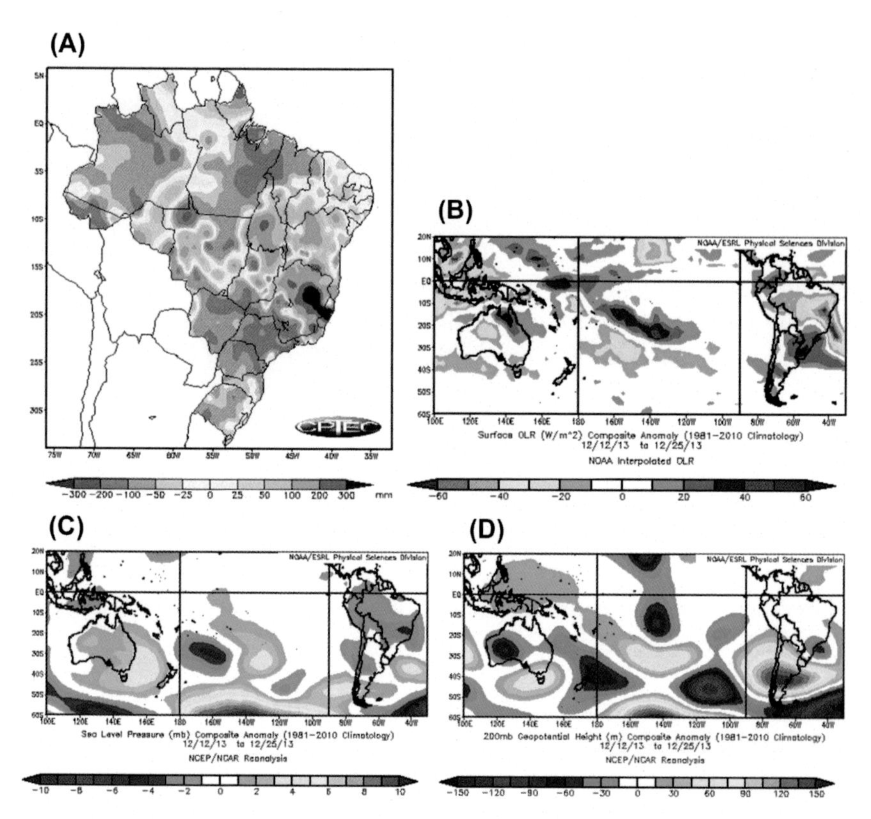

Figure 3.12 (A) Anomalous precipitation in December 2013; (B) anomalous outgoing longwave radiation; (C) anomalous sea level pressure; and (D) anomalous 200 hPa geopotential height during the rainiest period (December 12–December 25, 2013). *((A) From CPTEC/INPE; (D) From ESRL/PSD/NOAA.)*

Although there was no MJO in phases 8 or 1, which are favorable for excess rainfall in the SACZ (it was in phases 3, 4, 5), there were positive convective anomalies (negative OLR anomalies) in the subtropical southern Pacific, east of the date line, which are not usual in these MJO phases (but are common in phases 8 and 1), and that are able to trigger circulation anomalies that can produce enhanced rainfall in the SACZ (Grimm and Silva Dias, 1995; Grimm, 2018). The atmospheric fields in these extreme events are similar to those composited by Hirata and Grimm (2015) in southeastern Pacific and SA (cf. Figs. 3.9C and 3.12C; Figs. 3.10C and 3.12D).

5.3.3 The Drought and the Energy Crisis in the 2000–01 Summer

One of the worst and most striking droughts in central and east Brazil happened in the summer 2000–01. The rainfall deficits during summer 2001 reduced the river flow in several regions (northeast, central-west, and southeast Brazil), strongly affecting the hydroelectric power production. The government imposed energy conservation measures to avoid blackouts. This was a typical LN summer, in which the main circulation anomalies and impacts expected for this season were observed (see Section 4.2). In November 2000, there was above normal rainfall over central-east Brazil (Fig. 3.13, upper panel left), with low-level circulation anomalies fairly opposite to those in Fig. 3.5 (left, which depicts an EN situation), with cyclonic anomaly over central-east Brazil (not shown). Yet in January 2001, there was a strong drought (Fig. 3.13, upper panel right), with an anticyclonic anomaly over this region (Fig. 3.13, middle panel), fairly opposite to the anomaly in Fig. 3.5 (right, which depicts an EN situation), and an anomalous upper level cyclonic anomaly (Fig. 3.13, bottom panel). Therefore, the background circulation for drought was established over central and east SA, favoring synoptic features associated with droughts. A greater than normal activity of upper-level cyclonic vortices, with these systems penetrating deep within the continent, was observed in the summer 2000/01 (Cavalcanti, 2012).

5.3.4 The Dryness in Central South America in the 1960s and Early 1970s

During the period from 1960 to 1973, the level and extent of the Pantanal (central South American wetlands) in central-west Brazil, near Paraguay and Bolivia, showed anomalously low values (Hamilton et al., 1996, 2002). These prolonged dry conditions are related with interdecadal oscillations. The interdecadal modes with strongest influence in this region in summer (rainy season in the region, Fig. 3.1) are AMO and IPO (or its north Pacific component, Pacific Decadal Oscillation [PDO]) (Grimm and Saboia, 2015; Grimm et al., 2016). In this period the SST global interdecadal modes associated with IPO and PDO oscillations had negative factor scores, and the SST mode associated with the AMO had positive ones (Parker et al., 2007). This combination favored dry conditions in the region.

5.4 Middle-Lower Parana/La Plata Basin

The middle-lower Parana/La Plata basin, between 23°S and 38°S, has precipitation variability different and often opposite to that in the upper Parana/La Plata basin, which is situated in central-eastern SA. Although the middle-lower basin is not in the tropical belt, its regime is monsoonal, especially in its western part (Fig. 3.1), and therefore, this region is also mentioned here.

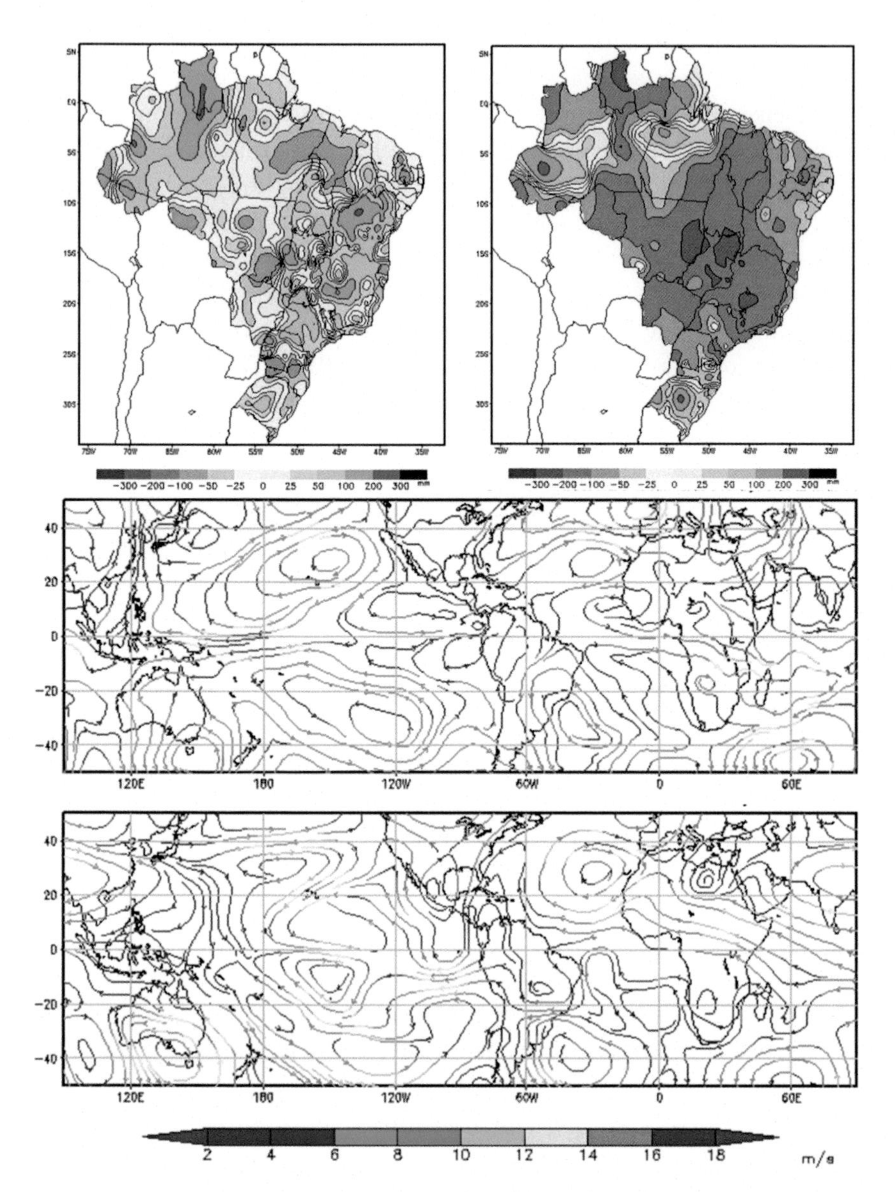

Figure 3.13 Upper panels: precipitation anomalies in November 2000 (left) and January 2001 (right). Lower panels: streamline anomalies at 850 hPa (middle) and 200 hPa (bottom) in January 2001. *(From CPTEC/INPE.)*

Although there is a dipole-like behavior in precipitation variability in all timescales between central-east and SESA, which includes the middle-lower Parana/La Plata basin, as described in previous sections, the systems causing extreme events in these regions are not the same. Frontal systems are important in both regions, but in central-east Brazil the SACZ is also very important.

Extreme events in the middle-lower Parana/La Plata River basin are even more strongly associated with extratropical wave activity/frontal systems (Teixeira and Satyamurty, 2007) than in the SACZ region, and the evolution of these events is also influenced by intraseasonal anomalies surrounding the subtropical jet (Hirata and Grimm, 2015). The intraseasonal variability influencing both regions may be connected with the MJO (period around 40 days) or another period (around 20 days) produced by a different mechanism (Nogues-Paegle et al., 2000). While for the SACZ region the MJO is more Influential, the higher frequency mode contributes more to the intraseasonal variability in the middle-lower Parana/La Plata basin.

A great portion of the excessive precipitation in the region is due to MCSs, especially its largest subclass, the mesoscale convective complexes (MCCs), which develop more frequently during the warm season over the region, and contributes to more than 40% of the rain (Rasmussen et al., 2016). According to the Tropical Rainfall Measuring Mission (TRMM) data, SESA contains the most extensive "hot spot" of the most intense thunderstorms on Earth (Zipser et al., 2006). This preference for the warm season is probably connected to favorable synoptic and mesoscale aspects (Velasco and Fritsch, 1987; Laing and Fritsch, 2000; Salio et al., 2007; Anabor et al., 2008; Durkee et al., 2009). Low static stability is also common during this season. Steep lapse rates develop from low-level heat and moisture supply from the Amazon basin favored by the Chaco low (especially in LLJ episodes) and cold-air advection from the upstream Andes. Also the mean position of the subtropical jet relative to the low-level heat and moisture supply is important. Upper-level disturbances in this jet provide (or prevent) enhanced ascent, and therefore, there are synoptic features associated with opposite phases of climate variability that can enhance or decrease the frequency of MCSs. For instance, the enhanced precipitation in the region during spring of EN episodes is due to the increased frequency of MCSs (Velasco and Fritsch, 1987).

5.4.1 The Floods in Southern Brazil, Southern Paraguay, and Northern Argentina in December 2015

The 2015–16 EN episode was a typical strong event with highest SST anomalies in the eastern Pacific, and its impacts can also be considered typical. Intense floods occurred in southern Brazil in the spring 2015, starting in October. The three states of southern Brazil, as well as parts of Paraguay, northern Argentina, and Uruguay were heavily affected and within the southernmost state of Brazil, Rio Grande do Sul, 43 cities were affected, 25 in emergency state, and almost 3000 families left their houses.

In Santa Catarina 30,000 people were affected. In Parana, besides harm to population during extreme events in the southwest region, the Binational Itaipu hydroelectric power plant warned against floods in west Parana and Paraguay and opened its spillways. About 9 million $L\,s^{-1}$ were released because of the strong precipitation in the Parana River basin. On the average, December was the month with more rainfall in the region (Fig. 3.14A). The upper-level circulation anomalies were very favorable to

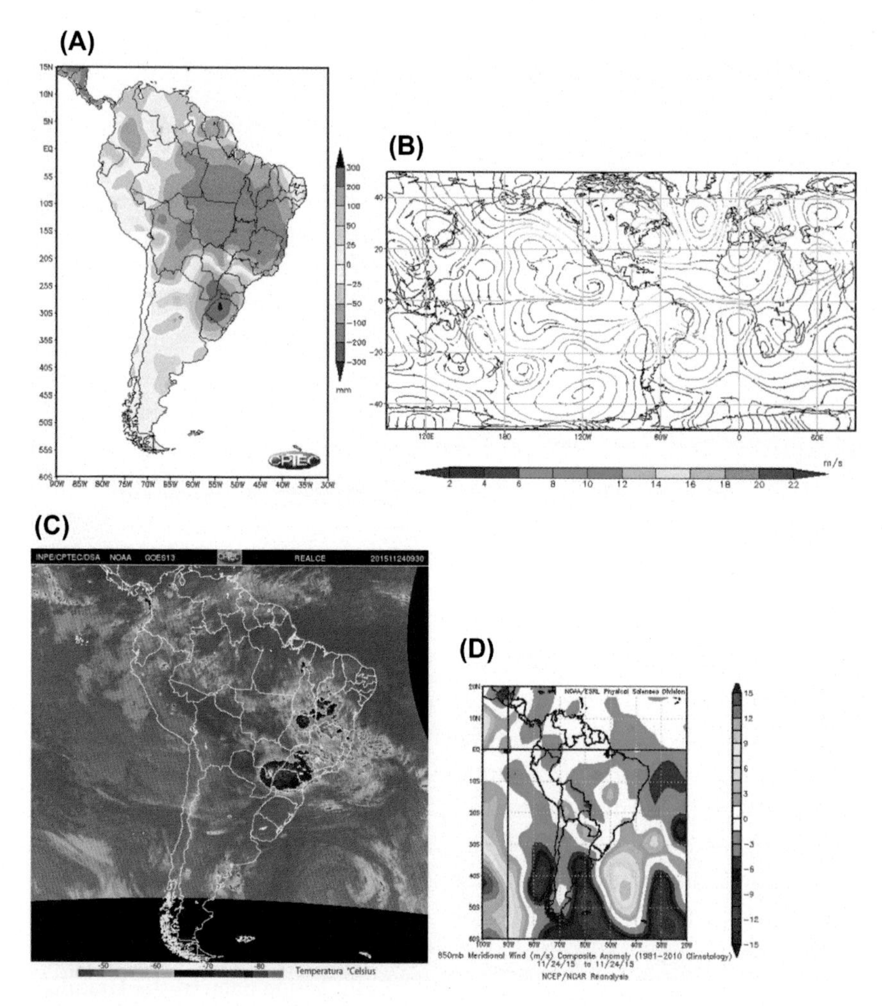

Figure 3.14 (A) Precipitation anomalies in December 2015; (B) streamline anomalies at 200 hPa in December 2015; (C) Infrared enhanced satellite image on November 24, 2015, 09:30 GMT (colors represent temperature). (D) 850 hPa meridional wind anomalies on November 24, 2015. *((C) From CPTEC/INPE; (D) From ESRL/PSD/NOAA.)*

enhanced precipitation over the middle-lower Parana/La Plata basin, displaying an anomalous pair cyclone/anticyclone over subtropical SA, as is typical during spring of EN (Fig. 3.14B).

During this season, there were many cases of MCCs, with great contribution of the LLJ. One of these cases happened on November 24, 2015 (Fig. 3.14C), when the low-level meridional southward wind east of the Andes showed strong anomalies (Fig. 3.14D), which resulted in daily precipitation above 85 mm day^{-1} in parts of central-west State of Parana.

6. OBSERVED TRENDS REGARDING EXTREMES

Several studies have been carried out on trends of climate extreme indices (CEI) using RClimDex/fclimdex.f software, recommended by the joint World Meteorological Organization Commission of Climatology and the Climate Variability and Predictability Project Expert Team on Climate Change Detection, Monitoring, and Indices. Here, only the indices related to precipitation are mentioned: CDD (annual maximum numbers of consecutive dry days), RX1day (annual maximum 1-day precipitation), RX5day (annual maximum consecutive 5-day precipitation), R20 mm (number of days per year with precipitation amount ≥20 mm), and R95p (sum of daily precipitation for a given period above the 95th percentile for the interval 1961–90). Although the first index indicates drought, the others measure abundance of high precipitation extreme events.

A pioneering study in this regard for SA, using rain gauge data (Haylock et al., 2006), showed that in the period 1960–2000 the number of extreme precipitation events increased in south and southeast Brazil, whereas in NE Brazil it was reduced. Similar results were obtained by Alexander et al. (2006) for 1951–2003. Positive tendencies in south and southeast Brazil were also detected by Marengo et al. (2010) and Penalba and Robledo (2010). However, Santos and Brito (2007) showed that the number of rainy days and extreme events increased in part of NE Brazil during 1935–2000.

Valverde and Marengo (2014) carried out a study for different Brazilian river basins, using the gridded daily data from the Climate Prediction Center (CPC/NOAA) and of some rain gauge stations, for the period 1979–2005. In several instances, the significant trends indicated by gridded data do not agree with those obtained from rain gauge data. In the Brazilian Amazon basin, the most consistent results are the increasing trends of RX1day, RX5day, R20 mm, and R95p over the northern part of the basin. In the Tocantins basin, in central Brazil, a positive trend of CDD appears in most of the basin's area, and the R95p shows a decrease in extreme rainfall in central and southern sectors of the basin. In the Sao Francisco basin, in central-east Brazil, RX5day and R95p tend to increase and CDD tends to decrease in the northern part of the basin, which is in southern NE Brazil, whereas the opposite happens in the southern part of the basin. In the Brazilian part of the Parana basin, the northern part showed increasing CDD and decreasing R95p with gridded data, although there are discrepancies with rain gauge data. Positive trends of R20 mm, RX1day, and RX5day were more significant in the southwestern sector, in the southern Paraguay subbasin.

There have been several studies on possible trends in the number of extreme events in southeast Brazil, a very populated region (e.g., Teixeira and Satyamurty, 2011; Dereczynski et al., 2013; Silva Dias et al., 2012; Zilli et al., 2016). Teixeira and Satyamurty (2011) did not find significant trends in annual frequency of heavy and extreme rainfall events in southeast Brazil, but detected them in south Brazil, in the period 1960–2004. In the city of Rio de Janeiro, an analysis of two rainfall time series with 30 and 33 years of data (within the periods 1967–2007 and 1964–2009), respectively, showed that among all the tested CEI related to extreme rainfall, only one (R95p) showed a positive significant trend for one of the stations (the one with 30 years data), which was is in a national park forest area (Dereczynski et al., 2013). Zilli et al. (2016) used data from 36 stations with at least 70 years of observations, including the period 1939–99. Positive trends in four indices related to intense rainfall during summer monsoon at 10% significance level were observed at six stations. The majority was located over the state of Rio de Janeiro (não including the city of Rio de Janeiro) and over urban areas of the state of Sao Paulo. In the city of Sao Paulo, a significant positive trend has been detected and correlated with several climatic indices. The one that explains most of the variance is that associated with PDO (Silva Dias et al., 2012). According to Grimm et al. (2016), the SACZ region is affected by the first mode of interdecadal precipitation variability, which is strongly connected to IPO, of which the PDO is the North Pacific component. Silva Dias et al. (2012) suggest that in the wet season the growth of the urban heat island and the role of air pollution in cloud microphysics may play a relevant role in the increasing extreme events over São Paulo.

The main problem in the interpretation of most trend analyses performed for SA is the relatively short period they cover, due to the scarceness of long daily precipitation series. Because interdecadal variability is significant in SA and it affects the extreme events (Grimm and Saboia, 2015;

Grimm et al., 2016), it is difficult to detect anthropogenic climate change with trends in periods shorter than 50 years. For instance, some studies point out rainfall trends in the Amazon, but the study with the longest timeseries does not show that the rainfall in the Brazilian Amazon basin is experiencing a significant change, except at a few individual stations, and subregions with upward trends are interspersed with subregions displaying downward trends (Satyamurty et al., 2009). Thus, over SA, the impact of anthropogenic climate change on precipitation is not as conclusive as it is on temperature.

REFERENCES

Alexander, L.V., Zhang, X., Peterson, T.C., Caesar, J., Gleason, B., Klein Tank, A., Haylock, M., Collins, D., Trewin, B., Rahimzadeh, F., Tagipour, A., Ambenje, P., Kumar, K.R., Revadekar, J., Griffiths, G., Vincent, L., Stephenson, D., Burn, J., Aguilar, E., Brunet, M., Taylor, M., New, M., Zhai, P., Rusticucci, M., Vazquez-Aguirre, J.L., 2006. Global observed changes in daily climate extremes of temperature and precipitation. J. Geophys. Res. Atmos. 111, D05109. https://doi.org/10.1029/2005JD006290.

Alvarez, M.S., Vera, C.S., Kiladis, G.N., Liebmann, B., 2016. Influence of the Madden Julian oscillation on precipitation and surface air temperature in South America. Clim. Dyn. 46, 245–262. https://doi.org/10.1007/s00382-015-2581-6.

Anabor, V., Stensrud, D.J., Moraes, O.L.L., 2008. Serial upstream-propagating mesoscale convective system events over Southeastern South America. Mon. Weather Rev. 136, 3087–3105.

Byerle, L.A., Paegle, J., 2002. Description of the seasonal cycle of low-level flows flanking the Andes and their interannual variability. Meteorologica 27, 71–88.

Carvalho, L.M.V., Jones, C., Silva Dias, M.A.F., 2002. Intraseasonal large-scale circulations and mesoscale convective activity in tropical South America during the TRMM-LBA campaign. J. Geophys. Res. 107 (D20), 8042. https://doi.org/10.1029/2001JD000745.

Carvalho, L.M.V., Jones, C., Liebmann, B., 2004. The South Atlantic Convergence Zone: intensity, form, persistence, relationships with intraseasonal to interannual activity and extreme rainfall. J. Clim. 17, 88–108.

Cavalcanti, I.F.A., 2012. Large scale and synoptic features associated with extreme precipitation over South America: a review and case studies for the first decade of the 21st century. Atmos. Res. 118, 27–40.

Cazes-Boezio, G., Robertson, A.W., Mechoso, C.R., 2003. Seasonal dependence of ENSO teleconnections over South America and relationships with precipitation in Uruguay. J. Clim. 16, 1159–1176.

Chan, S.C., Behera, S.K., Yamagata, T., 2008. Indian Ocean Dipole influence on South American rainfall. Geophys. Res. Lett. 35, L14S12.

Chaves, R.R., Nobre, P., 2004. Interactions between sea surface temperature over the South Atlantic Ocean and the South Atlantic Convergence Zone. Geophys. Res. Lett. 31, L03204. https://doi.org/10.1029/2003GL018647.

Cohen, J., Silva Dias, M.A., Nobre, C.A., 1995. Environmental conditions associated with Amazonian squall lines: a case study. Mon. Weather Rev. 123, 3163–3174.

Covey, D., Hastenrath, S., 1978. The Pacific El Niño phenomenon and the Atlantic circulation. Mon. Weather Rev. 106, 1280–1287.

Cunningham, C.C., Cavalcanti, I.F.A., 2006. Intraseasonal modes of variability affecting the South Atlantic Convergence Zone. Int. J. Climatol. 26, 1165–1180.

De Almeida, R.A.F., Nobre, P.R., Haarsma, J., Campos, E.J.D., 2007. Negative ocean-atmosphere feedback in the South Atlantic Convergence Zone. Geophys. Res. Lett. 34, L18809. https://doi.org/10.1029/2007GL030401.

Dereczynski, C., Silva, W.L., Marengo, J.A., 2013. Detection and projections of climate change in Rio de Janeiro, Brazil. Am. J. Clim. Change 2013 (2), 25–33.

Doyle, M.E., Barros, V.R., 2002. Midsummer low-level circulation and precipitation in subtropical South America and related sea surface temperature anomalies in the South Atlantic. J. Clim. 15, 3394–3410.

Durkee, J.D., Mote, T.L., Shepherd, M., 2009. The contribution of mesoscale convective complexes to rainfall across subtropical South America. J. Clim. 22, 4590–4605.

Ferraz, S.E.T., 2004. Intraseasonal variations of the summer precipitation over South America. PhD Thesis. Department of Atmospheric Sciences, University of São Paulo.

Fu, R., Zhu, B., Dickinson, R.E., 1999. How do atmosphere and land surface influence seasonal changes of convection in the tropical Amazon? J. Clim. 12, 1306–1321.

Gan, M.A., Kousky, V.E., Ropelewski, C.F., 2004. The South America monsoon circulation and its relationship to rainfall over West-Central Brasil. J. Clim. 17, 47–66.

Garreaud, R.D., Wallace, J.M., 1997. The diurnal march of convective cloudiness over the Americas. Mon. Weather Rev. 125, 3157–3171.

Garreaud, R.D., Wallace, J.M., 1998. Summertime incursions of midlatitude air into subtropical and tropical South America. Mon. Weather Rev. 126, 2713–2733.

Grimm, A.M., Silva Dias, P.L., 1995. Analysis of tropical-extratropical interactions with influence functions of a barotropic model. J. Atmos. Sci. 52, 3538–3555.

Grimm, A.M., 2003. The El Niño impact on the summer monsoon in Brazil: regional processes versus remote influences. J. Clim. 16, 263–280.

Grimm, A.M., 2004. How do La Niña events disturb the summer monsoon system in Brazil? Clim. Dyn. 22, 123–138.

Grimm, A.M., Vera, C.S., Mechoso, C.R., 2005. The South American monsoon system. In: Chang, C.-P., Wang, B., Lau, N.-C.G. (Eds.), The Global Monsoon System: Research and Forecast, pp. 219–238 WMO/TD 1266-TMRP 70. www.wmo.int/pages/prog/arep/tmrp/documents/global_monsoon_system_IWM3.pdf.

Grimm, A.M., Pal, J., Giorgi, F., 2007. Connection between spring conditions and peak summer monsoon rainfall in South America: role of soil moisture, surface temperature, and topography in eastern Brazil. J. Clim. 20, 5929–5945.

Grimm, A.M., Tedeschi, R.G., 2009. ENSO and extreme rainfall events in South America. J. Clim. 22, 1589–1609.

Grimm, A.M., Zilli, M.T., 2009. Interannual variability and seasonal evolution of summer monsoon rainfall in South America. J. Clim. 22, 2257–2275. https://doi.org/10.1175/2008JCLI2345.1.

Grimm, A.M., Ambrizzi, T., 2009. Teleconnections into South America from the tropics and extratropics on interannual and intraseasonal timescales. In: Vimeux, F., Sylvestre, F., Khodri, M. (Eds.), Past Climate Variability in South America and Surrounding Regions, first ed. Springer, Netherlands, pp. 159–191.

Grimm, A.M., 2011. Interannual climate variability in South America: impacts on seasonal precipitation, extreme events, and possible effects of climate change. Stoch. Environ. Res. Risk Assess. 25, 537–554. https://doi.org/10.1007/s00477-010-0420-1.

Grimm, A.M., Silva Dias, M.A.F., 2011. Synoptic and mesoscale processes in the South American monsoon. In: Chang, C.P., Ding, Y., Lau, N.-C., Johnson, R.H., Wang, B., Yasunari, T. (Eds.), The Global Monsoon System: Research and Forecast, World Scientific Series on Asia-Pacific Weather and Climate, vol. 5. World Scientific Publishing Company, Singapore, 606 p., Chapter 14, pp. 239–256, ISBN: 13 978-981-4343-40-4; ISBN: 10 981-4343-40-4.

Grimm, A.M., Cavalcanti, I.F.A., Berbery, E.H., 2015. The South American monsoon. CLIVAR Exch. 19 (1), 6–9. www.clivar.org/sites/default/files/documents/Exchanges_No_66.pdf.

Grimm, A.M., Saboia, J.P.J., 2015. Interdecadal variability of the South American precipitation in the monsoon season. J. Clim. 28, 755–775. https://doi.org/10.1175/JCLI-D-14-00046.1.

Grimm, A.M., Laureanti, N.C., Rodakoviski, R.B., Gama, C.B., 2016. Interdecadal variability and extreme precipitation events in South America during the monsoon season. Clim. Res. 68, 277–294. https://doi.org/10.3354/cr01375.

Grimm, A.M., 2018. Madden-Julian Oscillation Impacts on South America Summer Monsoon Season: Precipitation Anomalies, Extreme Events, and Teleconnections (Submitted).

Hamilton, S.K., Sippel, S.J., Melack, J.M., 1996. Inundation patterns in the Pantanal wetland of South America determined from passive microwave remote sensing. Arch. Hydrobiol. 137, 1–23.

Hamilton, S.K., Sippel, S.J., Melack, J.M., 2002. Comparison of inundation patterns among major South American floodplains. J. Geophys. Res. 107, 8038. https://doi.org/10.1029/2000JD000306.

Hastenrath, S., Heller, L., 1977. Dynamics of climatic hazards in northeast Brazil. Q. J. R. Meteorol. Soc. 103, 77–92.

Haylock, M.R., Peterson, T.C., Alves, L.M., Ambrizzi, T., Anunciação, Y.M.T., Baez, J., Barros, V.R., Berlato, M.A., Bidegain, M., Coronel, G., Corradi, V., Garcia, V.J., Grimm, A.M., Karoly, D., Marengo, J.A., Marino, M.B., Moncunill, D.F., Nechet, D., Quintana, J., Rebello, E., Rusticucci, M., Santos, J.L., Trebejo, I., Vincent, L.A., 2006. Trends in total and extreme South American rainfall in 1960–2000 and links with sea surface temperature. J. Clim. 19, 1490–1512.

Herdies, D.L., Da Silva, A., Silva Dias, M.A., Nieto-Ferreira, R., 2002. Moisture budget of the bimodal pattern of the summer circulation over South America. J. Geophys. Res. 107, 42/1–42/10.

Hirata, F.E., Grimm, A.M., 2015. The role of synoptic and intraseasonal anomalies in the life cycle of summer rainfall extremes over South America. Clim. Dyn. https://doi.org/10.1007/s00382-015-2751-6.

Jones, C., Carvalho, L.M.V., 2002. Active and break phases in the South American monsoon system. J. Clim. 15, 905–914.

Jones, C., Carvalho, L.M.V., Lau, K.M., Stern, W., 2004. Global occurrences of extreme precipitation events and the Madden–Julian oscillation: observations and predictability. J. Clim. 17, 4575–4589.

Kitoh, A., Endo, H., Krishna Kumar, K., Cavalcanti, I.F.A., Goswami, P., Zhou, T., 2013. Monsoons in a changing world: A regional perspective in a global context. J. Geophys. Res. Atmos. 118, 3053–3065.

Kousky, V., 1988. Pentad outgoing longwave radiation climatology for the South American sector. Revista Brasileira de Meteorologia 3, 217–231.

Laing, A.G., Fritsch, J.M., 2000. The large-scale environments of the global populations of mesoscale convective complexes. Mon. Weather Rev. 128, 2756–2776.

Li, W., Fu, R., 2004. Transition of the large-scale atmospheric and land surface conditions from the dry to the wet season over Amazonia as diagnosed by the ECMWF Re-Analysis. J. Clim. 17, 2637–2651.

Liebmann, B., Marengo, J.A., 2001. Interannual variability of the rainy season and rainfall in the Brazilian Amazon basin. J. Clim. 14, 4308–4318.

Liebmann, B., Mechoso, C.R., 2011. The South American monsoon system. In: Chang, C.-P., Ding, Y., Lau, N.-C., Johnson, R.H., Wang, B., Yasunari, T. (Eds.), The Global Monsoon System: Research and Forecast, second ed. World Scientific, pp. 137–157.

Lima, K.C., Satyamurty, P., Fernandez, J.P.R., 2010. Large-scale atmospheric conditions associated with heavy rainfall episodes in southeast Brazil. Theor. Appl. Climatol. 101, 121–135.

Machado, L.A.T., Rossow, W.B., Guedes, R.L., Walker, A.W., 1998. Life cycle variations of mesoscale convective systems over the Americas. Mon. Weather Rev. 126, 1630–1654.

Marengo, J.A., Liebmann, B., Kousky, V., Filizola, N., Wainer, I., 2001. On the onset and end of the rainy season in the Brazilian Amazon Basin. J. Clim. 14, 833–852.

Marengo, J.A., Soares, W.R., Saulo, C., Nicolini, M., 2004. Climatology of the low-level jet east of the Andes as derived from the NCEP reanalyses. J. Clim. 17, 2261–2280.

Marengo, J.A., Rusticucci, M., Penalba, O., Renom, M., 2010. An intercomparison of observed and simulated extreme rainfall and temperature events during the last half of the twentieth century: part 2: historical trends. Clim. Change 98, 509–529.

Marengo, J.A., Liebmann, B., Grimm, A.M., Misra, V., Silva Dias, P.L., Cavalcanti, I.F.A., Carvalho, L.M.V., Berbery, E.H., Ambrizzi, T., Vera, C.S., Saulo, A.C., Nogués-Paegle, J., Zipser, E., Seth, A., Alves, L.M., 2012. Recent developments on the South American monsoon system. Int. J. Climatol. https://doi.org/10.1002/joc.2254.

Marengo, J.A., Alves, L.M., 2012. The 2011 intense rainfall and floods in Rio de Janeiro. In: Blunden, J., Arndt, D.S. (Eds.), State of the Climate in 2011 Bull. Am. Meteorol. Soc. 93 (7), S176.

Marengo, J.A., Alves, L.M., Soares, W.R., Rodriguez, D.A., Camargo, H., Riveros, M.P., Pabló, A.D., 2013. Two contrasting severe seasonal extremes in tropical South America in 2012: flood in Amazonia and drought in northeast Brazil. J. Clim. 26, 9137–9154.

Moura, A.D., Shukla, J., 1981. On the dynamics of droughts in Northeast Brazil: observations, theory, and numerical experiments with a general circulation model. J. Atmos. Sci. 38, 2653–2675.

Muza, M.N., Carvalho, L.M.V., Jones, C., 2009. Intraseasonal and interannual variability of extreme dry and wet events over Southeastern South America and subtropical Atlantic during the austral summer. J. Clim. 22, 1682–1699.

Nicolini, M., Saulo, A.C., Torres, J.C., Salio, P., 2002. Enhanced precipitation over southeastern South America related to strong low-level jet events during austral warm season. Meteorologica 27, 59–69.

Nieto Ferreira, R., Rickenbach, T.M., Herdies, D.L., Carvalho, L.M.V., 2003. Variability of South American convective cloud systems and tropospheric circulation during January-March 1998 and 1999. Mon. Weather Rev. 131, 961–973.

Nieto-Ferreira, R., Rickenbach, T.M., Wright, E.A., 2011. The role of cold fronts in the onset of the South American monsoon. Q. J. R. Meteorol. Soc. 137 (Part B), 908–922. https://doi.org/10.1002/qj.810.

Nobre, P., Shukla, J., 1996. Variations of sea surface temperature, wind stress, and rainfall over the tropical Atlantic and South America. J. Clim. 10, 2464–2479.

Nobre, P., De Almeida, R., Malagutti, M., Giarolla, E., 2012. Coupled ocean–atmosphere variations over the South Atlantic Ocean. J. Clim. 25, 6349–6358. https://doi.org/10.1175/JCLI-D-11-00444.1.

Nogues-Paegle, J., Mechoso, C.R., Fu, R., Berbery, E.H., Chao, W.C., Chen, T., Cook, K., Diaz, A.F., Enfield, D., Ferreira, R., Grimm, A.M., Kousky, V., Liebmann, B., Marengo, J.A., Mo, K., Neelin, J.D., Paegle, J., Robertson, A.W., Seth, A., Vera, C.S., Zhou, J., 2002. Progress in Pan American CLIVAR research: understanding the South American monsoon. Meteorologica 1–2, 3–32.

Nogues-Paegle, J., Mo, K.C., 1997. Alternating wet and dry conditions over South America during summer. Mon. Weather Rev. 125, 279–291.

Nogues-Paegle, J., Byerle, L.A., Mo, K.C., 2000. Intraseasonal modulation of South American summer precipitation. Mon. Weather Rev. 128, 837–850.

Parker, D., Folland, C., Scaife, A., Knight, J., Colman, A., Baines, P., Dong, B., 2007. Decadal to multidecadal variability and the climate change background. J. Geophys. Res. 112, D18115. https://doi.org/10.1029/2007JD008411.

Penalba, O.C., Robledo, F., 2010. Spatial and temporal variability of the frequency of extreme daily rainfall regime in the La Plata Basin during the 20th century. Clim. Change 98 (3), 531–550. https://doi.org/10.1007/s10584-009-9743-7).

Rasmussen, K.L., Chaplin, M.M., Zuluaga, M.D., Houze Jr., R.A., 2016. Contribution of extreme convective storms to rainfall in South America. J. Hydrometeorol. 17, 353–367.

Robertson, A.W., Mechoso, C.R., 2000. Interannual and interdecadal variability of the South Atlantic Convergence Zone. Mon. Weather Rev. 128, 2947–2957.

Robertson, A.W., Farrara, J., Mechoso, C.R., 2003. Simulations of the atmospheric response to South Atlantic sea surface temperature anomalies. J. Clim. 16, 2540–2551.

Salio, P., Nicolini, M., Zipser, E.J., 2007. Mesoscale convective systems over Southeastern South America and their relationship with the South American low level jet. Mon. Weather Rev. 135, 1290–1309.

Santos, C.A., Brito, J.I.B., 2007. Análise dos índices de extremos para o semi-árido do Brasil e suas relações com TSM e IVDN. Revista Brasileira de Meteorologia 22, 303–312.

Satyamurty, P., Castro, A.A., Tota, J., Gularte, L.E.S., Manzi, A.O., 2009. Rainfall trends in the Brazilian Amazon Basin in the past eight decades. Theor. Appl. Climatol. https://doi.org/10.1007/s00704-009-0133-x.

Seluchi, M., Marengo, J.A., 2000. Tropical-mid latitude exchange of air masses during summer and winter in south America: climatic aspects and extreme events. Int. J. Climatol. 20, 1167–1190.

Seluchi, M.E., Chou, S.C., 2009. Synoptic patterns associated with landslide events in the Serra do Mar, Brazil. Theor. Appl. Climatol. 98, 67–77.

Silva, V.B.S., Kousky, V.E., 2012. The South American monsoon system: climatology and variability. In: Wang, S.-Y. (Ed.), Modern Climatology. InTech. ISBN: 978-953-51-0095-9.

Silva Dias, M.A.F., Dias, J., Carvalho, L.M.V., Freitas, E.D., Silva Dias, P.L., 2012. Changes in extreme daily rainfall for São Paulo, Brazil. Clim. Change. https://doi.org/10.1007/s10584-012-0504-7.

Siqueira, J.R., Machado, L.A.T., 2004. Influence of the frontal systems on the day-to-day convection variability over South America. J. Clim. 17, 1754–1766.

Tedeschi, R.G., Grimm, A.M., Cavalcanti, I.F.A., 2015. Influence of Central and East ENSO on extreme events of precipitation in South America during austral spring and summer. Int. J. Climatol. 35, 2045–2064. https://doi.org/10.1002/joc.4106.

Tedeschi, R.G., Grimm, A.M., Cavalcanti, I.F.A., 2016. Influence of Central and East ENSO on precipitation and its extreme events in South America during austral autumn and winter. Int. J. Climatol. https://doi.org/10.1002/joc.4670.

Teixeira, M.S., Satyamurty, P., 2007. Dynamical and synoptic characteristics of heavy rainfall episodes in Southern Brazil. Mon. Weather Rev. 135, 598–617.

Teixeira, M.S., Satyamurty, P., 2011. Trends in the frequency of intense precipitation events in Southern and Southeastern Brazil during 1960–2004. J. Clim. 24, 1913–1921. https://doi.org/10.1175/2011JCLI3511.1.

Valverde, M.C., Marengo, J.A., 2014. Extreme rainfall indices in the hydrographic basins of Brazil. Open J. Mod. Hydrol. 4, 10–26.

Vasconcellos, F.C., Cavalcanti, I.F.A., 2010. Extreme precipitation over Southeastern Brazil in the austral summer and relations with the Southern Hemisphere annular mode. Atmos. Sci. Lett. 11, 21–26.

Velasco, I., Fritsch, J.M., 1987. Mesoscale convective complexes in the Americas. J. Geophys. Res. 92, 9591–9613.

Vera, C., Higgins, W., Amador, J., Ambrizzi, T., Garreaud, R., Gochis, D., Gutzler, D., Lettenmaier, D., Marengo, J., Mechoso, C.R., Nogues-Paegle, J., Silva Dias, P.L., Zhang, C., 2006. Toward a unified view of the American monsoon system. J. Clim. 19, 4977–5000.
Virji, H., 1981. A preliminary study of summertime tropospheric circulation patterns over South America estimated from cloud winds. Mon. Weather Rev. 109, 599–610.
Webster, P.J., Chang, H.-R., 1988. Energy accumulation and emanation regions at low latitudes: impacts of a zonally varying basic state. J. Atmos. Sci. 45, 803–829.
Wheeler, M.C., Hendon, H.H., 2004. An all-season real-time multivariate MJO index: development of an index for monitoring and prediction. Mon. Weather Rev. 132, 1917–1932.
Williams, E., Dall'Antonia, A., Dall'Antonia, V., Almeida, J.M., Suarez, F., Liebmann, B., Malhado, A.C.M., 2005. The drought of the century in the Amazon basin: an analysis of the regional variation of rainfall in South America in 1926. Acta Amazon. 35, 231–238.
Yoon, J.H., Zeng, N., 2010. An Atlantic influence on Amazon rainfall. Clim. Dyn. 34, 249–264. https://doi.org/10.1007/s00382-009-0551-6.
Zeng, N., Yoon, Y., Marengo, J.A., Subrmanaiam, A., Nobre, C., Mariotti, A., 2008. Causes and impacts of the 2005 Amazon drought. Environ. Res. Lett. 3 (1):014002. https://doi.org/10.1088/1748-9326/3/1/014002.
Zhou, J., Lau, W.K.M., 1998. Does a monsoon climate exist over South America? J. Clim. 11, 1020–1040.
Zilli, M.T., Carvalho, L.M.V., Liebmann, B., Silva Dias, M.A., 2016. A comprehensive analysis of trends in extreme precipitation over southeastern coast of Brazil. Int. J. Climatol. https://doi.org/10.1002/joc.4840.
Zipser, E.J., Cecil, D.J., Liu, C., Nesbitt, S.W., Yorty, D.P., 2006. Where are the most intense thunderstorms on earth? Bull. Am. Meteorol. Soc. 87, 1058–1071.

CHAPTER 4

Precipitation Extremes in the West African Sahel: Recent Evolution and Physical Mechanisms

Théo Vischel[1], Gérémy Panthou[1], Philippe Peyrillé[2], Romain Roehrig[2], Guillaume Quantin[1], Thierry Lebel[1], Catherine Wilcox[1], Florent Beucher[2], Maria Budiarti[3]

[1]Univ. Grenoble Alpes, IRD, CNRS, IGE, Grenoble, France; [2]Centre National de Recherches Météorologiques, Météo-France & CNRS, Toulouse, France; [3]Badan Meteorologi Klimatologi dan Geofisika, Jakarta Pusat, Indonesia

Contents

Tropical Extremes: Natural Variability and Trends
ISBN 978-0-12-809248-4
https://doi.org/10.1016/B978-0-12-809248-4.00004-2

1. INTRODUCTION

The West African rainfall regime is driven by a monsoon system known as the West African Monsoon (WAM). As is the case for other significant monsoon systems (most notably, the Asian and South American monsoons), the WAM plays a major role in the water and energy redistributions in the intertropical belt. WAM's main climatological patterns are a negative rainfall gradient from the Guinean coast to the Saharan desert and a well-marked succession of dry and rainy seasons, both controlled by the migration of the Intertropical Convergence Zone (ITCZ). A high rainfall variability over a large range of space and timescales is superimposed on these very consistent climatological features. This has major impact on populations that are notoriously vulnerable to climate variability and change.

The Sahelian part of West Africa (roughly extending from 10°N to 15°N) is one region in the world where the meaning of "extreme climate" is truly expressed. Over the past century, Sahelian rainfall has undergone a very strong decadal variability, including the most intense and longest drought ever recorded in the world that lasted from 1970 to the middle of the 1990s (Hulme, 2001; Dai et al., 2004). Despite the emergence of a slight trend toward wetter conditions since the 1990s, often controversially considered as a recovery period (Ozer et al., 2003; Nicholson, 2005), the mean annual rainfall level remains lower than the average observed over the entire 20th century (Lebel and Ali, 2009; Gallego et al., 2015).

Many causes have been identified and investigated to explain the interdecadal variability of Sahelian rainfall, including land–atmosphere interactions (e.g., Charney, 1975), tropical, and global sea surface temperature (SST) anomalies (e.g., Folland et al., 1986; Lamb and Peppler, 1992; Giannini et al., 2003) and more recently greenhouse gases (GHG) (e.g., Dong and Sutton, 2015). However, the question of attribution remains challenging because of the multiple competing processes that act and interact at various ranges of space and timescales. The dominance of one mechanism or another also varies with location within the Sahel and with the period (Nicholson, 2013).

It is worth noting that while the persistent rainfall deficit over the Sahel perpetuates a significant risk of food shortages, as observed in 2005 and 2010 in Niger, the past 20 years have also seen an exponential increase of flood-related damages and fatalities (Tarhule, 2005; Di Baldassarre et al., 2010). The increasing vulnerability of populations (Di Baldassarre et al., 2010) and the changes in land use/cover (Descroix et al., 2009) can partly explain the growth of flood damages, but rainfall hazard has been shown to remain a major factor of flood increase in the recent years (Aich et al., 2015; Casse et al., 2016).

This relates to the emerging issue in the hydroclimate community of the possibility that global warming might generate increasing occurrences of extreme rainfall. However, because extreme events are rare by definition, detecting any statistically significant changes in their frequency is technically difficult. First of all, sufficiently long time series of observations—whether in situ or from satellite—are often lacking; this is especially true in West Africa where the national rain gauge networks are degrading and where it is often difficult to access the most recent data (Tarhule and Woo, 1998; Ali et al., 2005a). Furthermore, general circulation models (GCMs) still do not provide an appropriate alternative to observations for diagnosing the evolution of extreme rainfall because they have some biases in monsoon drivers such as ITCZ positions and SSTs (Siongco et al., 2015; Wang et al., 2014) and are likely to produce errors in trends (e.g., Saha et al., 2014). GCMs also do not capture the convection scales driving the production of intense storms. This is especially a concern in tropical regions where rainfall is essentially of convective origin: in the Sahel, more than 90% of the annual rain (Mathon et al., 2002) is produced by mesoscale convective systems (MCSs). The strong concentration of intense MCSs occurring during the 4 months of the core of the Sahelian rainy season (June–September) is deemed as unique on Earth by Zipser et al. (2006).

In situ data and GCM issues mainly explain why West Africa (like many other tropical regions) is underrepresented in studies evaluating rainfall intensification at the global scale either from GCM simulations (e.g., Min et al., 2011) or from ground-based rainfall measurements (e.g., Alexander et al., 2006; Westra et al., 2013). It also explains the lack of West Africa–specific studies on precipitation extremes.

Beyond the issue of the availability of suitable data for studying extreme rainfall evolutions in the long run, one may wonder how such evolutions might be related to changes in the organization of the convection. Because convective rainfall overarchingly predominates in the region, any change in the extreme precipitation characteristics could logically be traced back to changes in the characteristics—occurrence, intensity, and size—of the MCS.

Several major issues must thus be considered to advance our understanding of how modifications of extreme rainfall occurrences relate to the overall rainfall variability in the Sahel. The first one relates to the dependence of precipitation extremes on space and timescales. As convective rainfall is naturally highly variable in space and time, precipitation extremes are dependent on the spatiotemporal resolution at which the precipitation is examined. This dependence and the way it may change under nonstationary climate conditions are crucial to assessing the impacts of extreme precipitation. Another issue relates to the evolution of extreme value distributions at a given scale and how it might be linked to changes in convective system size and/or organization. This is not only important for impacts but also for providing clues on potential changes in the mesoscale atmospheric processes responsible for extreme rainfall. The final issue concerns the identification of the synoptic atmospheric conditions that could favor extreme precipitation events. This may help in understanding the reasons for the past variability and provide elements for making use of climate model scenarios to foresee the evolution of the extreme regime in an increasingly warmer world.

This chapter reviews the most recent results concerning the abovementioned issues, in the light of the recent literature and some original studies. The next section provides an overview of some key elements of understanding the evolution of precipitation in the Sahel and the physical mechanisms involved. This is analyzed from the perspective of changes at the large annual scale and the storm scale (mesoscale), with specific discussions about extreme rainfall events. Section 3 describes some essential features of the space–time structure of rainfall extremes, mainly addressed via the development of intensity–duration–area–frequency (IDAF) curves. In Section 4, recent methodological developments used to analyze recent trends in rainfall extremes are presented and applied to document the evolution of extreme precipitations in the Sahel from 1950 to 2015. A case study is presented in Section 5, based on a composite of extreme precipitation events that occurred over the last 20 years in Ouagadougou. It provides an insight into the atmospheric environment that may favor the occurrence of extreme storms in the Sahel. The most noteworthy results are summarized in the last section, giving some guidance for future research.

2. RECENT EVOLUTION OF RAINFALL REGIME IN WEST AFRICA

2.1 Rainfall Evolution at Annual Scale

Although the main climatic patterns (mean seasonal cycle and latitudinal gradients) of the rainfall regime in the Sahel have long been documented (Rodier, 1964), the way rainfall has evolved over the last century has only been addressed quite recently. Studies emerged with the first signs of drought at the end of the 1960s, especially after the extremely dry years of 1972 and 1973 that led to great famines, devastating lives and livelihoods and forcing communities to migrate. The situation caused a great stir in public opinion and aroused the interest and involvement of the scientific community.

2.1.1 The Great Sahelian Drought

First works in the 1970s on rainfall evolution evidenced a strong interannual variability of seasonal totals in the region and attempted to find signs of cyclicity to explain it, though without success (e.g., Bunting et al., 1976). Then many studies focused on the establishment of the drought and the causes of its persistence (e.g., Lamb, 1982; Hulme, 1992; Nicholson, 2001 for a few)—the latter of which has been progressively validated with the year-by-year update of rainfall measurements. Since the 1990s, there is abundant evidence in scientific literature that (1) the drought started at the end of the 1960s, after two decades of relatively wet years (Hubert and Carbonnel, 1987; Demarée, 1990; Le Barbé and Lebel, 1997; Tarhule and Woo, 1998) (see also an illustration for the central Sahel in Fig. 4.1), (2) the drought extended regionally all over West Africa with a gradual reduction of the rainfall deficit from 50% in the northern Sahel (around 16°N) to 10% in the southern West Africa (<10°N) (Lebel et al., 2003), and (3) the deficit mainly affected the rainfall amounts at the core of the rainy season without significantly changing the season duration (Lebel and Ali, 2009).

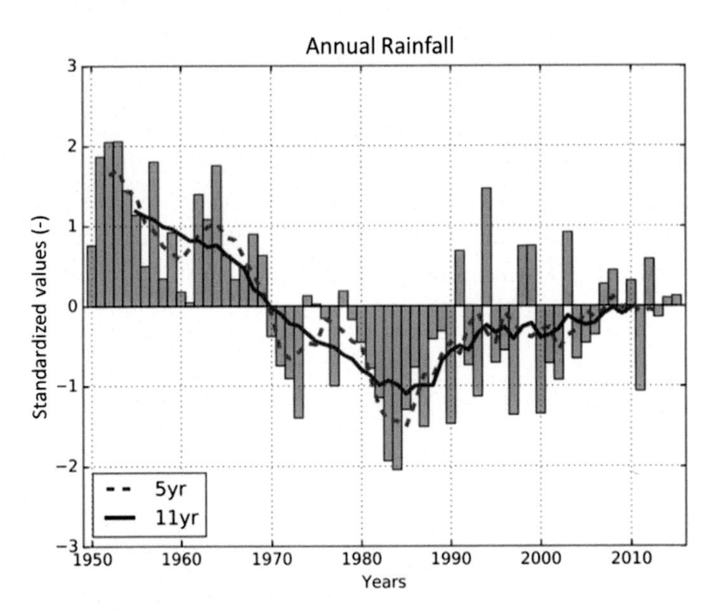

Figure 4.1 Annual standardized precipitation index in the central Sahel (computed from 43 stations in a box defined by longitudes 5°W–7°E and latitudes 9.5°N–15.5°N).

The low-frequency signals in global SSTs are now recognized as major drivers of the Sahelian drought (e.g., Giannini et al., 2003). The drought is indeed mainly attributed to a combined effect of (1) a global warming of tropical oceans and a positive phase of the Interdecadal Pacific Oscillation that both tend to inhibit the tropical convection, added to (2) a negative phase of the Atlantic Mutlidecadal Oscillation that limits the northward displacement of the ITCZ (Mohino et al., 2011).

2.1.2 The Question of Rainfall "Recovery"

After suffering new peaks of extreme dry years in 1983 and 1984 and the resulting devastating famines, the mean level of annual rainfall began to increase slightly at the end of the 1980s, causing a debate in the literature about a possible end of the drought (Nicholson, 2013). Speculating on annual rainfall anomalies (L'Hote et al., 2002), some studies considered that dry conditions predominated up into the early 21st century, whereas others suspected the end of the great Sahelian drought (Ozer et al., 2003). Some 15 years later, an upward trend of annual rainfall has been confirmed in the Sahel (Fig. 4.1). Its causes are not yet fully understood. The role of tropical SSTs seems to be of lesser influence when compared with other identified drivers such as the intensification of the Saharan heat low (SHL; Lavaysse et al., 2015), the

Mediterranean SST warming (Park et al., 2016), and the increase in atmospheric concentration of GHG (Dong and Sutton, 2015). Yet the varying sensitivity of GCMs to GHG versus SSTs (Gaetani et al., 2016) makes the attribution still debatable and calls for a more process-based assessment of climate simulations, including the regional circulation patterns that modulate the WAM (Janicot et al., 2015).

Moreover, the relative increase of rainfall has to be examined with a certain level of awareness for the following reasons: (1) the upward trend is only visible when smoothed over time at a pluriannual scale; (2) it actually masks a strong interannual variability with interspersed very wet and very dry years, with the latter forming the majority (Frappart et al., 2009; Ibrahim et al., 2012; Lodoun et al., 2013; Tarhule et al., 2014); (3) the trend is much smaller over the last decade, reaching a plateau with mean levels much lower than those of the two decades preceding the drought (Lebel and Ali, 2009; Ibrahim et al., 2012; Lodoun et al., 2013; Sanogo et al., 2015); and (4) regional contrasts exist within the Sahel, with a later onset of the trend in the western Sahel than in the central Sahel (Lebel and Ali, 2009; Mahé and Paturel, 2009).

Despite this, the term "recovery" is often taken for granted in the recent literature to describe the recent evolution of rainfall in the Sahel (e.g., Park et al., 2016; Dong and Sutton, 2015). This oversimplification should not make one lose sight of the abovementioned features and the other key processes of the monsoon variability occurring at intra-annual scales that have great impact on societies.

2.2 Rainfall Evolution at Mesoscale

2.2.1 Importance of Mesoscale Rainfall Variability for Climatological Studies

Until the 1990s, most of the studies on rainfall evolution in West Africa only considered changes in annual or monthly total precipitation, thus often ignored the role of the organized storms (MCSs), which are the primary elements of rainfall production in the region. MCSs are the result of many processes acting at various space and timescales from synoptic to mesoscale. They are responsible for the strong rainfall variability initiated at the storm scales, then propagated at much larger scales (up to decadal scales according to Balme et al., 2006a). Documenting their initiation and life cycle and how they change over time is crucial to understanding how these multiscale processes interact and modulate the long-term monsoon variability.

2.2.2 *Importance of Rainfall Variability at Mesoscale for Hydrological Impact Studies*

MCSs have major impacts on populations, as their occurrence, intensity, and size influences both agricultural yields (e.g., Guan et al., 2015) and the hydrological response of catchments (Vischel and Lebel, 2007; Vischel et al., 2009). An illustration of the importance of rainfall variability at mesoscale for hydrology is given in Fig. 4.2. A 13-year reference runoff series was produced by forcing a hydrological model representative of a 5000 km^2 domain in southern Niger (Massuel et al., 2011) with stochastically simulated reference rainfall series at mesoscale (Vischel et al., 2009). In the reference rainfall series, mean annual occurrence, intensity, and size of MCSs are consistent with MCS characteristics measured in situ over the AMMA-CATCH observatory (African Monsoon Multidisciplinary Analysis–Coupling the Tropical Atmosphere and the Hydrological Cycle, Lebel et al., 2009, see also Section 3.2) rain gauge network from years 1990 to 2012. From the rainfall series, drought scenarios were generated by gradually decreasing the mean annual umber of MCS in three different ways. The first way (#1 in Fig. 4.2) considers that rainfall decrease is due to a random reduction of the number of MCS at the core of the rainy season (from July to August). The second one (#2 in Fig. 4.2) is based on a gradual reduction of the number of MCSs from the less intense to the most intense. The third one (#3 in Fig. 4.2) is based on a gradual reduction of the umber of MCSs from the

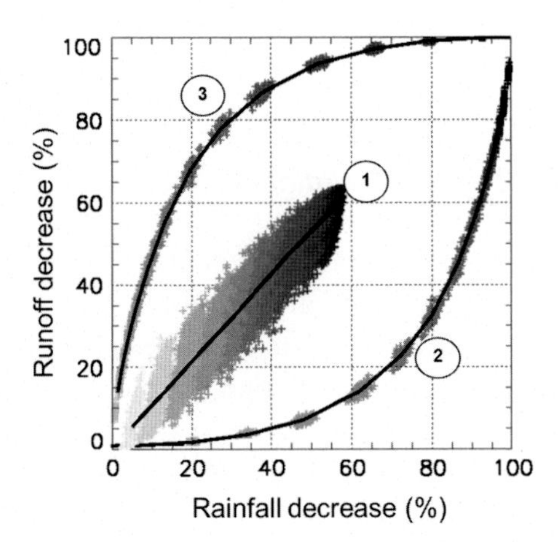

Figure 4.2 Simulated changes in annual simulated runoff of a Sahelian 5000 km^2 hydrosystem in the southern Niger for three different drought scenarios: (1) random decrease of the number of mesoscale convective systems (MCSs) in July and August, (2) decrease of the less intense MCS first, (3) decrease of the most intense MCS first.

most intense to the less intense. The three scenarios were used to force the hydrological model to assess their impacts on simulated runoff compared to the reference simulated series. In Fig. 4.2, mean annual changes in rainfall are reported on the x-axis and the associated mean annual changes in runoff on the y-axis. This purely numerical experiment shows that the three rainfall scenarios have a completely different impact on the simulated runoff. For instance, for a given decrease of 40% of mean annual rainfall, mean annual runoff decreased by a maximum of 5% for scenario #2, between 30% and 50% for scenario #1 and by more than 85% for scenario #3. Because of the nonlinearity of the rainfall–runoff relationship, a decrease in mean annual rainfall driven by a decreasing occurrence of intense events has much more impact on runoff than a similar decrease in rainfall driven by a decreasing number of small events. This result highlights the importance of documenting rainfall at mesoscale to assessing the hydrological impacts of climate change.

2.2.3 Recent Evolution of Mesoscale Rainfall Characteristics

As demonstrated above, the characterization of long-term changes in the rainfall regime at mesoscale is an important task for both climate and hydrological sciences. The first study related to this subject was conducted by Le Barbé and Lebel (1997). They proposed a statistical distribution model able to decompose point daily rainfall series into the product of the occurrence and the intensity of rainfall events. Comparing the distribution parameters between the contrasted wet (1950–69) and dry (1970–89) periods in Niger, they found that the decrease in the number of rain events at the core of the rainy season was the most important factor responsible for the drought. Their results have been generalized by Le Barbé et al. (2002) over a major part of West Africa and confirmed later by several other studies at the country scale (Moron et al., 2006; Frappart et al., 2009; Ibrahim et al., 2012; Lodoun et al., 2013). Another insight on decadal variability of the rainfall regime at mesoscale has been provided by Bell and Lamb (2006) in a study attempting to detect the paths of MCSs over multiple daily rain gauges contained in 4° × 4° regional boxes distributed over Niger, Burkina, Mali, and Senegal. Calibrated on the basis of 4 years of infrared satellite data (TAMSAT Product), a criterion of MCS detection was defined and used to analyze the evolution of MCS occurrence, intensity, and size/organization over the period 1950–98. Bell and Lamb (2006) found that the Sahelian drought was mainly associated with unusually disorganized and low-intensity MCSs. Although a lower spatial extension of MCSs is compatible with a local decrease in rainfall event occurrence, as evidenced by Le Barbé et al. (2002), the reduction of MCS intensity during the drought contradicts the other studies carried out in the region.

2.3 Extreme Rainfall

Tarhule (2005) was the first to stress the lack of attention received by extreme rainfall and flood events in the Sahel despite their considerable impacts on populations. With the update of rainfall data through the early 2000s, some studies noticed an increase in mean daily/event rainfall totals (Lebel and Ali, 2009; Frappart et al., 2009; Lodoun et al., 2013), as early as the end of the 1990s. Although this could suggest a possible change in Sahelian extreme rainfall regime, these preliminary results were not supported by specific studies on extreme precipitation.

In the middle of the 2000s, the joint CCl/CLIVAR/JCOMM Expert Team on Climate Change Detection and Indices (ETCCDI) proposed climate indices relevant for climate change detection. The indices for rainfall are applicable to daily series and are, for a few of them, dedicated to intense rainfall: Rx1day the monthly or annual maximum 1-day precipitation, R95pTOT and R99pTOT the sum of the 95th, and 99th percentile of daily precipitation. This initiative relayed by the project Climdex, which provided an R package for computing the indices, has likely facilitated the emergence of several studies about rainfall variability in West Africa, including an analysis of intense precipitation.

New et al. (2006) made the first application of the ETCCDI indices in Africa by analyzing six series of daily rainfall over the period 1950–2006 (two in Gambia and four in Nigeria). No significant changes were detected through extreme rainfall indices except a significant increasing trend for one station in Nigeria. Sarr et al. (2013, 2014) analyzed rainfall series in Senegal from 1950 to 2007 but did not find any clear signal of changes in extreme rainfall intensities. Sanogo et al. (2015) found a strong positive trend in the annual rainfall sums of days with rainfall totals exceeding the 95th percentile over the entire Sahel and the Guinean Coast over the period 1980–2010.

The dispersion of the results in the abovementioned studies makes it difficult to diagnose the real evolution of extreme rainfall in the region. Although the studies obviously differ in terms of region of study, length, and number of studied series, another aspect that can explain the lack of consensus is the impact of the low signal-to-noise ratio of extreme rainfall series, which makes potential trends very difficult to detect. This detection issue justifies the need for robust methods to detect trends in extreme rainfall series, such as those proposed by Panthou et al. (2012, 2013, 2014a) whose works will be described in further detail in Section 4.

2.4 Drivers of Rainfall Variability

Identification of the factors controlling rainfall variability in West Africa is a great challenge because the WAM encompasses multiple physical processes acting at different space and timescales.

2.4.1 Basic Description of West African Monsoon

A major feature of the WAM is the large meridional gradient of climate with hot and dry conditions north of the continent, defined as the SHL, where the wind usually blows northerly and advects dry air to the south. On the opposite side, the Guinean coast is moist and cooler and experiences the southwesterly monsoon wind, which transports moisture from the equatorial Atlantic Ocean to the continent. In between these two regions lies the Sahel that can be defined as the region where the gradient of temperature and water vapor is the greatest. The large meridional gradient of temperature maintains a mid-level jet, called the African easterly jet, which is the main stream for convective development and for the growth of the major synoptic scale disturbances over West Africa, namely the African easterly waves (AEW), detailed hereafter.

2.4.2 From Large-Scale Variability to Synoptic Disturbances

The WAM system is highly variable at different space and timescales. The interannual variability of the WAM rainfall is driven by the oceanic variability over the equatorial Atlantic Ocean; somewhat directly, because it represents the main source of humidity for the WAM; however, teleconnections with the Pacific Ocean are also significant (Janicot et al., 1996). The phases of ENSO can significantly modulate the WAM-driven rainfall, with the negative phase of ENSO associated with positive anomaly of precipitation over the Sahel (Joly and Voldoire, 2009; Joly et al., 2007). For the extreme precipitations that occurred in 2007 (Paeth et al., 2011), showed that the La Niña phase was influential on the occurrence of AEW and the depth of the monsoon westerlies. The rainfall patterns displayed during wet or dry years have been categorized by Nicholson (2008, 2009), depending on whether the rainfall anomaly over the Sahel and over the Guinean coast has opposite (dipole pattern) or common (nondipole pattern) signs. The dipole pattern is associated with a shift of the rain band, whereas the nondipole pattern corresponds to a general weakening or strengthening of rainfall within the rain band. During the annual cycle, a part of the variability of rainfall takes place (1) for periods of 15–30 days, corresponding to propagating modes that modulate the convection directly (Janicot et al., 2011) or through a pathway by the SHL (Roehrig et al., 2011) or (2) at synoptic timescales (2–10 day periods), for which AEWs are the leading drivers of convective activity.

2.4.3 Summary of the Key Factors for Convective Organization

The processes that lead to the formation of MCSs have been studied over the Sahel using radar and ground observations (Chong et al., 1987; Redelsperger et al., 2002; Barthe et al., 2010 among others) and convection-permitting numerical simulations (Diongue et al., 2002; Barthe et al., 2010). It is now recognized that a sufficiently intense African easterly jet, providing the background flow for the development of AEW, can assist with the organization of convection. The vertical wind shear is a key parameter as well, giving the leading direction of the MCS propagation and helping to structure and expand the convective systems. It is to be noted that surface heterogeneities can help initiate convection, but they seem to play a secondary role when an AEW is already formed (Taylor et al., 2011).

2.4.4 Basic Behavior of MCSs and AEWs

In this context, the availability of water vapor is by far the limiting factor of convective activity over the Sahel. This explains why MCSs develop over the Sahel during the summer season when sufficient moisture is available in the region. The humidity is mainly brought in the low atmospheric levels (~0–2 km above the ground) by the southwesterly monsoon flow. At the synoptic scale, the MCSs are embedded within AEWs, synoptic perturbations of 3000–4000 km wavelength that propagate westward at 6–7 $m\,s^{-1}$, and feed the MCSs with vorticity (Berry and Thorncroft, 2012; Schwendike and Jones, 2010) and humidity (Poan et al., 2015). Southerly wind anomalies form the trough of an AEW and are usually associated with humidification, whereas northerly wind anomalies form the ridge and result in a drying of the atmosphere. Depending on the latitude at which the MCS–AEW is initiated, the maximum rainfall is not necessarily coincident with the trough. For AEWs developing north of 10°N, rainfall is enhanced to the east of the trough, whereas for AEWs arriving south of 10°N, in a moister environment, the rainfall anomaly is greater to the west or ahead of the trough (Poan et al., 2015).

2.4.5 Looking Back at the Extremes

Although the processes at play in the formation of an MCS have been documented, relatively little attention has been paid to the dynamics of extreme events. A fairly recent work carried out by Lafore et al. (2017) analyzed the time sequence leading to the extreme event in Ouagadougou, on September 1, 2009. They showed that a combination of factors of different scales allowed the MCS–AEW system to reach a dramatic daily rainfall amount of 260 mm in Ouagadougou. Three successive AEWs associated with an active Kelvin wave together with a large moist anomaly over the Sahel were the leading elements of the atmospheric context of this extreme event. However, we do not know to what extent this specific event is representative of other extreme events in Ouagadougou or over other areas of the Sahel.

3. SPACE–TIME STRUCTURE OF PRECIPITATION EXTREMES

The characterization of rainfall variability at a wide range of scales has long been a pertinent research subject in various regions of the world. In particular, the estimation of extreme rainfall probabilities at different durations and spatial area is of major interest for operational hydrology because of the need to design flood control structures. Multiscale extreme rainfall frequency distribution can also be used to evaluate the typical space and timescales at which a given heavy rainfall event can be considered as the most severe (Ramos et al., 2005; Ceresetti et al., 2012). The description of the space–time structure of extreme storms is also useful in evaluating the atmospheric model simulations and more precisely the space–time resolutions at which they are able (or not) to reproduce the storm structures. The main difficulty to that end is the complex structure embedded in rainfall, which results from the interaction of several processes occurring at scales varying from the drop size to the synoptic scales, in addition to the difficulty in measuring this variability.

3.1 Intensity–Duration–Area–Frequency Curves in West Africa

The most usual representations of scale dependence of extreme rainfall distribution are IDAF curves. IDAF depict how rainfall quantiles (in intensity or depth) evolve with time and space aggregation for different probabilities of exceedance (most often expressed in terms of return periods). They are usually composed of combined intensity–duration–frequency curves (IDF), which document the way the point rainfall distribution changes with time resolution, and areal reduction factors (ARF) which is, for a given duration and return period, the ratio between areal and point rainfall.

The assessment of IDAFs involves several requirements: (1) rainfall records at resolutions compatible with the scales of internal storm variability, typically spanning from local scale to 100 km in space and between minutes to a day in time; (2) sufficiently long series of rainfall to get a robust estimation of extreme value distributions; and (3) a suitable methodological framework allowing for a reliable assessment of extreme rainfall distributions and scaling effects. In West Africa, national weather services may provide long rainfall time series but most often at a daily time step and low spatial densities (lower than one rain gauge per 3500 km^2 Ali et al., 2005b), which do not cover the space–time scales of MCS internal structures. Several studies have attempted to estimate IDF from daily data sets in the region, but using strong unverified assumptions on rainfall time-scaling properties (e.g., De Paola et al., 2014; Data et al., 2016). Some authors have estimated IDF from satellite precipitation products (Endreny and Imbeah, 2009; Paeth et al., 2010; Awadallah and Awadallah, 2013), but the reliability of their results is questionable because of the short length of the satellite time series, not to mention their uncertainties due to indirect rainfall estimates. In some rare cases, IDFs have been estimated from subdaily data in Nigeria (Oyebande, 1982; Oyegoke and Oyebande, 2008), Congo (Mohymont and Demarée, 2010), and the Ivory Coast (Soro et al., 2010), but these studies often use empirical methods consisting of separately adjusting a frequency distribution model at each rainfall duration and then fitting an empirical scaling formulation. Although it has the advantage of being very easy to implement, this approach has been shown to present a risk of producing gross errors in IDF relationships such as parasitic oscillations or nonphysical curve intersections (Koutsoyiannis et al., 1998).

3.2 Recent Advances Based on Fractal Rainfall Properties

The most significant advances in the understanding of West African MCS internal structures have been made through the AMMA project (Redelsperger et al., 2006) and particularly its mesoscale observing system AMMA-CATCH (Lebel et al., 2009). AMMA-CATCH is dedicated to long-term monitoring of climatological, hydrological, and ecological changes in West Africa. It gathers data from three densely instrumented sites with areas varying from 10,000 to 16,000 km^2, located at different latitudes in Mali, Niger, and Benin, so as to sample the characteristic ecoclimatic gradients of the region. It is, to date, the only network in West Africa to provide densely distributed rainfall data at subdaily time steps, operating since 1990 in Niger, 1999 in Benin, and 2005 in Mali. AMMA-CATCH rainfall data sets have helped in documenting many features of MCS rainfall such as the rain intensity distribution (Balme et al., 2006b), the statistical spatial structure (Guillot and Lebel, 1999; Ali et al., 2003), and the rainfall kinematics (Depraetere et al., 2009; Vischel et al., 2011).

During the Special Observing Period in 2006, a Doppler RONSARD radar was settled in the Benin site and recorded rainfall during 3 months. By analyzing the radar rain rate maps, Verrier et al. (2010) and de Montera et al. (2010) have shown that rain storms typical of the region have multifractal properties. Despite being based on short-term data and not directly focused on extreme events, these two studies have demonstrated the potential of scale invariant models based on the fractal theory to describe the space–time storm structures in the region. From longer subdaily rainfall series provided by the AMMA-CATCH recording rain gauge networks, Panthou et al. (2014b) in the Niger site and Agbazo et al. (2016) in the Benin site have shown that the scale dependence of maximum rainfall values was satisfactorily represented by a fractal model based on a simple scaling relationship. The combination of the scaling relationship with the generalized extreme value (GEV) distribution[1] used to fit the distribution of rainfall maxima (as proposed by Menabde et al., 1999) led these two authors to propose a robust estimate of IDF curves in their regions of study. The building of spatial rain fields from recording rain gauges (Vischel et al., 2011) also allowed Panthou et al. (2014b) to successfully test a dynamic scaling model for ARF and then provide the only IDAF product available to date in West Africa.

Fig. 4.3 shows for the region of Niamey (Niger) how an extreme rainfall distribution (represented through return levels plots) changes with rainfall spatial resolution (ranging from point measurement to 2500 km^2) for three different rainfall durations (1, 6, and 24 hours). From the IDAF model (solid lines in Fig. 4.3), it can be noted that at a given spatial (respectively temporal) resolution and a given frequency, the rainfall intensity decreases with time (respectively spatial) resolution. This is an expected behavior coming from the smoothing effects of space and time aggregation on rainfall return levels. Another noticeable pattern is the diminution of rainfall return levels with spatial resolution, which is more pronounced for short than for long durations over a given range of spatial resolutions. This results from the

[1] Generalized extreme value distribution. See Section 4.1 "Dealing with detection issues" for further description.

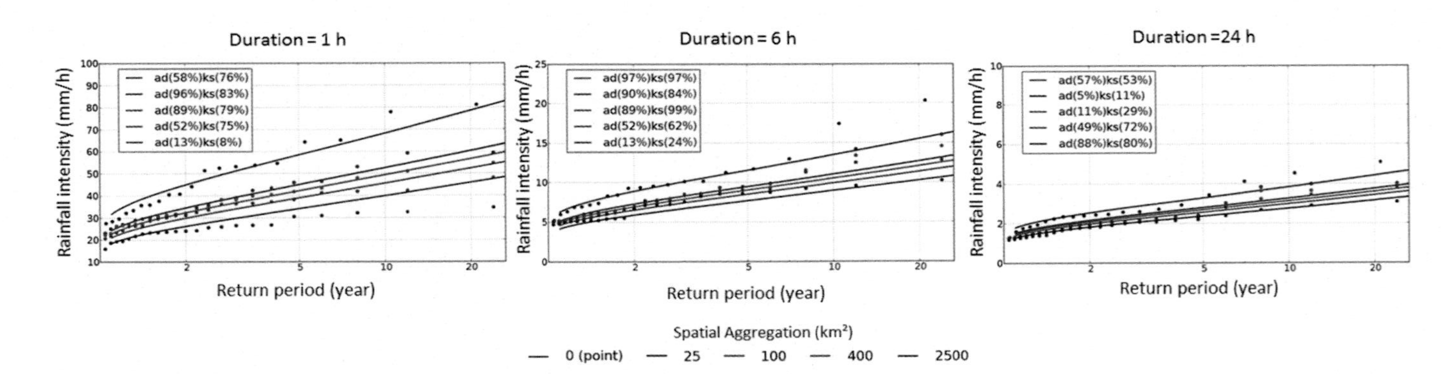

Figure 4.3 Empirical return level plots obtained at the Niamey Airport station in comparison with the global intensity–duration–area–frequency model for different durations (1, 6, and 24 hours from left to right) and different spatial aggregations (from point to 2500 km^2). *(Adapted from Panthou, G., Vischel, T., Lebel, T., Quantin, G., Molinié, G., 2014b. Characterising the space–time structure of rainfall in the Sahel with a view to estimating IDAF curves. Hydrol. Earth Syst. Sci. 18, 5093–5107.* https://doi.org/10.5194/hess-18-5093-2014.*)*

propagation of the storms over the study area, which increases the spatial extension of rainy areas for long durations and then reduces the smoothing effects of the spatial aggregation.

The quality of the proposed IDAF model is assessed by testing the adequacy of the selected distribution with empirical rainfall values, illustrated in Fig. 4.3 at the station of Niamey Airport. In each panel of Fig. 4.3, the *P*-values of two goodness of fit tests (ad for Anderson Darling and ks for Kolmogorov-Smirnov) are reported. They show good adequacy of the model with levels of significance always greater than 10% except for the smallest time step (1 hour) and the largest area (2500 km^2). At this time step, rainfall is highly intermittent in space at the scale of the study area. Consequently, the short duration rainfall samples mix both positive and zero rainfall values, which make the scaling properties more complex and the simple scaling assumptions more difficult to verify.

The IDF and IDAF models proposed by Agbazo et al. (2016) and Panthou et al. (2014b) are of great interest for better documenting the space and time structures of extreme storms in the region. They also provide quantitative information about rainfall return levels for hydrological engineers. However, these products are based on strong assumptions about the temporal stationarity of rainfall distributions, which is obviously questionable considering the strong past decadal variability of rainfall regime in West Africa. The associated limitations raise a general question of the relevance of assigning a return period to a rainfall event in a changing climate, an issue which is far from being purely a West African concern. Some recent studies have proposed methodologies to take into account rainfall distribution nonstationarities in IDF curves. Similar approaches would be probably worth developing in a West African context. By assessing the evolution of daily rainfall extremes over the recent decades, the results presented in the following section are a first step in that direction.

4. TRENDS IN PRECIPITATION EXTREMES

4.1 Dealing With Detection Issues

4.1.1 Requirements for Analyzing Trends in Extremes

In Section 2.3 we mentioned the difficulty of drawing clear conclusions on the recent evolution in extreme rainfall regime in West Africa from the few papers dealing with this aspect in the literature. It is likely that these studies are actually missing some of the main requirements that must be met when analyzing trends in precipitation extremes.

First, as already mentioned in the introduction, studies must be based on sufficiently long periods of records to perform robust statistical analyses on extremes, which are rare by definition. The constraint of record length explains why most studies on extreme rainfall trends in different regions of the world are based on the analysis of daily values recorded by rain gauges. To date, rain gauges are the only instruments to provide long-term rainfall observations, satellite observations being still too recent in that respect.

Second, meaningful statistical tests must be used to detect significant trends in series characterized by a strong natural variability both in space and time. Morin (2011) showed how the chance of detecting a significant existing trend in precipitation data, even of high magnitude, can be reduced in the case of series with low signal-to-noise ratio. Because point series of rainfall extremes are inherently very noisy and do not properly sample the spatial variability of extremes, their individual analysis can mask changes taking place at regional scales and can make trends very difficult to detect (Groisman et al., 2005). One recommended way to overcome the detection problem is to collectively analyze data from several geographical locations to optimally combine the temporal and spatial information provided by point rainfall series covering a climatic region (Morin, 2011; Khaliq et al., 2009).

4.1.2 Variability of Extreme Rainfall Series and Impact on Trend Detection Illustrated in the Sahel

The high variability of rainfall extreme series is illustrated in Fig. 4.4 for three Sahelian daily rainfall time series measured over the period 1950–90 by rain gauges located in Niger. The three stations are separated from one another by less than 60 km, and their environment characteristics are comparable, with similar vegetation cover (shrubby savanna) and topography (relatively flat in the region). Climatological features are thus expected to be the same, which is confirmed by the interannual averaged values of annual cumulative rainfall (gray lines Fig. 4.4A) and mean extreme rainfall (defined here by annual daily maxima, gray lines Fig. 4.4B). However, although Fig. 4.4A shows a quite consistent evolution of annual rainfall between the three stations (notably with a fairly visible contrast between the wet and dry periods, respectively, before and after the year 1970), series of rainfall extremes display very distinctive and erratic patterns without any obvious common trends or break points.

In light of this qualitative illustration, Panthou et al. (2013) applied several tests of nonstationarity (assuming linear trends and/or change points in the series) individually at 126 daily rainfall stations spread over Mali, Burkina,

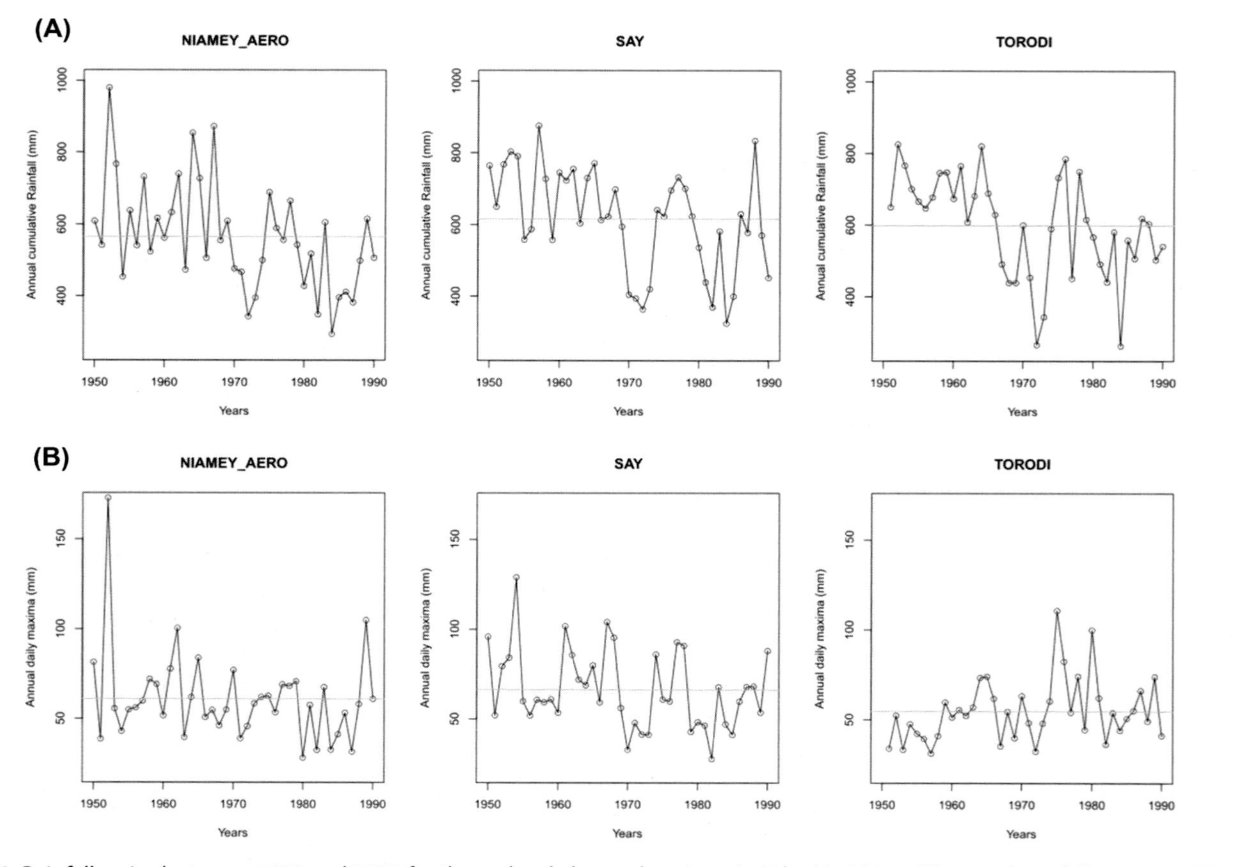

Figure 4.4 Rainfall series between 1950 and 1990 for three closely located stations (<60 km) in Niger: (A) annual rainfall amounts, (B) annual daily maximum rainfall.

Niger, and Benin. For series of cumulative annual rainfall values, Panthou et al. (2013) confirmed that all tests consistently detect nonstationarities over a majority (more than 90%) of the stations at a high level of significance (<5%), whereas only a few stations were shown to be nonstationary when these tests were applied to annual daily maxima series. Obviously, a significant trend at annual scale does not ensure the existence of a trend in extreme values and conversely an absence of trend at the annual scale does not prevent from detecting a trend in extremes (Groisman et al., 2005). However, Panthou et al. (2013) showed that the use of a regional statistical approach, integrating the spatial information of the entire rainfall networks, leads to decidedly different conclusions as significant nonstationarities can be then detected in the studied extreme rainfall series.

4.1.3 A Regional Approach to Analyze and Detect Nonstationarities in Extreme Rainfall Series

The regional statistical approach proposed by Panthou et al. (2013) for testing the stationarity of rainfall extreme series is based on the extreme value theory (EVT). The EVT provides a set of statistical distributions specifically dedicated to model the distribution of extreme values, including the GEV distribution function, commonly used to estimate the distribution of block maxima (most often annual daily maxima in case of hydrometeorological data) and the generalized Pareto distribution used to model the distribution of peaks over a threshold.

While often used for hydrological applications, mostly on individual stations for pointwise estimates of return levels, these distributions can also be estimated using several stations jointly. More precisely, the distribution parameters can be formulated as depending on spatially varying covariates able to represent the variation in space of the extreme distribution over a given region. The regional rainfall data set can then be pooled into one single sample from which the regional extreme distribution can be directly inferred by the maximum likelihood method (e.g., Blanchet and Lehning, 2010). Applied on the 126 stations in the central Sahel, Panthou et al. (2012) showed that a regional GEV distribution with the mean interannual annual rainfall over the period of study 1950–90 taken as a linear covariate of the location parameter was much more efficient than a classical pointwise approach for estimating the rainfall return levels.

Similarly, time-varying covariates can be included in the formulation of the extreme distribution parameters. Panthou et al. (2013) used time-dependent parameters to detect regionally significant trends and/or break points in series of annual maximum daily rainfall. To do so they compared the capability of (1) a pointwise stationary GEV distribution versus a time-dependent GEV distribution and (2) a regional stationary GEV distribution versus a regional time-dependent GEV distribution in their ability to model the distribution of extreme rainfall time series. Both the pointwise and regional nonstationary GEV models were shown to significantly outperform the stationary models and demonstrated that the distribution of daily rainfall maxima underwent a significant breakpoint in 1969–70.

An important methodological result emerging from Panthou et al. (2012, 2013) is the good capacity of the proposed regional statistical approach to reduce the effect of record length and noise of extreme rainfall series on trend detection.

4.2 Trends in Daily Precipitation Extremes in the Central Sahel (1950–2015)

The methodological developments proposed by Panthou et al. (2012, 2013) opened the possibility for Panthou et al. (2014a) to analyze the evolution of precipitation extremes in the Sahel in light of regional WAM multidecadal variability. To accomplish this, Panthou et al. (2014a) used daily rainfall data from 43 stations located between 9.5°N and 15.5°N, 5°W and 7°E, and having operated continuously from 1950 to 2010. For the purpose of this book chapter, the main results obtained by Panthou et al. (2014a) have been extended for 5 years after an update (up to 2015) of the entire data set.

4.2.1 Evolution of Annual Daily Maxima

The use of a regional GEV model adjusted to the 43 stations allows for the computation of a regional standardized index of annual daily rainfall maxima. The index reported in Fig. 4.5 is also smoothed by moving averages over 5 and 11 years, to better highlight the decadal variability. Fig. 4.5 can be compared with Fig. 4.1, which similarly represents standardized values of annual rainfall amounts over the same period and for the same stations.

The evolution of annual rainfall amounts (Fig. 4.1) and annual daily maxima (Fig. 4.5) curves is quite coherent until the beginning of the 1990s. Both annual totals and annual maxima mark a negative break at the end of the 1960s, already evidenced by Panthou et al. (2013) and start an increase in the middle of the 1980s. However, although the mean annual rainfall stabilizes at the middle of 1990s below the average levels for the period 1950–2015, the daily maxima continue to increase with a marked rise over the period 2000–15, reaching levels as high as those observed during the wet period preceding the drought.

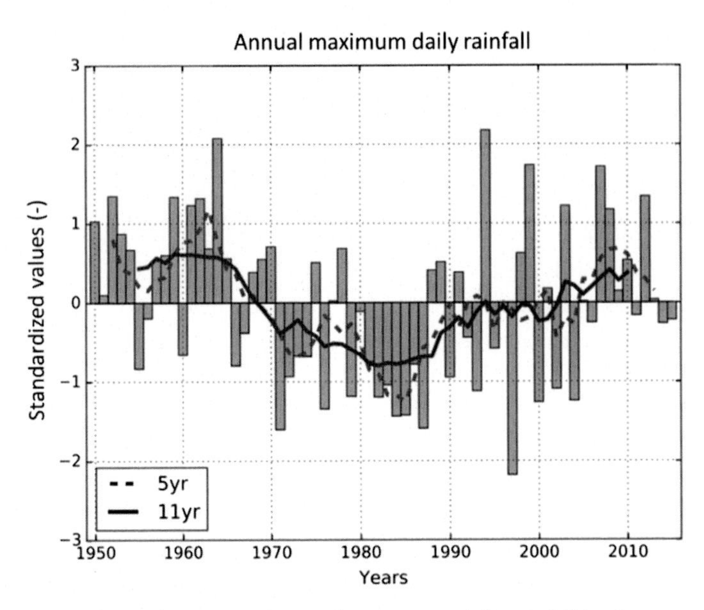

Figure 4.5 Standardized values of annual maximum daily rainfall in the central Sahel (computed from 43 stations in a box defined by longitudes 5°W–7°E and latitudes 9.5°N–15.5°N). *(Updated after Panthou, G., Vischel, T., Lebel, T., 2014a. Recent trends in the regime of extreme rainfall in the Central Sahel. Int. J. Climatol. 34, 3998–4006.* https://doi.org/10.1002/joc.3984.*)*

4.2.2 Evolution of Values Over Threshold

As maxima analysis is only based on one value per station per year and as it does not provide information about the occurrence of extreme events, Panthou et al. (2014a) added supplementary analyses based on a peak over threshold methods. Using again the regional statistical approach, they defined at each station (1) extreme events as daily values over a threshold being exceeded only twice a year on average over the period of study and (2) intense rainfall events as daily values over a threshold being exceeded only 10 times a year on average over the period of study. From north to south, the threshold values vary between 30 and 60 mm for extreme events and between 12 and 32 mm for intense events.

Two main results can be drawn from Fig. 4.6, which describes the changes over the period 1950–2015 in mean annual number of extreme and intense events (Fig. 4.6A) and their relative contribution to annual rainfall amounts (Fig. 4.6B). First, the occurrence of extreme events follows a trend similar to annual daily maxima. Their evolution is marked by a significant rise over the last 15 years, to a level close to those of the wet period

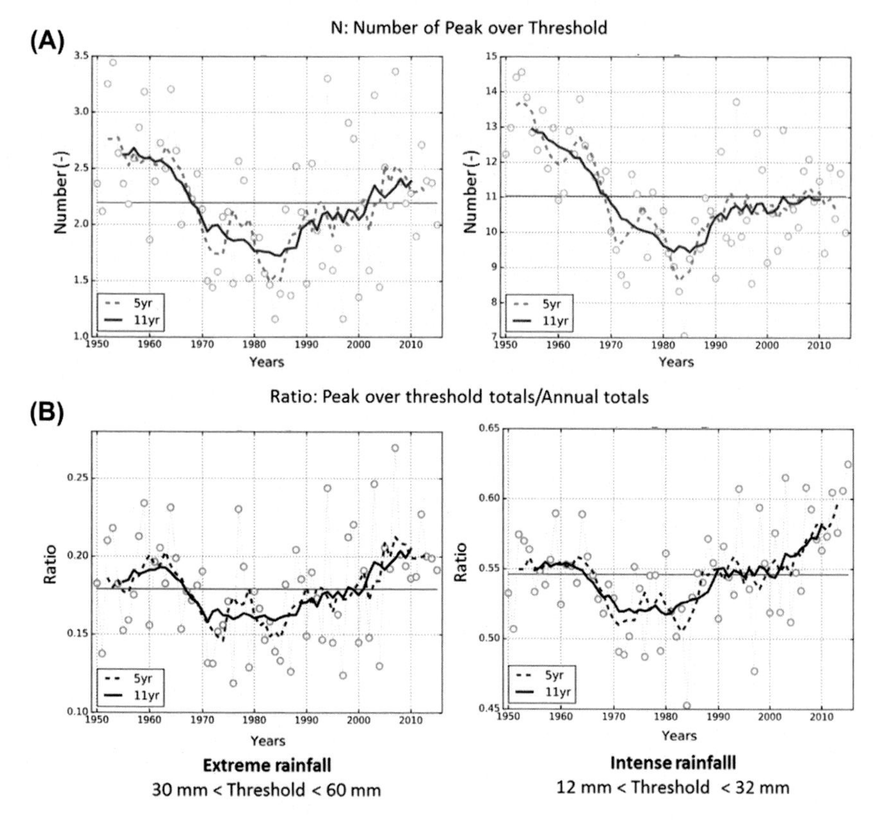

Figure 4.6 (A) Evolution of the number of extreme (left) and intense (right) daily rainfall events, (B) Evolution of extreme (left) and intense (right) daily rainfall events expressed as a ratio of total rainfall. *(Updated after Panthou, G., Vischel, T., Lebel, T., 2014a. Recent trends in the regime of extreme rainfall in the Central Sahel. Int. J. Climatol. 34, 3998–4006.* https://doi.org/10.1002/joc.3984.*)*

before 1970, whereas the occurrence of intense events stays under the interannual mean level. Second, the relative contribution of both extreme and intense events to the annual totals has sharply increased since 2000 to unprecedented levels. Over the study period 1950–2015 and despite the presence of two very wet decades in the 1950s and 1960s, the importance of intense and extreme events in the rainfall regime has never been as high as in the last 15 years.

The notable increase of extreme rainfall without any clear trend in annual rainfall confirms the findings of Panthou et al. (2014a) that an important change in rainfall regime occurred at the turn of this century, with extreme events becoming more pronounced.

4.3 Link With Mesoscale Convective System Spatial Extent

The recent intensification of the Sahelian rainfall regime raises the question of potential changes in the spatial extent and organization of MCSs. It is of particular interest here because the propensity of convection to aggregate and organize may contribute to changes in extreme rainfall events (Bony et al., 2015).

Although IR data from geostationary satellites are highly relevant to documenting the life cycle and properties of MCSs (e.g., Mathon et al., 2002; Fiolleau and Roca, 2013), they only provide data over the last 30 years. Alternatively, Bell and Lamb (2006) showed that indices based on daily gauge-based rain measurements could be defined to document MCS properties in relation with the decadal variability. Their study covered four 4° × 4° square boxes in Niger, Burkina Faso, Mali, and Senegal over the period 1950–98. Here, we propose to update their results related to the spatial extent of MCSs by using the daily data set of Panthou et al. (2014a) that covers the boxes in Niger (11°N–15°N, 0°E–4°E) and Burkina Faso (10°N–14°N, 0°W–4°W) up to 2015.

As suggested by Bell and Lamb (2006), the degree of convection organization of MCSs is assessed through the computation of a daily disturbance extent index (DDEI) defined as the proportion of stations that received rainfall on a given day. The DDEI thus represents the horizontal extent of MCSs. The evolution of the mean annual DDEI values in Niger and Burkina Faso is reported in Fig. 4.7 over the period 1950–2015.

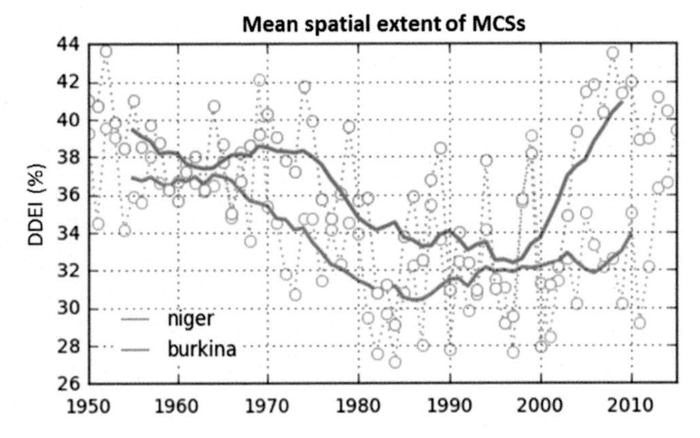

Figure 4.7 Evolution over the period 1950–2015 of the mean annual daily disturbance extent index (DDEI) representing the spatial extent of mesoscale convective systems in Niger and Burkina Faso areas. *Plain lines* represent the 11-year moving averages.

As already indicated by Bell and Lamb (2006) (Fig. 15 in their paper), Fig. 4.7 shows that the mean MCS extents significantly decrease (by a maximum of 16% in Niger and 18% in Burkina Faso) at the end of the 1960s in conjunction with drought conditions. In the middle of the 1980s, MCS spatial extent started a slight but steady increase in Niger, reaching the average levels of the period. In Burkina Faso, the decrease in mean MCS extent lasted up to the end of the 1990s but then sharply increased, exceeding all prior levels as early as 2005.

By updating one of the results of the earlier study of Bell and Lamb (2006) in light of more recent rainfall data, we thus show a trend toward bigger rainy systems over the last 15 years in the central Sahel.

5. LARGE-SCALE ATMOSPHERIC ENVIRONMENT OF EXTREME PRECIPITATING EVENTS OVER THE CENTRAL SAHEL: THE CASE STUDY OF OUAGADOUGOU, BURKINA FASO

Despite being at an early stage of research, valuable knowledge has been gained in recent years about trends in extreme rainfall in the Sahel. However, the associated atmospheric environment has not yet been well documented and understood. This section provides for the first time an outline of a few major atmospheric features associated with the occurrence of extreme rainfall over Ouagadougou, Burkina Faso. Ouagadougou has been chosen as it allows to extend the study of Lafore et al. (2017), but the following results seem to be representative of extreme rainfall events occurring over the central Sahel region (work in progress, not shown). Extreme precipitating events are detected following the work of Panthou et al. (2014a) then a composite of their large-scale environment is used to extract their common features. Data and methodology are described in Section 5.1. Section 5.2 highlights the annual and daily cycle of the occurrence of the extreme events over Ouagadougou. Section 5.3 provides insight on the mechanisms at play during the average composite extreme event occurring over Ouagadougou.

5.1 Methodology and Data

5.1.1 Rainfall and Atmospheric Data

Extreme events are detected at the daily timescale over the period 1950–2010, following the methodology of Panthou et al. (2014a) previously described in Section 4.2. From the data set of Panthou et al. (2014a), we focus only on the events occurring over Ouagadougou (except for the annual cycle of the frequency of occurrence of extreme events for which we use all the stations—Fig. 4.8). Although this data set is suitable for analyzing extreme events from a local perspective (ground station), its time resolution is not sufficient to accurately sample atmospheric processes, such as AEWs or MCSs, whose scales range from a few hours to a few days. Therefore, 6-hourly or even 3-hourly data are more suitable. We thus use rainfall estimates from the Tropical Rainfall Measuring Mission (TRMM) satellite to refine the time occurrence of the extreme events detected from daily frequency values by the approach of Panthou et al. (2014a) to the 6-hour frequency. The 3B42 version 7 of TRMM rainfall estimates (Huffman et al., 2007) is available on a $0.25° \times 0.25°$ spatial grid at a 3-hour frequency for the period 1998–present. Note that these rainfall estimates are not used quantitatively. They only provide an assessment of the time when each event reaches its maximum intensity over a few TRMM pixels containing Ouagadougou (see below).

The state of the atmosphere is described using the European Centre for Medium-Range Weather Forecasts (ECMWF) Interim Reanalysis (ERA-Interim), available from 1 January 1979 to present. Outputs are available every 6 hours with a $0.75° \times 0.75°$ horizontal resolution and on 36 vertical levels. We use wind and column water vapor (CWV) for describing the atmospheric dynamics and thermodynamics associated with extreme events.

Considering the available period of each of these data sets, the present analysis focuses on the period 1998–2010. This period samples 20 extreme events occurring over Ouagadougou. The composite sequence of the

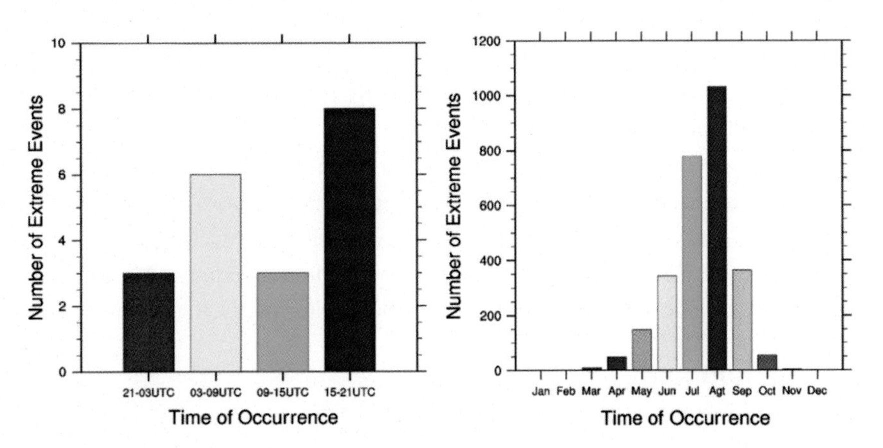

Figure 4.8 Left: Mean annual cycle of extreme events over the central Sahel over the period 1979–2010. Right: Average time of arrival of extreme events in Ouagadougou over the period of JJAS 1998–2010 (defined by combined analysis of daily extremes at the Ouagadougou station defined by Panthou et al., 2014a and Tropical Rainfall Measuring Mission 3B42 3-hourly rainfall estimates).

atmosphere structure associated with these extreme events is computed at a 6-hourly resolution.

5.1.2 Refinement of the Dates of Extreme Events

As the work of Panthou et al. (2014a) is based on daily accumulated rainfall amount at ground stations, the extreme events are detected at a daily time resolution. Here, we further assess the time of occurrence of these events using that of the maximum of TRMM 6-hourly rainfall estimates falling in the area of $1° \times 1°$ centered on Ouagadougou within the 48 hours, which encompass Panthou's date (midnight). The area of $1° \times 1°$ is chosen to capture the large-scale signature of extreme precipitating events. As previously mentioned, this methodology has been developed to describe the extreme events from a large-scale perspective. It is therefore more suitable to describe the atmospheric conditions than a local analysis based on ground station observations.

5.1.3 Building of the Extreme Rainfall Event Composite

The composite structure of these extreme events is then computed as the average of the atmospheric state associated with each of the 20 events occurring over Ouagadougou. A composite sequence is also computed from 5 days prior to the events (Lag −5) to the date of the events (Lag 0). To better emphasize the dynamical and thermodynamical structure associated with these events, and especially how they can depart from the slowly varying state of the WAM, we compute composites of both raw and anomalous atmospheric variables. Anomalies are defined as departure from the mean 1998–2010 annual cycle (see e.g., Poan et al., 2013 for further details).

5.2 Seasonal and Diurnal Cycles of Extreme Events

Fig. 4.8 shows the annual cycle of the frequency of occurrence of extreme events of the central Sahel and the daily cycle of the frequency of occurrence of extreme events occurring over Ouagadougou. Extreme events are more frequent over the Sahel during the monsoon season from June to September, when the monsoon flow provides enough moisture (and thus instability) over the region. A peak in August is observed, when the Sahel is known to be the moistest (Sultan and Janicot, 2003). With respect to the diurnal cycle, TRMM data indicate that the maximum of events in Ouagadougou is obtained in the late afternoon (15:21 UTC, 8 events) and in the early morning (3:9 UTC). Consistent with the dynamics of MCSs over the region and the study Mathon et al. (2002), the late afternoon peak of extreme events probably corresponds to the local development of convection, whereas the early morning peak is consistent with the passage of mature systems formed further to the east.

5.3 Large-Scale Atmospheric Conditions Leading to Extreme Events

Fig. 4.9 displays the atmospheric conditions that occurred on average when Ouagadougou is impacted by an extreme event. As expected, TRMM rainfall shows a maximum over Ouagadougou reaching a magnitude of 3.5 mm hour^{-1} (Fig. 4.9A). The composite event occurs within a northeasterly wind shear corresponding to the southern branch of the 700 hPa large-scale anticyclonic circulation. The maximum precipitation in Ouagadougou covers a large band of rainfall, extending up to 20°N and in good accordance with the monsoon flow northern penetration (Fig. 4.9C). The southwesterly monsoon flow is key for moistening the 5°N–20°N latitudinal band, in particular the region east of Ouagadougou, where CWV reaches values greater than 50 mm, a threshold usually required to obtain MCS over the Sahel.

The maximum precipitation in Ouagadougou is associated with a cyclonic vortex (C2) impacting the deep layer between 925 and 700 hPa (Fig. 4.9B and D). It is embedded within the trough of an AEW on the southern flank of the African easterly jet (green contour). Note that west of the event, around 10°W, a huge ridge extends from the equator to 30°N, largely exceeding the typical scale of an AEW. The CWV anomaly shows at the same time an unusual pattern: a moist anomaly exists all around the cyclonic vortex extending from 0°E to 15°E (blue area surrounding C2), and a secondary moist anomaly is located further to the north at the Mauritania–Mali border (A1). The pattern of CWV anomaly is large compared to the one expected within an AEW, which is usually confined within the southerly wind anomaly (Poan et al., 2013). Here the moist pattern of half wavelength covers almost 2500 km, from 10°W to 15°E. It is noteworthy that this is a very specific configuration because northerly wind usually transports drier air from the north to the south. Here the large-scale moist anomaly can fuel the MCS with moisture from mostly all directions, including from the north.

The features emphasized by Lafore et al. (2017) appear to be common to most of the extreme events occurring over Ouagadougou: the MCS leading to extreme precipitation is embedded within an AEW trough and is associated with a strong wind shear and a large-scale moist anomaly taking place over the continent.

The time sequence of CWV and 925 hPa wind anomalies is displayed in Fig. 4.10 to assess the origin of the atmospheric drivers. The large-scale

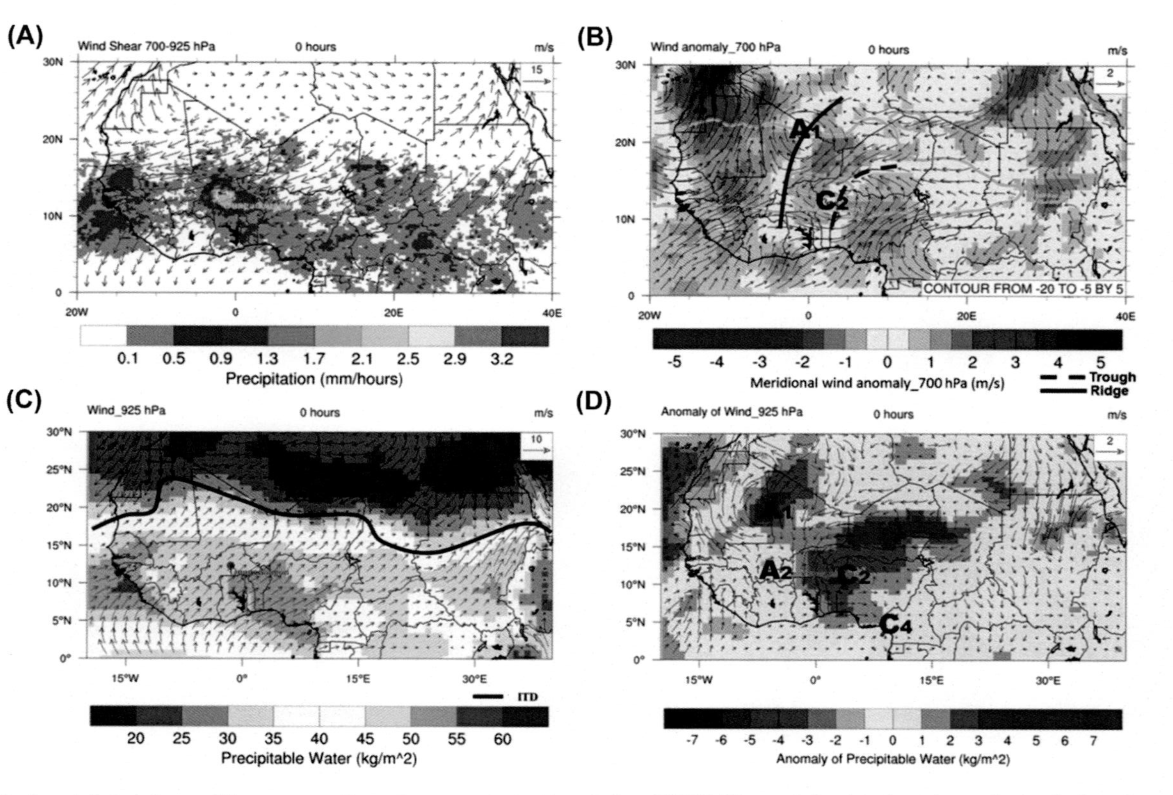

Figure 4.9 Horizontal structure of the composite extreme event at Lag 0 for: (A) TRMM precipitation (mm hour^{-1}, *shaded*) and 700–925 hPa wind shear (m s^{-1}, vectors); (B) 700 hPa wind anomaly (ms^{-1}, vectors) and 700 hPa meridional wind anomaly (ms^{-1}, *shaded*); (C) CWV (kg m^{-2}, *shaded*) and 925 hPa wind (m s^{-1}, vectors); (D) CWV anomaly (kg m^{-2}, *shaded*) and 925 hPa wind (m s^{-1}, vectors). In (C) the *black line* indicates the position of zero zonal wind. In (B) the *green line* represents the AEJ core, the *solid (dashed) line* indicates the through (ridge) of the AEW. In (B and D) the symbols A and C indicate the anticyclonic and cyclonic circulations. *AEW*, African easterly waves; *CWV*, column water vapor; TRMM, Tropical Rainfall Measuring Mission.

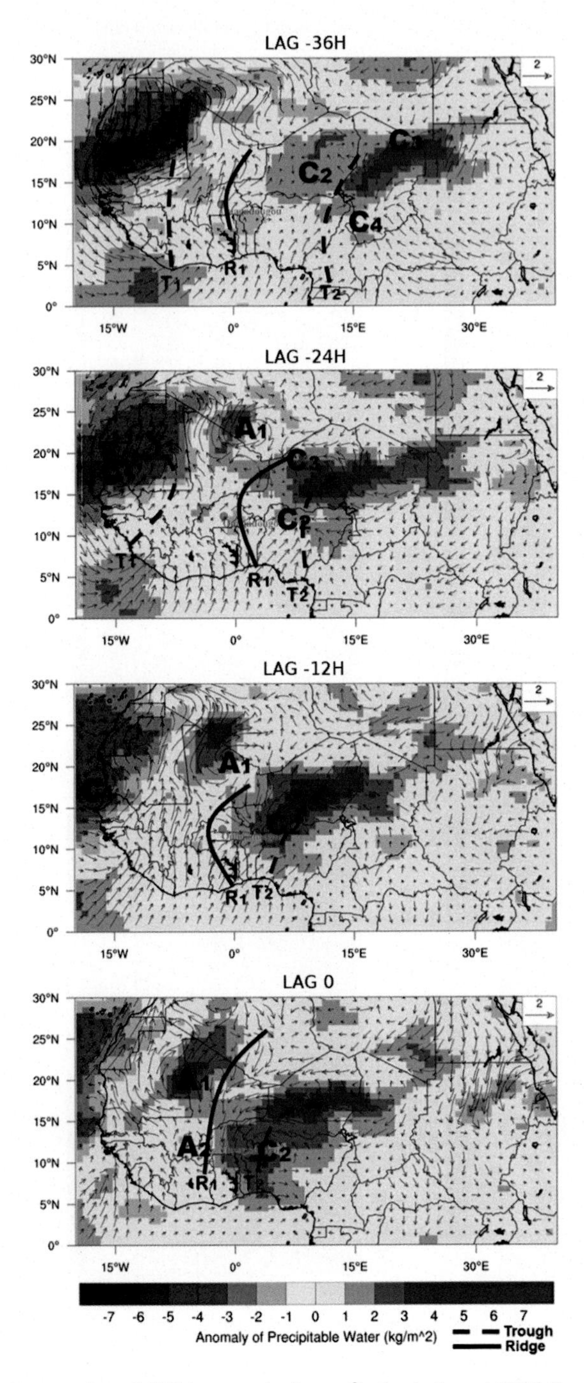

Figure 4.10 Composite of CWV anomaly (kg m^{-2}, *shaded*) and 925 hPa wind anomaly m s^{-1}, vectors) for Lag −36, −24, −12 hours and Lag 0. The *solid (dashed) line* indicates the trough (ridge) of the AEW. A and C indicate the anticyclonic and cyclonic circulations. T2 is the trough associated with the arrival of the extreme event in Ouagadougou. *AEW*, African easterly waves; *CWV*, column water vapor.

moist anomaly can be detected 36 hours before the occurrence of the events (Lag −36 hours) around 15°E (top), whereas a dry anomaly still prevails on the western part of the continent (C1). The trough and the cyclonic vortex leading to the extreme event (T2, C2) can also already be detected at 15°E. A noticeable feature is the southerly wind anomaly extending from the equator to 15°N, west of trough T2, contributing to the moistening of the atmosphere over the continent.

Over time these structures move gradually westward with the moist anomaly splitting into two subregions at Lag −12 hours, whereas the dry anomaly heads to and reaches the Atlantic Ocean. Note that from Lag −24 hours to Lag 0, the northerly wind west of C2 brings moist air into the cyclonic circulation, allowing supplementary moisture to arrive within the MCS. This is rather unusual in the context of AEWs (e.g., Poan et al., 2013). There again the large-scale moist anomaly is key to the continuous moistening of the MCS from the north.

A synthetic view is illustrated in Fig. 4.11, showing a lag–longitude diagram of the 700-hPa meridional wind and CWV anomalies.

The moist large-scale anomaly can easily be seen at Lag −5 days around 30°E with a wavelength of ~7000 km propagating at 6–8 ms^{-1}. Within the large-scale moist/dry pattern, a first AEW (T1/R1) is initiated at Lag −3 days during a dry phase. It is followed by a moister AEW (T2) in which the extreme event forms (see also Fig. 4.10). There are thus several scales at play during the composite extreme event, with a large-scale moist anomaly moving westward at about the same velocity as the AEW but with a wavelength twice as big. Note the AEW trough (T2) propagates faster after the extreme event (Lag 0, 5°E), likely indicating some interaction between the MCS, the AEW, and the large-scale moist anomaly. It is also noteworthy that 5 days prior to the event, the large-scale pattern can already be tracked, suggesting some potential predictability for such extreme events.

5.4 Discussion

The features underlined by Lafore et al. (2017) appear to be at work over Ouagadougou for the composite extreme precipitating event, and in fact for most of the extreme precipitating events. Indeed, a visual analysis of each event does reveal that the mean composite is representative of the atmospheric conditions associated with each of the 20 events. These conditions include patterns of different scales such as an AEW and the large-scale moist anomaly.

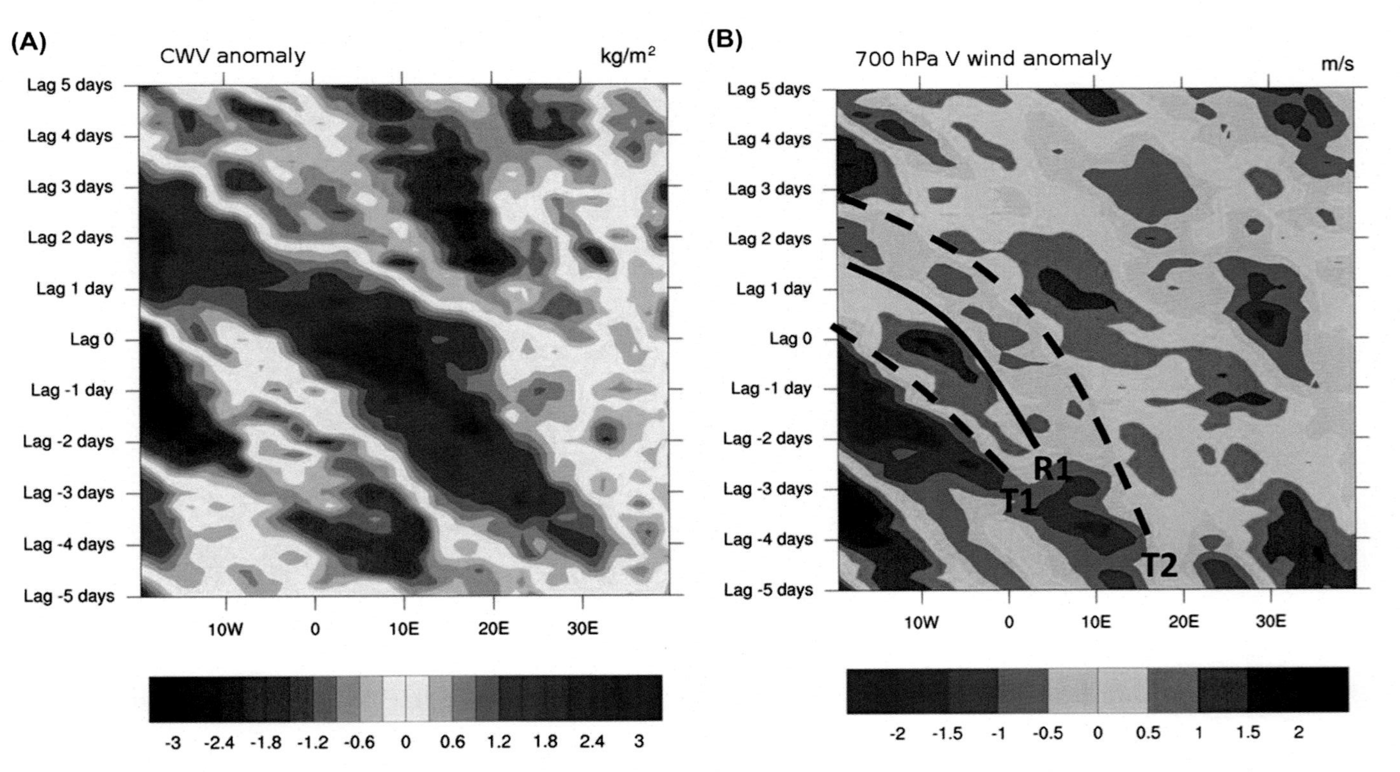

Figure 4.11 Lag–longitude composite structure averaged over 12°N–16°N of (A) column water vapor (CWV) anomaly (kg m^{-2}, *shaded*) and (B) 700 hPa meridional wind anomaly (m s^{-1}).

The nature of the large-scale moist anomaly remains unclear but appears consistent with the intraseasonal variability of the WAM (Janicot et al., 2011 and references therein) and especially with that of the SHL (Chauvin et al., 2010; Roehrig et al., 2011). Further analysis is necessary to fully understand this moist anomaly. Here we consider a single set covering all selected extreme events in Ouagadougou, but it might be insightful to separate these events depending on their arrival time (late afternoon vs. early morning) to better understand the interaction between diurnal and MCS life cycles.

6. CONCLUSIONS AND PERSPECTIVES

While long involved in documenting the reasons for the great Sahelian drought that struck the region at the end of the 1960s for almost 30 years, climate scientists are now challenged with the understanding of the new turn taken by rainfall since the end of the 1990s: a return to wetter conditions, though still dryer than the two decades preceding the drought; and a strong interannual variability of annual totals. The last 25 years have experienced both low agricultural yields and an increasing number of damaging floods, raising the question of a more extreme climate with more dry spells and more precipitations extremes.

Tarhule (2005) was the first to signal the growing societal importance of floods in the Sahel. Intuitively, he suggested that "[i]t is reasonable to expect [...] that more frequent high-intensity rainfall would characterize a return to wetter conditions, with a concomitant increase in flood risk." Eleven years after, are there robust scientific arguments to confirm this intuition? More generally, what do we know about rainfall extremes in the Sahel? This chapter summarizes the state of knowledge in that regard.

We show that some advances have been made in characterizing the space–time structure of extreme Sahelian storms. In particular, scale invariance properties of extreme rainfall have been demonstrated in the framework of the fractal theory. This helped in characterizing the dependence of extreme rainfall distribution on time duration (between 1 and 24 hours) and spatial area (from point to 10,000 km^2). The rainfall scaling properties allowed for the generation of IDAF curves for the surroundings of Niamey (Niger), which are very useful tools for hydrological engineering and for assessing the severity of rainfall events in observations or atmospheric model simulations.

Although the recent evolution of the annual rainfall regime has been well documented in the Sahel, there are only a few papers dealing with the recent changes in Sahelian MSC properties, and even fewer that consider the most intense of them. MCSs are, however, key elements for understanding the hydroclimatic variability in the Sahel as they produce the great majority of seasonal rainfall. As shown in this chapter, their occurrence and intensity directly influence the hydrological response of catchments. Studies dealing with trends in extreme rainfall in the Sahel often suffer from the difficulty of (1) obtaining time series with long enough periods of records and/or (2) identifying robust methodologies to detect significant trends from notoriously noisy time series of extremes. Dealing with the detection issues, Panthou et al. (2012, 2013) proposed a robust regional statistical methodology based on the EVT. This allowed Panthou et al. (2014a) to assess the recent trends in extreme Sahelian rainfall. They showed that the recent return to higher annual rainfall in the Central Sahel has happened in a context of a persisting deficit of the number of rain events, compensated by a larger share of strong rainfall events. This new rainfall regime is typical of a more extreme climate characterized by harsher dry spells during the rainy season and more precipitation extremes. This trend seems to be accompanied by an increasing mean size of MCSs. This is coherent with the fact that bigger MCSs are more likely to produce higher rainfall amounts (e.g., Roca et al., 2014).

This chapter provided an original analysis of the atmospheric processes responsible for the 20 most extreme events that have happened in Ouagadougou since 1997. A composite analysis of these 20 events revealed the presence of consistent atmospheric features. First, favorable conditions for the development and maintenance of convection are provided by a large-scale moist anomaly, already visible in the eastern Sahel 3 days before the event and arriving at Ouagadougou in phase with a marked AEW. The concurrence of synoptic (AEW) and large-scale (moist anomaly) atmospheric drivers is a key feature of the occurrence of extreme rainfall events in Ouagadougou. It allows for the moistening associated with the well-developed vortex to be found in the composite with a northerly transport of moist air to the west of the MCS and advection by the monsoon flow on its eastern flank.

According to the recent evolution of the rainfall regime shown in this chapter, we recommend using with caution the term "recovery," often used to qualify the rainfall of the last two decades. This word only reflects the evolution of mean annual rainfall levels, and it masks a strong interannual variability and a significant intensification of the rainfall regime at the daily scale. When used without caution it can also inadvertently downplay the societal impacts of climate extremes that remain in force in the Sahel. Dry years have continued to affect crop yields and water resources, and heavy rainfall has more than ever destroyed crops and increased the risk of flooding. By combining an increase in extreme precipitation and in dry spells, the Sahelian rainfall regime actually exhibits all the signs and symptoms of a "higher hydroclimatic intensity" as defined by Giorgi et al. (2011). This term or "period of rainfall intensification" is preferred to "recovery" when describing the Sahelian climate of the last two decades.

Despite great advances, several questions related to extreme rainfall are still open and deserve to be clarified through further research.

In particular, some studies recently showed that during the last two decades, the mean annual rainfall levels remained much longer in deficit over the western Sahel than in the eastern and central Sahel (Lebel and Ali, 2009; Mahé and Paturel, 2009). This east–west dipole is seen in the recent CMIP simulations and seems to be a dominating pattern for the decades to come (Monerie et al., 2012). An important issue is thus the way this regional contrast is reflected in the evolution of the mean and extreme rainfall regimes at mesoscale. More generally, the question of regional contrasts must also be examined in the southern West Africa (Sudanian and Guinean regions) to highlight the impact of the regional north–south gradient and the potential differences between inland and coastal precipitation.

Further research is also needed to identify the mechanisms involved in the evolution of extreme events in the context of climate change. The role played by the atmospheric features shown to be responsible for extreme events in Ouagadougou has to be confirmed at the regional scale and possibly on longer timescales. Then it will be possible to analyze how the occurrence of these features has changed in the recent past and how it will change in the future. Climate models will be helpful in that respect. Although they still struggle in representing convection processes, they most often provide a good representation of synoptic patterns driving rainfall over West Africa.

Another task is to more accurately characterize the typology of MCSs that locally produce extreme rainfall amounts. The increasing mean size of MCSs presented in this chapter suggests that bigger systems might be more frequent over the last 10 years. This preliminary result must be confirmed by deeper analyses of the interaction between local intense rainfall and associated MCSs. In the short term, it is planned to realize a combined analysis of rain gauge–based measurements and infrared data from geostationary satellites available since the beginning of the 1980s. This should allow for assessing changes in size and duration of MCSs and detecting the life cycle stages at which they are the most prone to triggering extremes.

A related issue is improving documentation on the internal structure of extreme rainy systems. Although most of the studies dealing with extreme rainfall evolution are based on daily values, it is also very important to document how rainfall intensities have evolved at subdaily scales. Fig. 4.12 shows 5-min time step records of the two biggest events available in the AMMA-CATCH Niger database over the period 1990–2016 (Lebel et al., 2009). Both events occurred in the morning and accumulated more than 200 mm of rainfall in less than 6 hours (return period estimated around 3000 years). Interestingly, these two extreme events display very distinctive temporal structures characterized for first event (event of Koure 2003) by two main convective cells with maximum intensities around 10 mm/5 min, and for the other (event of Dantiandou 2014) by one single very intense convective cell with a maximum intensity peaking at 14.5 mm/5 min. These time structures probably come from a different degree of convection organization within the MCSs that produced these events. Moreover, it is likely that these two types of hyetographs, if occurring over a similar catchment, would have a very different hydrological impact. Runoff production being very sensitive to short timescale rainfall intensities, especially in the Sahel where soils have very low infiltration capacities, it is likely that an event such as that of Dantiandou 2014 would produce more runoff than the event of Koure 2003.

This illustration raises the question of how the intensification demonstrated at a daily basis has been translated into subdaily rainfall intensities. Are convective cells in extreme events more intense or more numerous, or both? How has the temporal structure of extreme events evolved over time? Investigations based on the high-resolution rainfall data set provided by the Observation Service AMMA-CATCH are ongoing on this subject, mainly with aim of assessing recent trends in maximal short time rainfall intensities. This will contribute to the development of nonstationary IDAF curves that are of high interest for infrastructure design in a changing climate (e.g., Cheng and AghaKouchak, 2014). This will also be of great value for better evaluating the relative contribution of rainfall and land-use changes in the recent increase in floods in the Sahel (Descroix et al., 2012; Aich et al., 2015; Casse et al., 2016).

Finally, it is worth noting that the most significant advances presented in this chapter are the result of combined statistical and physical analyses of both ground-based and satellite observations. They have been fostered by international scientific programs such as the AMMA (Redelsperger et al., 2006)

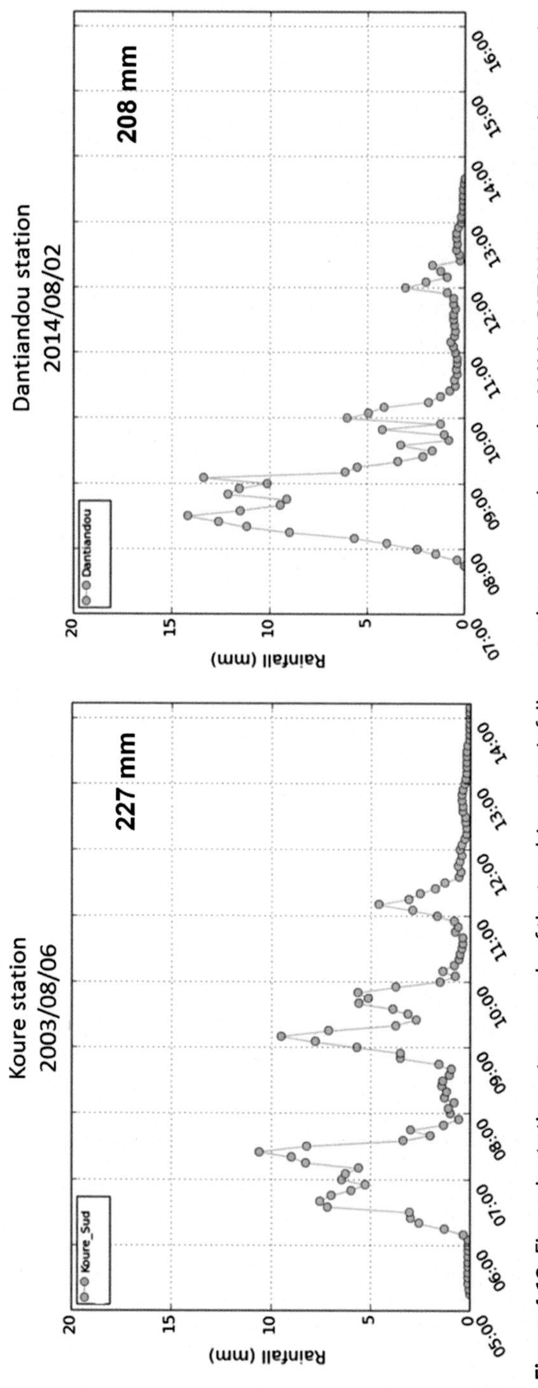

Figure 4.12 Five-minute time step records of the two biggest rainfall events that occurred over the AMMA-CATCH Niger network since it has operated in 1990.

project and its current successor AMMA-2050 (www.amma2050.org), as well as by projects dedicated to climate observations such as the AMMA-CATCH Observing System (Lebel et al., 2009, www.amma-catch.org) and the satellite mission Megha-Tropiques (Roca et al., 2015, meghatropiques.ipsl.polytechnique.fr). All these initiatives help to reinforce two essential pillars required if one wants to respond to the challenges posed by the ongoing global changes: (1) continuing long-term climate observations; and (2) favoring the exchange of scientific expertise beyond disciplinary barriers.

ACKNOWLEDGMENTS

The research leading to these results has received funding from the NERC/DFID Future Climate For Africa programme under the AMMA-2050 project, grant numbers NE/M020428/1, NE/M019950/1. This work was also supported by the French national programme EC2CO-LEFE "Recent evolution of hydroclimatic hazards in the Sahel: detection and attribution."

REFERENCES

Agbazo, M.N., Koto N'Gobi, G., Kounouhewa, B., Alamou, E., Afouda, A., Akpo, A., 2016. Estimation of IDF curves of extreme rainfall by simple scaling in northern Oueme Valley, Benin Republic (West Africa). Earth Sci. Res. J. 20, 1–7. https://doi.org/10.15446/esrj.v20n1.49405.

Aich, V., Liersch, S., Vetter, T., Andersson, J., Müller, E., Hattermann, F., 2015. Climate or land use?—Attribution of changes in river flooding in the Sahel zone. Water 7, 2796–2820. https://doi.org/10.3390/w7062796.

Alexander, L.V., et al., 2006. Global observed changes in daily climate extremes of temperature and precipitation. J. Geophys. Res. 111, 22. https://doi.org/10.1029/2005JD006290.

Ali, A., Lebel, T., Amani, A., 2003. Invariance in the spatial structure of Sahelian rain fields at climatological scales. J. Hydrometeorol. 4, 996–1011.

Ali, A., Amani, A., Diedhiou, Lebel, T., 2005a. Rainfall estimation in the Sahel. Part II: evaluation of rain gauge networks in the CILSS countries and objective intercomparison of rainfall products. J. Appl. Meteorol. 44, 1707–1722.

Ali, A., Lebel, T., Amani, A., 2005b. Rainfall estimation in the Sahel. Part I: error function. J. Appl. Meteorol. 44, 1691–1706.

Awadallah, A.G., Awadallah, N.A., 2013. A novel approach for the joint use of rainfall monthly and daily ground station data with TRMM data to generate IDF estimates in a poorly gauged arid region. Open J. Mod. Hydrol. 03, 1–7. https://doi.org/10.4236/ojmh.2013.31001.

Balme, M., Lebel, T., Amani, A., 2006a. Années sèches et années humides au Sahel: quo vadimus? J. Sci. Hydrol. 51 (2), 254–271.

Balme, M., Vischel, T., Lebel, T., Peugeot, C., Galle, S., 2006b. Assessing the water balance in the Sahel: impact of small scale rainfall variability on runoff: Part 1: rainfall variability analysis. J. Hydrol. 331, 336–348.

Barthe, C., Asencio, N., Lafore, J.-P., Chong, M., Campistron, B., Cazenave, F., 2010. Multiscale analysis of the 25–27 July 2006 convective period over Niamey: comparison between Doppler radar observations and simulations. Q. J. R. Meteorol. Soc. 136, 190–208.

Bell, M.A., Lamb, P.J., 2006. Integration of weather system variability to multidecadal regional climate change: the West African Sudan-Sahel zone, 1951-98. J. Clim. 19, 5343–5365.

Berry, G.J., Thorncroft, C.D., 2012. African easterly wave dynamics in a mesoscale numerical model: the upscale role of convection. J. Atmos. Sci. 69, 1267–1283.

Blanchet, J., Lehning, M., 2010. Mapping snow depth return levels: smooth spatial modeling versus station interpolation. Hydrol. Earth Syst. Sci. 14, 2527–2544. https://doi.org/10.5194/hess-14-2527-2010.

Bony, S., et al., 2015. Clouds, circulation and climate sensitivity. Nat. Geosci. 8, 261–268. https://doi.org/10.1038/ngeo2398.

Bunting, A.H., Dennett, M.D., Elston, J., Milford, J.R., 1976. Rainfall trends in the West African Sahel. Q. J. R. Meteorol. Soc. 102, 59–64. https://doi.org/10.1002/qj.49710243105.

Casse, C., Gosset, M., Vischel, T., Quantin, G., Tanimoun, B.A., 2016. Model-based study of the role of rainfall and land use–land cover in the changes in the occurrence and intensity of Niger red floods in Niamey between 1953 and 2012. Hydrol. Earth Syst. Sci. 20, 2841–2859. https://doi.org/10.5194/hess-20-2841-2016.

Ceresetti, D., Anquetin, S., Molini, G., Leblois, E., Creutin, J.-D., 2012. Multiscale evaluation of extreme rainfall event predictions using severity diagrams. Weather Forecast. 27, 174–188. https://doi.org/10.1175/WAF-D-11-00003.1.

Charney, J.G., 1975. Dynamics of deserts and drought in the Sahel. Q. J. R. Meteorol. Soc. 101, 193–202. https://doi.org/10.1002/qj.49710142802.

Chauvin, F., Roehrig, R., Lafore, J.-P., 2010. Intraseasonal variability of the Saharan heat low and its link with midlatitudes. J. Clim. 23, 2544–2561.

Cheng, L., AghaKouchak, A., 2014. Nonstationary precipitation intensity-duration-frequency curves for infrastructure design in a changing climate. Sci. Rep. 4. https://doi.org/10.1038/srep07093. http://www.ncbi.nlm.nih.gov/pmc/articles/PMC4235283/.

Chong, M., Amayenc, P., Scialom, G., Testud, J., 1987. A tropical squall line observed during the COPT 81 experiment in West Africa. Part 1: kinematic structure inferred from dual-Doppler radar data. Mon. Weather Rev. 115, 670–694.

Dai, A., Lamb, P.J., Trenberth, K.E., Hulme, M., Jones, P.D., Xie, P., 2004. The recent Sahel drought is real. Int. J. Climatol. 24, 1323–1331. https://doi.org/10.1002/joc.1083.

Data, R., Sule, B.F., Ige, T., 2016. Synthesis of isopluvial maps for Nigeria. Using IDF equations derived from daily rainfall data. J. Sci. Eng. Res. 3, 505–514.

de Montera, L., Verrier, S., Mallet, C., Barthès, L., 2010. A passive scalar-like model for rain applicable up to storm scale. Atmos. Res. 98, 140–147. https://doi.org/10.1016/j.atmosres.2010.06.012.

De Paola, F., Giugni, M., Topa, M.E., Bucchignani, E., 2014. Intensity-duration-frequency (IDF) rainfall curves, for data series and climate projection in African cities. SpringerPlus 3, 133.

Demarée, G.R., 1990. An indication of climatic change as seen from the rainfall data of a mauritanian station. Theor. Appl. Climatol. 42, 139–147. https://doi.org/10.1007/BF00866869.

Depraetere, C., Gosset, M., Ploix, S., Laurent, H., 2009. The organization and kinematics of tropical rainfall systems ground tracked at mesoscale with gages: first results from the campaigns 1999–2006 on the Upper Ouémé Valley (Benin). J. Hydrol. 375, 143–160. https://doi.org/10.1016/j.jhydrol.2009.01.011.

Descroix, L., et al., 2009. Spatio-temporal variability of hydrological regimes around the boundaries between Sahelian and Sudanian areas of West Africa: a synthesis. J. Hydrol. 375, 90–102.

Descroix, L., Genthon, P., Amogu, O., Rajot, J.-L., Sighomnou, D., Vauclin, M., 2012. Change in Sahelian Rivers hydrograph: the case of recent red floods of the Niger River in the Niamey region. Glob. Planet. Change 98–99, 18–30. https://doi.org/10.1016/j.gloplacha.2012.07.009.

Di Baldassarre, G., Montanari, A., Lins, H., Koutsoyiannis, D., Brandimarte, L., Blöschl, G., 2010. Flood fatalities in Africa: from diagnosis to mitigation. Geophys. Res. Lett. 37. https://doi.org/10.1029/2010GL045467.

Diongue, A., LaFore, J.-P., Redelsperger, J.-L., Roca, R., 2002. Numerical study of a Sahelian synoptic weather system: initiation and mature stages of convection and its interactions with the large-scale dynamics. Q. J. R. Meteorol. Soc. 128, 1899–1927.

Dong, B., Sutton, R., 2015. Dominant role of greenhouse-gas forcing in the recovery of Sahel rainfall. Nat. Clim. Change 5, 757–760. https://doi.org/10.1038/nclimate2664.

Endreny, T.A., Imbeah, N., 2009. Generating robust rainfall intensity–duration–frequency estimates with short-record satellite data. J. Hydrol. 371, 182–191. https://doi.org/10.1016/j.jhydrol.2009.03.027.

Fiolleau, T., Roca, R., 2013. Composite life cycle of tropical mesoscale convective systems from geostationary and low Earth orbit satellite observations: method and sampling considerations. Q. J. R. Meteorol. Soc. 139, 941–953. https://doi.org/10.1002/qj.2174.

Folland, C., Palmer, T., Parker, D., 1986. Sahel rainfall and worldwide sea temperatures, 1901–85. Nature 320, 602–607.

Frappart, F., et al., 2009. Rainfall regime across the Sahel band in the Gourma region, Mali. J. Hydrol. 375, 128–142.

Gaetani, M., Flamant, C., Bastin, S., Janicot, S., Lavaysse, C., Hourdin, F., Braconnot, P., Bony, S., 2016. West African monsoon dynamics and precipitation: the competition between global SST warming and CO_2 increase in CMIP5 idealized simulations. Clim. Dyn. 1–21. https://doi.org/10.1007/s00382-016-3146-z.

Gallego, D., Ordóñez, P., Ribera, P., Peña-Ortiz, C., García-Herrera, R., 2015. An instrumental index of the West African monsoon back to the nineteenth century: an instrumental index of the West African monsoon. Q. J. R. Meteorol. Soc. 141, 3166–3176. https://doi.org/10.1002/qj.2601.

Giannini, A., Saravanan, R., Chang, P., 2003. Oceanic forcing of Sahel rainfall on interannual to interdecadal time scales. Science. 30. http://www.agu.org/pubs/crossref/2003/2003GL017371.shtml.

Giorgi, F., Im, E.-S., Coppola, E., Diffenbaugh, N.S., Gao, X.J., Mariotti, L., Shi, Y., 2011. Higher hydroclimatic intensity with global warming. J. Clim. 24, 5309–5324. https://doi.org/10.1175/2011JCLI3979.1.

Groisman, P.Y., Knight, R.W., Easterling, D.R., Karl, T.R., Hegerl, G.C., Razuvaev, V.N., 2005. Trends in intense precipitation in the climate record. J. Clim. 18, 1326–1350.

Guan, K., Sultan, B., Biasutti, M., Baron, C., Lobell, D.B., 2015. What aspects of future rainfall changes matter for crop yields in West Africa? Geophys. Res. Lett. http://onlinelibrary.wiley.com/doi/10.1002/2015GL063877/full.

Guillot, G., Lebel, T., 1999. Disaggregation of Sahelian mesoscale convective system rain fields further developments and validation. J. Geophys. Res. 104, 31533–31551.

Hubert, P., Carbonnel, J.P., 1987. Approche statistique de l'aridification de l'Afrique de l'Ouest. J. Hydrol. 95, 165–183.

Huffman, G.J., et al., 2007. The TRMM multisatellite precipitation analysis (TMPA): quasi-global, multiyear, combined-sensor precipitation estimates at fine scales. J. Hydrometeorol. 8, 38. https://doi.org/10.1175/JHM560.1.

Hulme, M., 1992. Rainfall changes in Africa: 1931–1960 to 1961–1990. Int. J. Climatol. 12, 685–699. https://doi.org/10.1002/joc.3370120703.

Hulme, M., 2001. Climatic perspectives on Sahelian desiccation: 1973–1998. Glob. Environ. Change 11, 19–29. https://doi.org/10.1016/S0959-3780(00)00042-X.

Ibrahim, B., Polcher, J., Karambiri, H., Rockel, B., 2012. Characterization of the rainy season in Burkina Faso and it's representation by regional climate models. Clim. Dyn. 39, 1287–1302. https://doi.org/10.1007/s00382-011-1276-x.

Janicot, S., Moron, V., Fontaine, B., 1996. Sahel droughts and ENSO dynamics. Geophys. Res. Lett. 23, 515–518.

Janicot, S., et al., 2011. Intraseasonal variability of the West African monsoon. Atmos. Sci. Lett. 12, 58–66.

Janicot, S., et al., 2015. The recent partial recovery in Sahel rainfall: a fingerprint of greenhouse gases forcing? Gewex News 27.

Joly, M., Voldoire, A., 2009. Influence of ENSO on the West African monsoon: temporal aspects and atmospheric processes. J. Clim. 22, 3193–3210.

Joly, M., Voldoire, A., Douville, H., Terray, P., Royer, J.-F., 2007. African monsoon teleconnections with tropical SSTs: validation and evolution in a set of IPCC4 simulations. Clim. Dyn. 29, 1–20.

Khaliq, M.N., Ouarda, T.B.M.J., Gachon, P., Sushama, L., St-Hilaire, A., 2009. Identification of hydrological trends in the presence of serial and cross correlations: a review of selected methods and their application to annual flow regimes of Canadian rivers. J. Hydrol. 368, 117–130. https://doi.org/10.1016/j.jhydrol.2009.01.035.

Koutsoyiannis, D., Kozonis, D., Manetas, A., 1998. A mathematical framework for studying rainfall intensity-duration-frequency relationships. J. Hydrol. 206, 118–135.

Lafore, J.-P., et al., 2017. A Multi-Scale Analysis of the Extreme Rain Event of Ouagadougou in 2009: High-Impact Weather System, West Africa, African Easterly Waves. Q. J. R. Meteorol. Soc. 149, 3094–3109. https://doi.org/10.1002/qj.3165.

Lamb, P.J., 1982. Persistence of Subsaharan drought. Nature 299, 46–48. https://doi.org/10.1038/299046a0.

Lamb, P.J., Peppler, R.A., 1992. Further case studies of tropical Atlantic surface atmospheric and oceanic patterns associated with Sub-Saharan drought. J. Clim. 5, 476–488. https://doi.org/10.1175/1520-0442(1992)005<0476:FCSOTA>2.0.CO;2.

Lavaysse, C., Flamant, C., Evan, A., Janicot, S., Gaetani, M., 2015. Recent climatological trend of the Saharan heat low and its impact on the West African climate. Clim. Dyn. 1–20. https://doi.org/10.1007/s00382-015-2847-z.

Le Barbé, L., Lebel, T., 1997. Rainfall climatology of the HAPEX-Sahel region during the years 1950-1990. J. Hydrol. 188–189, 43–73.

Le Barbé, L., Lebel, T., Tapsoba, D., 2002. Rainfall variability in West Africa during the years 1950-90. J. Clim. 15, 187–202.

Lebel, T., Ali, A., 2009. Recent trends in the central and Western Sahel rainfall regime (1990-2007). J. Hydrol. 375, 52–64.

Lebel, T., Diedhiou, A., Laurent, H., 2003. Seasonal cycle and interannual variability of the Sahelian rainfall at hydrological scales. J. Geophys. Res. 108. https://doi.org/10.1029/2001JD001580.

Lebel, T., et al., 2009. AMMA-CATCH studies in the Sahelian region of West-Africa: an overview. J. Hydrol. 375, 3–13.

Lodoun, T., Giannini, A., Traoré, P.S., Somé, L., Sanon, M., Vaksmann, M., Rasolodimby, J.M., 2013. Changes in seasonal descriptors of precipitation in Burkina Faso associated with late 20th century drought and recovery in West Africa. Environ. Dev. 5, 96–108. https://doi.org/10.1016/j.envdev.2012.11.010.

L'Hote, Y., Mahé, G., Some, B., Triboulet, J.P., 2002. Analysis of a Sahelian annual rainfall index from 1896 to 2000; the drought continues. Hydrol. Sci. 47 (4), 563–572.

Mahé, G., Paturel, J.-E., 2009. 1896-2006 Sahelian annual rainfall variability and runoff increase of Sahelian Rivers. Surf. Geosci. Hydrol. Hydrogeol. 341, 538–546.

Massuel, S., Cappelaere, B., Favreau, G., Leduc, C., Lebel, T., Vischel, T., 2011. Integrated surface water–groundwater modelling in the context of increasing water reserves of a regional Sahelian aquifer. Hydrol. Sci. J. 56, 1242–1264. https://doi.org/10.1080/02626667.2011.609171.

Mathon, V., Laurent, H., Lebel, T., 2002. Mesoscale convective system rainfall in the Sahel. J. Appl. Meteorol. 41, 1081–1092.

Menabde, M., Seed, A., Pegram, G., 1999. A simple scaling model for extreme rainfall. Water Resour. Res. 35, 335–339. https://doi.org/10.1029/1998WR900012.

Min, S.-K., Zhang, X., Zwiers, F.W., Hegerl, G.C., 2011. Human contribution to more-intense precipitation extremes. Nature 470, 378–381. https://doi.org/10.1038/nature09763.
Mohino, E., Janicot, S., Bader, J., 2011. Sahel rainfall and decadal to multi-decadal sea surface temperature variability. Clim. Dyn. 37, 419–440. https://doi.org/10.1007/s00382-010-0867-2.
Mohymont, B., Demarée, G.R., 2010. Courbes intensité-durée-fréquence des précipitations à Yangambi, Congo, au moyen de différents modèles de type Montana. Hydrol. Sci. J. 51 (2), 239–253.
Monerie, P.-A., Fontaine, B., Roucou, P., 2012. Expected future changes in the African monsoon between 2030 and 2070 using some CMIP3 and CMIP5 models under a medium-low RCP scenario. J. Geophys. Res. Atmos. 117, n/a–n/a. https://doi.org/10.1029/2012JD017510.
Morin, E., 2011. To know what we cannot know: global mapping of minimal detectable absolute trends in annual precipitation: minimal detectable precipitation trends. Water Resour. Res. 47, n/a–n/a. https://doi.org/10.1029/2010WR009798.
Moron, V., Robertson, A.W., Ward, M.N., 2006. Seasonal predictability and spatial coherence of rainfall characteristics in the tropical setting of Senegal. Mon. Weather Rev. 134, 3248–3262. https://doi.org/10.1175/MWR3252.1.
New, M., et al., 2006. Evidence of trends in daily climate extremes over southern and west Africa. J. Geophys. Res. 111, D14102.
Nicholson, S.E., 2001. Climatic and environmental change in Africa during the last two centuries. Clim. Res. 17, 123–144.
Nicholson, S., 2005. On the question of the "recovery" of the rains in the West African Sahel. J. Arid Environ. 63, 615–641.
Nicholson, S.E., 2008. The intensity, location and structure of the tropical rainbelt over west Africa as factors in interannual variability. Int. J. Climatol. 28, 1775–1785.
Nicholson, S.E., 2009. A revised picture of the structure of the "monsoon" and land ITCZ over West Africa. Clim. Dyn. 32, 1155–1171.
Nicholson, S.E., 2013. the West African Sahel: a review of recent studies on the rainfall regime and its interannual variability. ISRN Meteorol. 2013, 1–32. https://doi.org/10.1155/2013/453521.
Oyebande, L., 1982. Deriving rainfall intensity-duration-frequency relationships and estimates for regions with inadequate data. Hydrol. Sci. J. 27, 353–367. https://doi.org/10.1080/02626668209491115.
Oyegoke, S.O., Oyebande, L., 2008. A new technique for analysis of extreme rainfall for Nigeria. Environ. Res. J. 2, 7–14.
Ozer, P., Erpicum, M., Demarée, G., Vandiepenbeeck, M., 2003. The Sahelian drought may have ended during the 1990s. Hydrol. Sci. J. 48, 489–492. https://doi.org/10.1623/hysj.48.3.489.45285.
Paeth, H., Fink, A.H., Pohle, S., Keis, F., Mächel, H., Samimi, C., 2010. Meteorological characteristics and potential causes of the 2007 flood in sub-Saharan Africa. Int. J. Climatol., n/a–n/a. https://doi.org/10.1002/joc.2199.
Paeth, H., Fink, A.H., Pohle, S., Keis, F., Mächel, H., Samimi, C., 2011. Meteorological characteristics and potential causes of the 2007 flood in sub-Saharan Africa. Int. J. Climatol. 31, 1908–1926.
Panthou, G., Vischel, T., Lebel, T., Blanchet, J., Quantin, G., Ali, A., 2012. Extreme rainfall in West Africa: a regional modeling. Water Resour. Res. 48, W08501. https://doi.org/10.1029/2012WR012052.
Panthou, G., Vischel, T., Lebel, T., Quantin, G., Pugin, A.-C.F., Blanchet, J., Ali, A., 2013. From pointwise testing to a regional vision: an integrated statistical approach to detect nonstationarity in extreme daily rainfall. Application to the Sahelian region. J. Geophys. Res. Atmos. 118, 8222–8237. https://doi.org/10.1002/jgrd.50340.

Panthou, G., Vischel, T., Lebel, T., 2014a. Recent trends in the regime of extreme rainfall in the Central Sahel. Int. J. Climatol. 34, 3998–4006. https://doi.org/10.1002/joc.3984.
Panthou, G., Vischel, T., Lebel, T., Quantin, G., Molinié, G., 2014b. Characterising the space–time structure of rainfall in the Sahel with a view to estimating IDAF curves. Hydrol. Earth Syst. Sci. 18, 5093–5107. https://doi.org/10.5194/hess-18-5093-2014.
Park, J., Bader, J., Matei, D., 2016. Anthropogenic Mediterranean warming essential driver for present and future Sahel rainfall. Nat. Clim. Change 6, 941–945. https://doi.org/10.1038/nclimate3065.
Poan, D.E., Roehrig, R., Couvreux, F., Lafore, J.-P., 2013. West African monsoon intraseasonal variability: a precipitable water perspective. J. Atmos. Sci. 70, 1035–1052.
Poan, D.E., Lafore, J.-P., Roehrig, R., Couvreux, F., 2015. Internal processes within the African easterly wave system. Q. J. R. Meteorol. Soc. 141, 1121–1136.
Ramos, M.H., Creutin, J.-D., Leblois, E., 2005. Visualization of storm severity. J. Hydrol. 315, 295–307. https://doi.org/10.1016/j.jhydrol.2005.04.007.
Redelsperger, J.-L., Diongue, A., Diedhiou, A., Ceron, J.-P., Diop, M., Gueremy, J.-F., Lafore, J.-P., 2002. Multi-scale description of a Sahelian synoptic weather system representative of the West African Monsoon. Q. J. R. Meteorol. Soc. 128, 1229–1257.
Redelsperger, J.-L., Thorncroft, C.D., Diedhiou, A., Lebel, T., Parker, D.J., Polcher, J., 2006. African monsoon multidisciplinary analysis: an international research project and field campaign. Bull. Am. Meteorol. Soc. 87, 1739–1746. https://doi.org/10.1175/BAMS-87-12-1739.
Roca, R., Aublanc, J., Chambon, P., Fiolleau, T., Viltard, N., 2014. Robust observational quantification of the contribution of mesoscale convective systems to rainfall in the tropics. J. Clim. 27, 4952–4958. https://doi.org/10.1175/JCLI-D-13-00628.1.
Roca, R., et al., 2015. The Megha-Tropiques mission: a review after three years in orbit. Front. Earth Sci. 3. https://doi.org/10.3389/feart.2015.00017. http://journal.frontiersin.org/article/10.3389/feart.2015.00017.
Rodier, J., 1964. Régimes hydrologiques de l'Afrique Noire à l'ouest du Congo. Mémoire Orstom, Paris.
Roehrig, R., Chauvin, F., Lafore, J.-P., 2011. 10–25-Day intraseasonal variability of convection over the Sahel: a role of the Saharan heat low and midlatitudes. J. Clim. 24, 5863–5878. https://doi.org/10.1175/2011JCLI3960.1.
Saha, A., Ghosh, S., Sahana, A.S., Rao, E.P., 2014. Failure of CMIP5 climate models in simulating post-1950 decreasing trend of Indian monsoon. Geophys. Res. Lett. 41, 7323–7330. https://doi.org/10.1002/2014GL061573.
Sanogo, S., Fink, A.H., Omotosho, J.A., Ba, A., Redl, R., Ermert, V., 2015. Spatio-temporal characteristics of the recent rainfall recovery in West Africa. Int. J. Climatol., n/a–n/a. https://doi.org/10.1002/joc.4309.
Sarr, M.A., Zoromé, M., Seidou, O., Bryant, C.R., Gachon, P., 2013. Recent trends in selected extreme precipitation indices in Senegal – a changepoint approach. J. Hydrol. 505, 326–334. https://doi.org/10.1016/j.jhydrol.2013.09.032.
Sarr, M.A., Gachon, P., Seidou, O., Bryant, C.R., Ndione, J.A., Comby, J., 2014. Inconsistent linear trends in Senegalese rainfall indices from 1950 to 2007. Hydrol. Sci. J. 1405221 53351007. https://doi.org/10.1080/02626667.2014.926364.
Schwendike, J., Jones, S.C., 2010. Convection in an African easterly wave over West Africa and the eastern Atlantic: a model case study of Helene (2006). Q. J. R. Meteorol. Soc. 136, 364–396.
Siongco, A.C., Hohenegger, C., Stevens, B., 2015. The Atlantic ITCZ bias in CMIP5 models. Clim. Dyn. 45, 1169–1180. https://doi.org/10.1007/s00382-014-2366-3.
Soro, G.E., Goula Bi, T.A., Kouassi, F.W., Srohourou, B., 2010. Update of intensity-duration-frequency curves for precipitation of short durations in tropical area of West Africa (Cote D'ivoire). J. Appl. Sci. 10, 704–715. https://doi.org/10.3923/jas.2010.704.715.

Sultan, B., Janicot, S., 2003. The West African monsoon dynamics. Part II: the "preonset" and "onset" of the summer monsoon. J. Clim. 16, 3389–3406.

Tarhule, A., 2005. Damaging rainfall and flooding: the other Sahel hazards. Clim. Change 72, 355–377. https://doi.org/10.1007/s10584-005-6792-4.

Tarhule, A., Woo, M.-K., 1998. Changes in rainfall characteristics in northern Nigeria. Int. J. Climatol. 18, 1261–1271. https://doi.org/10.1002/(SICI)1097-0088(199809)18:11.

Tarhule, A., et al., 2014. Exploring temporal hydroclimatic variability in the Niger Basin (1901-2006) using observed and gridded data. Int. J. Climatol., n/a-n/a. https://doi.org/10.1002/joc.3999.

Taylor, C.M., Gounou, A., Guichard, F., Harris, P.P., Ellis, R.J., Couvreux, F., De Kauwe, M., 2011. Frequency of Sahelian storm initiation enhanced over mesoscale soil-moisture patterns. Nat. Geosci. 4, 430–433.

Verrier, S., de Montera, L., Barthès, L., Mallet, C., 2010. Multifractal analysis of African monsoon rain fields, taking into account the zero rain-rate problem. J. Hydrol. 389, 111–120. https://doi.org/10.1016/j.jhydrol.2010.05.035.

Vischel, T., Lebel, T., 2007. Assessing the water balance in the Sahel: impact of small scale rainfall variability on runoff. Part 2: idealized modeling of runoff sensitivity. J. Hydrol. 333, 340–355.

Vischel, T., Lebel, T., Massuel, S., Cappelaere, B., 2009. Conditional simulation schemes of rain fields and their application to rainfall–runoff modeling studies in the Sahel. J. Hydrol. 375, 273–286. https://doi.org/10.1016/j.jhydrol.2009.02.028.

Vischel, T., Quantin, G., Lebel, T., Viarre, J., Gosset, M., Cazenave, F., Panthou, G., 2011. Generation of high resolution rainfields in West Africa: evaluation of dynamical interpolation methods. J. Hydrometeorol. 110426113802000. https://doi.org/10.1175/JHM-D-10-05015.1.

Wang, C., Zhang, L., Lee, S.-K., Wu, L., Mechoso, C.R., 2014. A global perspective on CMIP5 climate model biases. Nat. Clim. Change 4, 201–205. https://doi.org/10.1038/nclimate2118.

Westra, S., Alexander, L.V., Zwiers, F.W., 2013. Global increasing trends in annual maximum daily precipitation. J. Clim. 26, 3904–3918. https://doi.org/10.1175/JCLI-D-12-00502.1.

Zipser, E.J., Liu, C., Cecil, D.J., Nesbitt, S.W., Yorty, D.P., 2006. Where are the most intense thunderstorms on earth? Bull. Am. Meteorol. Soc. 87, 1057–1071. https://doi.org/10.1175/BAMS-87-8-1057.

CHAPTER 5

Evaluating Large-Scale Variability and Change in Tropical Rainfall and Its Extremes

Richard P. Allan, Chunlei Liu
Department of Meteorology, University of Reading, Reading, United Kingdom

Contents

1. INTRODUCTION

Extremes of tropical rainfall are determined by the complex nature of the atmospheric circulation and its thermodynamic structure and are commonly associated with societal damage through immediate responses such as flooding and landslides. Conversely a sustained dearth of rainfall accumulations can, among other factors, contribute to crippling drought over longer, multiannual timescales. Tropical depressions and cyclones often wreak severe damage on a day-to-day basis to low-lying, highly populated regions through intense rainfall combined with strong winds and storm surges or may lead to devastating landslides in mountainous terrain; the fickle nature of weather patterns is ultimately responsible for the location and timing of these weather extremes, most acute for associated convective cells, which operate at smaller spatial scales. Where longer-term accumulations matter, in cases such as multiple failed seasonal rains, atmospheric circulation patterns again play a dominant role, driven by the slower heartbeat of the ocean as it paces out the internal rhythm of El Niño Southern Oscillation (ENSO)

Tropical Extremes: Natural Variability and Trends
ISBN 978-0-12-809248-4
https://doi.org/10.1016/B978-0-12-809248-4.00005-4

and its decadal characteristics and manifestation across tropical arid zones. Yet despite the dominant role of chaotic atmospheric and oceanic fluctuations in meting out extremes of weather, there are a number of drivers at the largest terrestrial scales which, barely perceptibly, nudge the distribution of extremes away from the present-day patterns through global climatic changes that are linked to human activities and their influence on the planet's energy budget. Rising concentrations of greenhouse gases in the atmosphere are slowly but inexorably perturbing the climate system (IPCC, 2013), altering the ultimate fuel for rainfall extremes: water and energy. Although detailed computer simulations capture the physical fluid, thermodynamic essence of the atmosphere, the processes building up to rainfall extremes, and the fine-scale detail of convective storms, are not explicitly represented and must be approximated, albeit in a physically based way. Therefore the reliability of future projections out across many, progressively more distant decades relies on fundamental linkages between the skill in depicting essential physics at the tens to hundreds of kilometer scales and the unrepresented detail, which determines impacts.

The exploitation of observations in combination with global climate simulations of the present and the future offers a potential method of exploring the consistency and links between current variability in characteristics of rainfall extremes and the responses out into a substantially warmer future. Furthermore, monitoring of observed changes, within the context of large-scale physical drivers, offers a capacity for continual reassessment and update of existing knowledge. In the following discussions, current changes in tropical precipitation, its characteristics and determinant variables, are assessed from an observational perspective and linked, where practicable, to aspects of future responses. This offers just one tool among many in underpinning the evaluation of tropical extremes and how they will alter in the future, with anticipated significant consequences for human societies.

2. PHYSICAL CONSTRAINTS ON CHANGES IN TROPICAL PRECIPITATION

The links between Earth's energy budget and the hydrological cycle have long been recognized (e.g., Mitchell et al., 1987) and are discussed in detail by Allen and Ingram (2002) in relation to climate change. The way radiative forcings are initially partitioned between the atmosphere and the surface has recently received considerable attention (Myhre et al., 2017) because large adjustments in atmospheric circulation and precipitation (Bony et al., 2013) ultimately transfer this perturbation to the energy budget into the oceans where the bulk of the climate system's heat capacity resides. This potential to alter characteristics of the hydrological cycle through shifts in the large-scale weather patterns is complimented by a regional perspective on energy and moisture budgets (Muller and O'Gorman, 2011), which also influences the characteristics of precipitation extremes associated with their dynamical and thermodynamic response to climate change (Marvel and Bonfils, 2013).

Eqs. (5.1) and (5.2) illustrate the atmospheric energy and moisture budgets, which may apply regionally or tropics-wide:

$$LP = Q - S + H_t + \Delta H, \tag{5.1}$$

$$P = E - W_t - \Delta W, \tag{5.2}$$

where P is precipitation in $kgm^{-2}s^{-1}$, L is the latent heat of vaporization (ignoring any exchanges involving ice for simplicity), Q is the net longwave radiative cooling minus the heating from absorbed sunlight, S is the sensible heating of the atmosphere from the surface, H_t is the energy export out of the region, and ΔH is the energy change within the atmospheric column. Precipitation is further constrained by the water budget (Eq. 5.2) where E is surface evapotranspiration of water into the atmosphere, W_t is the transport of water (including ice and liquid but primarily vapor) out of the region by atmospheric winds, and ΔW represents the changes in water storage in the atmospheric column.

Eq. (5.1) has been extensively used to interpret global-scale changes in precipitation (Andrews et al., 2010), where $H_t = 0$, ΔH is considered small and S is often neglected yet can produce substantial deviations (e.g., Ming et al., 2010) from the resulting simplified perturbation budget, $L\Delta P \sim \Delta Q$: changes in precipitation (ΔP) are thus directly linked to changes in the net radiative cooling of the atmosphere (ΔQ), which more phyiscally applies to the free-troposphere above the mixing condensation level (O'Gorman et al., 2012). A complementary perspective may be applied to surface evaporation (Roderick et al., 2014) although in reality the system is coupled and the surface and atmospheric energy budgets will adjust to any change to restore radiative-dynamical balance. Recent advances (Andrews et al., 2010) have decomposed ΔQ into (1) fast adjustments to radiative forcings, such that direct heating by greenhouse gases, for example, suppresses precipitation through increased atmospheric stability or reduced tropical circulation (e.g., Bony et al., 2013; He and Soden, 2016) and through coupled processes reduce evaporation thereby effectively transferring the heat from the atmosphere to the heat sink of the vast oceans, and (2) a slower response dependent on the changes in ocean surface temperature that are a result of a response toward a new equilibrium temperature set by the radiative forcing and atmospheric feedbacks (see discussion in O'Gorman et al., 2012). This framework has proven useful in interpreting and deconstructing the simulated responses in precipitation into fast adjustments to radiative forcings and slower responses to surface temperature changes, both of which may be manifest regionally through complex responses in atmospheric circulation (Bony et al., 2013). While applying to large-scale mean precipitation responses, attempts have also been made to link to extremes of precipitation (Myhre et al., 2017), which are also constrained by energy and water budgets, yet the depiction of associated fine-scale detail by climate models is questionable.

Muller and O'Gorman (2011) find that local changes in P are well approximated by the global changes in P combined with the change in dry static energy export from the region which is, in turn, influenced by thermodynamic and dynamic components (e.g., Emori and Brown, 2005). Radiative forcing by greenhouse gases and aerosols can influence the global P changes through rapid adjustments in the energy budget (Andrews et al., 2010), but these adjustments are also manifest as responses in the atmospheric circulation (Bony et al., 2013), both factors determining how regional precipitation extremes respond to climate change. Lin et al. (2016) find that future responses of precipitation extremes are strongly influenced by the balance between greenhouse gas and aerosol radiative forcings. The further constraint of the moisture budget (Eq. 5.2) modifies the response and ensures a close coupling between E and P at the largest scales.

A further determinant of the tropical circulation, more particularly the rainy belt within the Intertropical Convergence Zone (ITCZ), relates to hemispheric energetics. Frierson et al. (2013) demonstrate how transport of heat to the northern hemisphere by the ocean explains the northward mean position of the ITCZ; perturbations to the interhemispheric energy balance through cooling of the northern hemisphere by aerosol radiative forcing have been implicated in the 1980s Sahel drought (Hwang et al., 2013), and subsequent recovery in mean and extreme Sahel rainfall has been linked to preferential heating of the northern hemisphere (or more particularly the Sahara heat low region) by greenhouse gas radiative forcing (Dong and Sutton, 2015; Taylor et al., 2017). Because climate models disagree on the sign of hemispheric precipitation differences and energy transports (Loeb et al., 2016), this may provide an important area of research in improving the simulation of the tropical rainy belt and extremes within (e.g., Hawcroft et al., 2016).

Much of the precipitation extremes within the tropical rainy belt are associated with monsoon circulation, which dominate societies across west Africa and south Asia. Based on an energy and moisture budget approach, Levermann et al. (2009) proposed a simplified depiction of the monsoon (although their use in linking to abrupt transitions has been questioned by Boos and Storelvmo, 2016): precipitation is determined by a balance between the energy transports (influenced by the land–ocean temperature contrast and the associated monsoonal wind), the net radiative cooling and further constrained by the moisture balance set by the land–ocean atmospheric specific humidity difference, and the monsoonal wind (both constrained by the land–ocean temperature contrast). Current changes in seasonal rainfall are strongly influenced by internal fluctuations of the climate system (Maidment et al., 2015; Sukhatme and Venugopal, 2016) yet greenhouse gas forcing is increasingly playing a role (Dong and Sutton, 2015; Gu et al., 2016).

Considering that changes in atmospheric water storage are small compared to the fluxes into and out of the column, it is simple to link regional P–E to atmospheric moisture transports into or out of a region from Eq. (5.2) such that $P-E \sim -W_t$. Because W_t depends on the atmospheric winds and moisture amount, assuming relatively small changes in the mean atmospheric circulation implies that W_t approximately scales with the Clausius–Clapeyron equation:

$$\frac{1}{W_t}\frac{dW_t}{dT} \approx \frac{1}{e_s}\frac{de_s}{dT} \approx \frac{d\ln(e_s)}{dT} = \frac{L}{RT^2}, \tag{5.3}$$

where e_s is the saturation water vapor pressure and R_v is the water vapor gas constant such that $L/R_vT^2 \sim 0.07\,K^{-1}$ at temperatures near to the surface (e.g., Allan, 2012). Assuming $\Delta W \sim 0$ in Eq. (5.2), $W_t \sim (P-E)$ and therefore $\Delta(P-E) \sim \Delta W_t \sim 0.07(P-E)\Delta T$. Applying this scaling implies an amplification of P−E patterns with warming, which appears sufficient to capture the zonal responses of P−E simulated under climate change (Held and Soden, 2006) and is corroborated by an observed amplification of salinity patterns (Durack et al., 2012), which are strongly related to the freshwater input at the ocean surface via P−E. It is therefore conceptually useful to separate the wet and dry portions of the tropical circulation to understand how extremes in tropical rainfall characteristics will change in response to a warming climate. However, relating these constraints to precipitation extremes is not straightforward, particularly with respect to drought over land in which deficiencies in commonly used metrics such as the Palmer Drought Severity

Index (PDSI) and conflation of variability with climate change inhibit attribution of current trends (Trenberth et al., 2014).

Held and Soden (2006) noted that evaporation varies more smoothly over space and time than precipitation and suggested that subtropical precipitation declines could be linked with decreasing P−E . However, additional moisture export from the subtropical evaporative oceans is to some extent compensated by enhanced evaporation, and the amplification of P−E patterns is also somewhat offset by a reduced tropical circulation (Chadwick et al., 2013). An anticipated decline in the strength of the tropical Walker circulation can be linked to reduced effective mass flux required to reconcile the large low-level moisture increases of around 7%/K with the "muted" precipitation changes constrained by the energy balance (Eq. 5.2) to increase at a lesser rate (around 1%–3%/K) although is additionally modulated by changes in temperature lapse rate with altitude set by the moist adiabat (Held and Soden, 2006). Furthermore, suppression of precipitation through the direct atmospheric heating by greenhouse gases (Bony et al., 2013) appears to explain remaining signals of subtropical precipitation decline (He and Soden, 2016).

Over tropical land, P or P−E provides an incomplete representation of aridity (Roderick et al., 2014), and the ratio of precipitation to potential evapotranspiration provides a more meaningful, impact-relevant quantity (Scheff and Frierson, 2015). Regardless, because P−E is positive for land over multiannual timescales (it must balance surface runoff), increased moisture transport into continents (Zahn and Allan, 2013b), as a consequence of increased atmospheric moisture, cannot explain continental drying since P−E should become more positive with warming (Greve et al., 2014). Local continental drying can nevertheless be explained by the location and change in spatial gradients in temperature and relative humidity (Byrne and O'Gorman, 2015). However, considering multiannual timescales over fixed geographical locations conflates the seasonal progression and interannual variability of the rainy belt: mixing the wet and dry components of the tropical circulation neglects the possibility for $P<E$ in the dry season and precludes simple isolation of recent drivers of P−E over land (Kumar et al., 2015). Because the precipitation responses to climate change constitute a complex mix relating to the stronger warming over land than ocean, the direct influence of radiative forcing of greenhouse gases on tropical circulation and the pattern of sea surface temperature (SST) changes (He and Soden, 2016), compounded by substantial internal variability in the atmospheric circulation (Gu et al., 2016), monitoring of the current wet and dry region changes in the context of these driving factors is crucial (Allan et al., 2014a).

Evaluation of extreme precipitation events, which evolve on small time and space scales, is a challenge both to observe (Wilcox and Donner, 2007) and more especially to simulate because the detailed structure cannot be explicitly represented in state-of-the-art climate models (e.g., Pendergrass et al., 2016). They are also subject to the same large-scale biases affecting mean precipitation relating to ITCZ location, energy, and moisture transports via the atmospheric circulation and land surface characteristics. Therefore understanding and attributing changes in precipitation extremes associated with complex atmospheric circulations such as the Indian and west African monsoons (Vittal et al., 2016; Taylor et al., 2017) require large-scale dynamical context in addition to appreciation of the local-scale thermodynamic and microphysical factors. Attempting a generalized theory of how precipitation extremes respond to climate change requires a unification of these competing factors: O'Gorman (2015) approximates the extreme surface precipitation (P_{ext}) as:

$$P_{ext} = \varepsilon \frac{1}{g} \int_{p_s}^{100\,\text{hPa}} \omega(p) \left. \frac{dq_s(T)}{dp} \right|_{\theta_e^*} dp, \tag{5.4}$$

where ε is an effective efficiency at which condensed water is converted to precipitation, ω is the vertical velocity in pressure coordinates, which acts on the vertical change in saturation specific humidity (q_s) with vertical pressure along a moist adiabat (constant equivalent potential temperature, θ_e^*), which are integrated through the troposphere, here approximated as between surface pressure, p_s, and 100 hPa. The changes in precipitation extremes in response to climate change are therefore determined by a microphysical component (ε), a dynamical component relating to ω, and a thermodynamic component relating to the q_s term, quite well approximated by the near-surface specific humidity that is constrained by the Clausius–Clapeyron equation at the largest scales to increase at around 7%/K (O'Gorman, 2015). This rate is therefore a reasonable starting point, with microphysical and dynamical factors altering the precipitation extreme response away from this scaling.

Because precipitation extremes are still governed by moisture and energy budgets, there can be a complex response across space and timescales. For example, latent heating can stabilize the atmospheric profile at larger scales yet invigorate ascent at the smallest scales, and the potential for substantial changes in the characteristics of rainfall is possible and very likely (Trenberth et al., 2003) yet inadequately represented by coarse resolution climate models in the tropics (Pendergrass et al., 2016). However, the use of observations to evaluate the response of precipitation extremes to year-to-year variability (Liu and Allan, 2012), usually in terms of intensity distribution (Pendergrass and Hartmann, 2014), and further relate this to climate change response (O'Gorman, 2012), offers a promising emergent constraint.

3. OBSERVED CURRENT CHANGES IN TROPICAL CLIMATE

3.1 Tropic-Wide Changes

Changes at the largest scales provide physical context for the current evolution of tropical climate extremes. Fig. 5.1 illustrates how tropical mean (30°S–30°N) climate has varied and changed over the period 1979–2016, comparing a range of satellite and surface-based observations with atmosphere-only climate model simulations where realistic radiative forcings

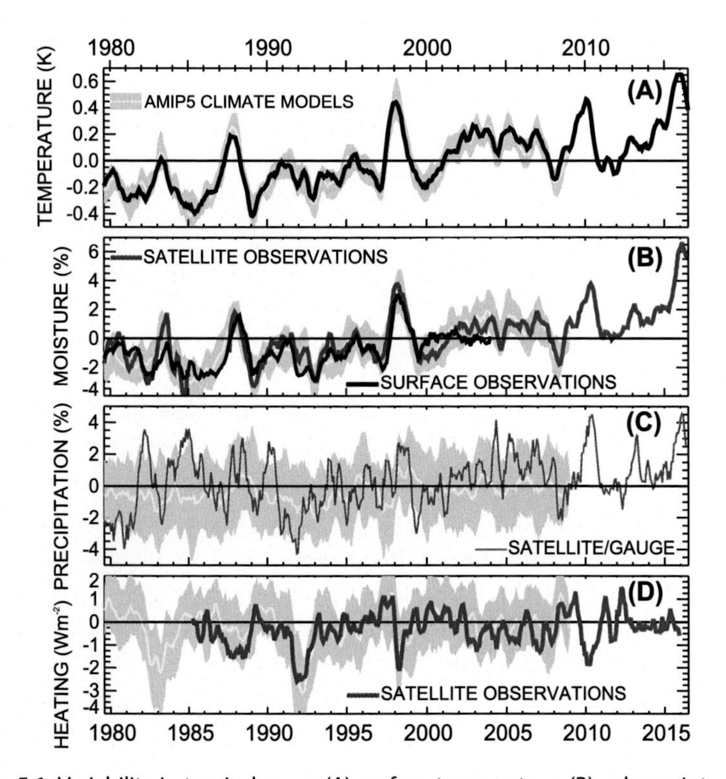

Figure 5.1 Variability in tropical mean (A) surface temperature, (B) column-integrated or near surface water vapor, (C) precipitation, and (D) top of atmosphere net radiation for atmosphere-only climate models (AMIP5) using prescribed observed sea surface temperature and sea ice and a combination of satellite-based and surface observations over the period 1980–2016 (5 month smoothing is applied).

and observed SST and sea ice conditions are prescribed as boundary conditions (these Atmospheric Model Intercomparison Project (AMIP) experiments form part the Coupled Model Intercomparison Project phase 5 CMIP5; Taylor et al., 2011). The tropics have warmed over the period, by 0.15 K/decade when considering deseasonalized monthly anomalies for 1988–2015 (this period was chosen to allow better comparison with other variables, which suffer from reliability issues prior to 1988) using the HadCRUT4 v4.5 data set; this trend is punctuated by warm El Niño events (e.g., 1997–98, 2015–16) and cool La Niña episodes (e.g., 1999, 2011). Atmospheric water vapor closely tracks the temperature changes and there is broad agreement between the range of surface-based, satellite-based, and AMIP simulations. The satellite ocean sampling estimates (here a combination of microwave measurements taken from the F08/F11/F13/F17 series of Defense Meteorological Satellite Program satellites; Wentz, 2013) combined with the European Centre for Medium-range Weather Forecasts ReAnalysis Interim product (ERA Interim combines a range of observations with a numerical model using data assimilation; Dee et al., 2011) over land indicate a moistening of 1.2%/decade over the period 1988–2015. Combining the temperature and water vapor trends provides a long-term estimate of how low-altitude moisture responds to decadal warming, $dW/dT = 1.2\%\,\text{decade}^{-1}/0.15\,\text{K}\,\text{decade}^{-1} = 8\%/\text{K}$, which is close to that expected from the Clausius–Clapeyron equation when applied to low-altitude tropical temperature (e.g., Allan, 2012), and assuming tropical mean relative humidity changes are small, which may be somewhat questionable for decadal variations over land (Simmons et al., 2010). This available atmospheric moisture essentially provides the fuel for intense precipitation events where water vapor is continually transported from surrounding regions and converged into storms (Trenberth et al., 2003).

Although atmospheric moisture is central for individual storm cells, the collective characteristics of longer timescale extremes tropics-wide are determined also by energy balance as discussed in Section 2. Tropical mean precipitation from the Global Precipitation Climatology Project (GPCP; Adler et al., 2018), which is a combination of satellite-based and land surface gauge-based estimates, appears to display greater relative month-to-month variability than temperature or water vapor; the covariability with these variables and similarity to AMIP5 simulations appears less coherent (Fig. 5.1C) with only marginally significant trends over the 1988–2015 period of 0.8%/decade. Insight into the causes of this poorer coupling and the higher frequency variability relates to nonlinearities, export of dry static energy from the tropics and complex balances associated with convection as discussed by Su and Neelin (2003). Because short-lived mesoscale systems contribute 75% of tropical precipitation (Roca et al., 2014), the link between extreme precipitation and total accumulations is obvious.

Finally, changes in net heating into the tropics, measured at the top of the atmosphere by satellites, show good agreement with the AMIP5 simulations (details in Allan et al., 2014b), indicating that radiative forcing and feedback response are well simulated when realistic ocean surface temperature is prescribed. Variability is dominated by reduced heating following the Pinatubo volcanic eruption in 1991, which increased planetary albedo through the introduction of reflective aerosol in the stratosphere, and enhanced heat loss during El Niño events in which a warmer atmosphere loses more energy to space through infrared emission. How this heat input and its changes are distributed at the surface and exported poleward (recall the transport term in Eq. 5.1) is less certain yet is crucial for regional scale responses in precipitation and its extremes.

3.2 Responses of Wet and Dry Portions of the Tropical Circulation

Based on the moisture budget considerations (Eq. 5.2), there is a strong physical basis for expecting contrasting precipitation responses in comparatively wet and dry portions of the atmospheric circulation as discussed in Section 2. A simple consequence of increases in atmospheric water vapor is the enhanced transport of moisture from the evaporative subtropical oceans into the wet convergence zones (e.g., Zahn and Allan, 2013a). The consequent amplification of P−E patterns over the oceans, combined with changes in runoff, lead to amplification of ocean salinity patterns (Durack et al., 2012). The observed changes are somewhat below the Clausius–Clapeyron rate (Skliris et al., 2016) and much of the signal in precipitation change may relate to circulation response to greenhouse gas forcings (He and Soden, 2016).

Over land, multiannual P−E is positive, being balanced by surface runoff, so increased atmospheric moisture and enhancement of existing moisture convergence over tropical land (Zahn and Allan, 2013b) will further increase P−E implying a wet gets wetter response. This is however not necessarily the case for the dry season where P−E may be negative as evaporation of the plentiful soil moisture, following the rainy season, dominates any dry season rainfall such that a more "thirsty" atmosphere is able to effectively diverge evaporated water following the wet season over land (Kumar et al., 2015). Furthermore, soil moisture feedbacks (Berg et al., 2016), changes in relative humidity (Byrne and O'Gorman, 2016), and consideration of temperature and moisture gradients complicate the signal over land (Byrne and O'Gorman, 2015). The anticipated diverging responses in wet and dry circulation regimes is difficult to detect from a fixed geographical frame of reference (Polson and Hegerl, 2016), which effectively integrates over contrasting atmospheric circulation regimes associated with a local wet and dry season. Therefore a dynamically determined wet and dry regime, which evolves over time, is conceptually useful.

Fig. 5.2 displays land precipitation integrated separately over the wet and dry components of the tropical circulation, based on the method of Liu and Allan (2013). As tropical SSTs increase through the 21st century in coupled simulations, precipitation increases in the wet regime/season and decreases in the dry regime/season, although changes in the 20th century for the wet regime are complex, relating to the mix of greenhouse gas and aerosol radiative forcings (Salzmann, 2016) combined with the seasonal position of the tropical rainy belt (Feng et al., 2013) and its associated monsoonal systems relative to the geographic coverage of land. Although observations, primarily based on rain gauge measurements, capture some aspects of the multidecadal changes, variability is larger than the model ensemble spread. By considering a long AMIP simulation conducted with the Goddard Institute for Space Studies (GISS) climate model, precipitation variability appears to be generally well explained by the substantial decadal variability in SSTs that is unforced and internal to the climate system. Thus, although the commonly quoted mechanism for "wet gets wetter, dry gets drier" does not apply in general over land, intriguing signals of amplification in the seasonality of rainfall (Chou et al., 2013) are apparent over tropical land with potentially serious consequences for ecosystem productivity (Murray-Tortarolo et al., 2016) yet are confounded by dominant signals of internal variability (Lintner et al., 2012) and by differences between observed and simulated climatological locations of wet and dry regimes and their spatial movement over time (Polson and Hegerl, 2016). The role of radiative forcings in determining wet/dry regime responses due to dynamical adjustments (He and Soden, 2016) and the influence on seasonal tropical rainfall extremes is of considerable importance to the evolving impacts of climate change and merits further scrutiny and continued monitoring.

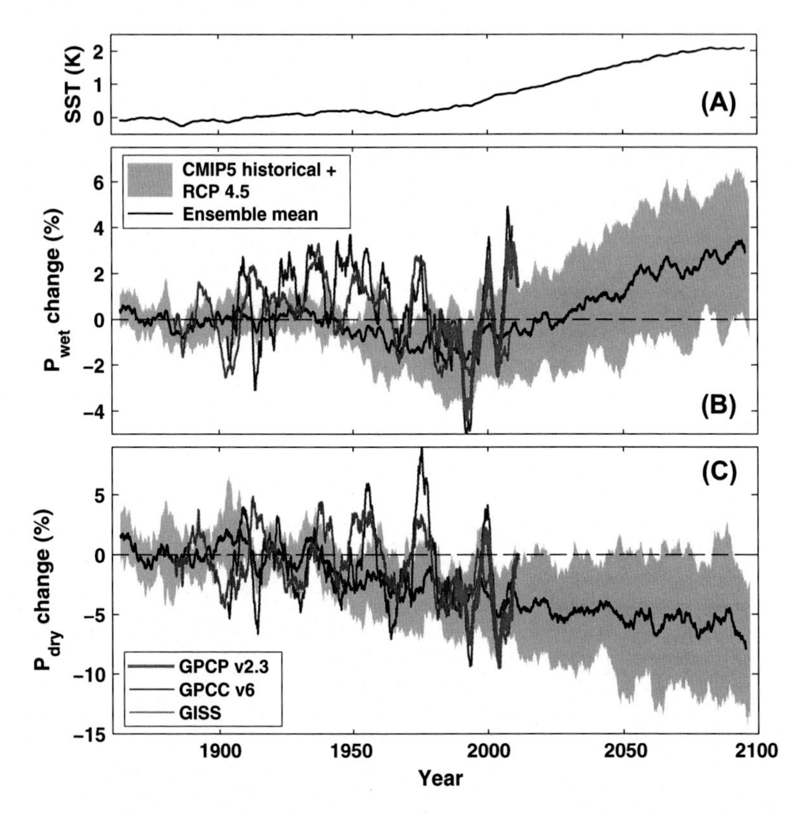

Figure 5.2 Historical and projected future changes in tropical (A) sea surface temperature and tropical precipitation over land for (B) the wettest 30% of the atmospheric circulation and (C) the driest 70% of the tropical circulation in fully coupled climate model simulations (CMIP5 historical and RCP4.5 experiments), an atmosphere-only climate model simulation (the Goddard Institute for Space Studies (GISS) climate model) and Global Precipitation Climatology Project (GPCP) and Global Precipitation Climatology Center (GPCC) observations.

4. EVALUATION OF CURRENT AND FUTURE CHANGES IN EXTREME RAINFALL

The "where and when" of tropical rainfall extremes is slave to the detailed dynamical and thermodynamic structure of the tropical atmosphere and in some circumstances its local interaction with the land surface (Taylor et al., 2012). Effective evaluation of rainfall extremes can also be conducted by considering the intensity distribution, independent of geographical region (e.g., Pendergrass and Hartmann, 2014). Fig. 5.3 illustrates how scaling of precipitation extremes with tropical mean surface temperature can be evaluated for present-day variability and further may be used to gauge the robust nature of future climate change projections.

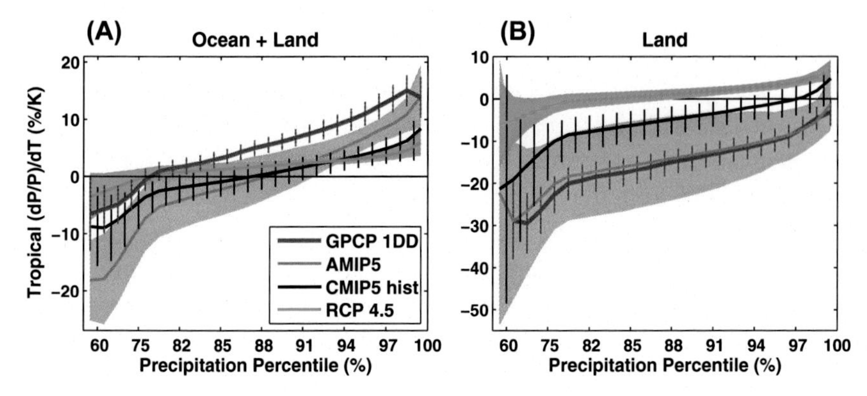

Figure 5.3 Responses in 5-day mean precipitation to tropic-wide mean temperature in % per Kelvin across percentiles of precipitation intensity over (A) the global tropics and (B) tropical land. Observed scalings for the present day (1998–2008) and regression uncertainty are shown for GPCP 1DD daily observations and atmosphere-only climate model simulations (AMIP). Also shown are scalings for the 1985–2005 present-day period for coupled climate model historical simulations (CMIP5 hist) and scalings estimated from the difference between future (2080–2100, CMIP5 RCP4.5) and present (1985–2005, CMIP5 hist) intensity distribution changes to illustrate the long-term climate change response. *CMIP5*, Coupled Model Intercomparison Project Phase 5; *GPCP*, Global Precipitation Climatology Project.

For GPCP observations, the intensity distribution is generated for 5-day mean rainfall accumulations, which were considered the limit at which satellite products covering multiple decades are able to consistently capture the scaling of rainfall events (Liu and Allan, 2012). Consistent with the analysis of monthly wet and dry regimes in Section 3, an amplification of the tropic-wide precipitation intensity distribution with warming is evident in both observations and simulations with rainfall increasing in high percentiles and decreasing in low percentiles (Fig. 5.3A). The GPCP observations indicate a more positive scaling than the ensemble mean of the AMIP simulations by up to about 10%/K although there is greater consistency for the most intense 99th percentile where observed rainfall increases by ~14%/K; this is somewhat larger than expected from the Clausius–Clapeyron equation but consistent with previous studies when considering simulated climate change responses (Sugiyama et al., 2010) and has been linked to a shift in the intensity distribution (Pendergrass and Hartmann, 2014).

The AMIP simulations provide the most consistent comparison with observations due to the realistic prescribed SSTs and radiative forcings, yet fully coupled models are necessary to project future changes. Therefore, the coupled model versions are also tested for the present day (CMIP5 hist experiment in Fig. 5.3A), and these show a somewhat reduced amplification of the precipitation intensity distribution compared with the AMIP simulations for the tropical land plus ocean: the scaling at the 99th percentile is around 8%/K. The scaling for the present-day coupled simulations can further be compared for consistency with the climate change response, where this is calculated as the precipitation changes in each intensity bin from the present-day simulations (1985–2005 from the hist experiment) to the future (2080–2100 for the RCP4.5 moderate emissions future scenario experiment). There is again broad consistency between present-day variability and future responses with slightly reduced magnitudes of scaling and a smaller 99th percentile response (~6%/K), slightly larger than the model median values calculated by O'Gorman (2012). The differences can be explained in terms of rather different spatial patterns of warming between present-day variability and future climate change that not only influence the climate sensitivity (Gregory and Andrews, 2016) but also are related to the suppression of rainfall at the largest scales by the direct atmospheric heating effect of greenhouse gases (e.g., Allan et al., 2014a), which increase in tandem with tropical temperature during climate change but do not for short-term interannual variability.

Although some consistency in precipitation responses is evident between observations and models across timescales for the entire tropics, it is important to assess changes over land where the bulk of the impacts from extreme rainfall events or multi-season rainfall deficits occur. Fig. 5.3B repeats the precipitation intensity scalings for tropical land. Consistency between AMIP simulations and observations, which over land are primarily determined by rain gauge measurements, is excellent. However, in contrast with the tropics-wide analysis in Fig. 5.3A, precipitation intensity declines with tropics-wide warming. This is explained by considering that warm El Niño events are associated with substantial spatial reorganization of tropical atmospheric circulation such that on average there is less rainfall over tropical land (e.g., Trenberth et al., 2014). A similar picture is evident for the coupled model simulations of the present day (CMIP5 hist) although the ensemble mean precipitation scaling is up to 10%/K more positive than the AMIP simulations, partially explained by the inaccurate representation of ENSO-like processes in coupled simulations that display a diversity of multidecadal variability (e.g., see Fig. S2 of Palmer and McNeall, 2014).

Importantly, the future responses in precipitation distribution over tropical land do not resemble the present-day changes yet closely match the tropics-wide responses shown in Fig. 5.3A. The warming pattern of future climate change response is dissimilar to present-day variability, and the climate change scaling is not significantly influenced by ENSO-related changes in atmospheric circulation, although will be influenced by long-term responses in the tropical hydrological cycle to changes in SST (Hegerl et al., 2015) and its spatial patterns (Xie et al., 2010). This also includes narrowing of the ITCZ (Byrne and Schneider, 2016) and widening of the entire tropical belt (Seidel et al., 2008) in response to global warming not only induced by increasing emissions of greenhouse gases by human activities but also influenced by aerosol pollution and tropospheric ozone (Allen et al., 2012). This underscores the limitations of inferring climate change responses in precipitation from interannual variability over land. Yet the consistency between climate change responses over land and ocean and between the tropics-wide present-day variability and climate change response offers a possibility to provide more meaningful evaluation of future responses.

Fig. 5.4 shows an example evaluation of future climate change response, concentrating on the 99th percentile of 5-day rainfall intensity over the entire tropics where consistency between present-day variability in AMIP simulations and future CMIP5 model simulation response is superior to lower precipitation percentiles. Consistent with O'Gorman (2012), a robust relationship between present-day variability and climate change response is present across models, although the scaling for variability is approximately twice the climate

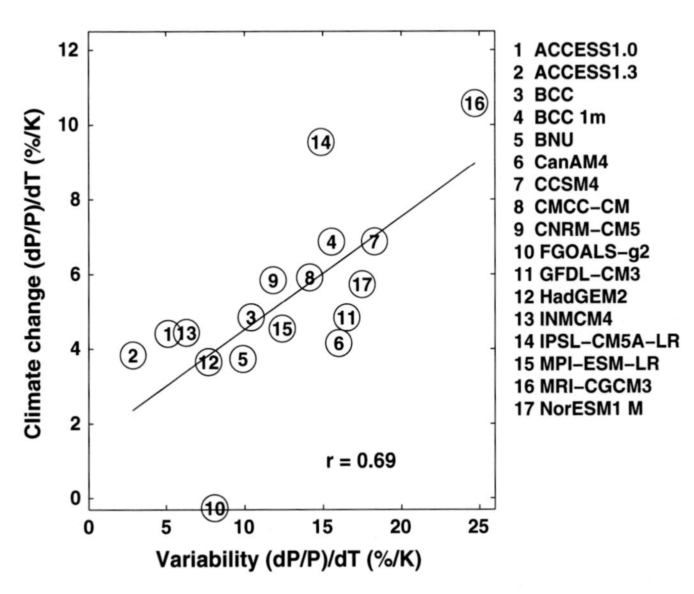

Figure 5.4 Links between present-day variability and future climate change response in tropical extreme (99th percentile) 5-day mean precipitation intensity response to tropical mean temperature changes across Coupled Model Intercomparison Project Phase 5 climate models.

change scaling. The observed 10%–17%/K extreme precipitation response diagnosed from variability is approximately equivalent to a 4%–7%/K climate change response. Although a more sophisticated statistical and physical representation of the most extreme rainfall events is required to fully understand future responses (Westra et al., 2014), this basic use of observations to evaluate present-day responses in precipitation extremes offers a potentially valuable emergent constraint on the simulated climate change response.

5. CONCLUSIONS

Extremes of tropical rainfall wreak damage on agriculture, infrastructure, and people. The where, when, and magnitude of drought and deluge are centered on the complex evolution of the atmosphere and ocean circulation. Therefore limiting impacts demands sufficient preparedness, which necessitates comprehensive prediction and monitoring strategies. Sophisticated numerical modeling and observing systems are vital in providing such a capability yet both are deficient to different degrees, and interpretation requires a physically based context to draw on their full potential. This is particularly acute when evaluating the current and future impact of climate changes on characteristics of rainfall extremes throughout the tropics (Hegerl et al., 2015). One strategy among many is to combine observations and simulations within an interpretive framework to more effectively gauge how tropical extremes are currently evolving and how they will alter in the future. This chapter has summarized some recent advances in understanding tropical rainfall extremes at the largest spatial scales and illustrates examples of how observations and simulations may be combined to monitor ongoing changes in tropical precipitation extremes and evaluate future responses to the warming of climate driven by continued emissions of greenhouse gases and aerosol pollution through intense human activities.

Water and energy are the fuel for tropical rainfall extremes and provide a powerful framework for interpreting current changes at the largest spatial scales. Water vapor fluctuates with ENSO-driven variability in tropical temperatures at the rate expected from the Clausius–Clapeyron equation, around 8%/K based on the observations employed. Warming trends over recent decades, linked to climate response to human activities, have resulted in a 1.2%/decade increase in low-altitude moisture. Climate models are able to capture the essential water vapor responses to climate variability and change, yet representing rainfall extremes is more challenging. The barely perceptible response of tropical mean precipitation to multidecadal warming is expected based on energy budget constraints at the surface (Roderick et al., 2014), in the atmosphere (Allen and Ingram, 2002), and their dynamical coupling (Bony et al., 2013), including complex adjustments to the mix of radiative forcings (Myhre et al., 2017). Signals are further obscured by the strong influence of internally generated variability in atmosphere and ocean circulation (Gu et al., 2016) and the more challenging measurement and sampling of highly heterogeneous precipitation fields in space and time by a disparate set of observing systems (Hegerl et al., 2015).

Despite these challenges in detecting and understanding changes in the hydrological cycle, contrasting responses of precipitation to tropical temperature are apparent in wet and dry portions of the atmospheric circulation (Liu and Allan, 2013) manifest particularly as an amplification of wet and dry seasons (Chou et al., 2013). This is confirmed by accounting for contrasting climatological locations of wet and dry regimes and their spatial movements over time using global data sets (Polson et al., 2016) and long island records (Polson and Hegerl, 2016). Although greater moisture transports from subtropical ocean source regions to wet convergence zones are expected in the tropics, a more complex explanation for the scaling of precipitation and its extremes is emerging: recent subtropical ocean precipitation decline is strongly linked to a dynamical response to the direct atmospheric heating by increased carbon dioxide, with moisture transports apparently playing a secondary role (He and Soden, 2016). Furthermore, precipitation responses over land are driven by a complex mix of these dynamical responses (Bony et al., 2013) modulated by the geographical position of mountains and regional gradients in temperature and humidity (Byrne and O'Gorman, 2015) and modified through land–atmosphere feedbacks (Berg et al., 2016). Fixed geographical locations sample both wet and dry portions of the atmospheric circulation, for example, wet and dry seasons; these contrasting meteorological regimes are expected to display disparate behaviors in response to a warming climate and are further subject to substantial influence from subtle changes in the position and characteristics of large-scale atmospheric circulation patterns (Kumar et al., 2015), for example, monsoon systems (Vittal et al., 2016; Taylor et al., 2017). Therefore detecting past changes and projecting future responses over land are problematic because of these multiple complex drivers and observational sampling deficiencies (Greve et al., 2014).

The links between precipitation and aridity are also questionable (Roderick et al., 2014; Scheff and Frierson, 2015; Greve et al., 2014) and further complicated by biophysical responses to carbon dioxide

(Milly and Dunne, 2016). Clear signals of increased rainfall and its intensity over wet regions are physically understandable and related to enhanced moisture convergence yet more research is required to monitor and understand declines in precipitation in the drier portion of the atmospheric circulation in observations and simulations (Liu and Allan, 2013) and attribute these changes to thermodynamic processes, circulation responses to radiative forcing or unforced internal variability, or feedback processes. Despite this complexity and the uncertainty in location of future spatial shifts in the tropical circulation, climate models consistently project large rainfall changes will occur (including both substantial increases or decreases depending on location) across considerable proportions of tropical land over the 21st century (Chadwick et al., 2015), and the intensification of rainfall events, where and when they occur, is an inevitable consequence of climate warming.

As for monthly precipitation, the heaviest rainfall events are slave to the locations of the tropical rainy belt and the convective organization within, as well as the positions of the tracks of tropical cyclones. Decadal climate variability, including the timing and nature of ENSO events, strongly determines the spatial rearrangement of tropical precipitation extremes (Sukhatme and Venugopal, 2017): both extremes and precipitation trends at larger space and timescales are determined by the complex mix of radiative forcing of climate and the decadal manifestation of ENSO or multidecadal ocean decadal variability (Gu et al., 2016). This makes the detection of changes in mean and extreme precipitation a challenge: more extreme rainfall in wet and dry regions was identified over higher latitude regions with a lack of clear signal in the tropics for the period 1950–80 (Donat et al., 2016) yet this period is unusual, with substantial changes in aerosol forcing, which alter the spatial pattern of wet and dry portions of the tropical circulation (Bollasina et al., 2011), and definition of reference periods can substantially influence the results (Sippel et al., 2018).

Examination of precipitation intensity distribution, independent of geographical location, offers a powerful way to explore and interpret the current and future changes in tropical extremes (e.g., Allen and Ingram, 2002). A consistent shift of the intensity distribution is apparent for 5-day accumulations for the available observations and simulations employed with intensification of the heaviest rainfall events at a rate close to or above that expected from the Clausius–Clapeyron equation. However, linking the interannual covariation with tropical warming to climate change is problematic, especially over land where warm El Niño years are associated with less rainfall across tropical continental regions. Nevertheless, scaling over the entire tropics, or at least the oceans, offers a promising emergent constraint on the heaviest rainfall events (O'Gorman, 2012).

A further challenge in projecting future responses in tropical rainfall extremes is the inability for models to represent mesoscale convection, including its spatial organization (Pendergrass et al., 2016). Diversity among model responses linked to the convection scheme used (Turner and Slingo, 2009) cast doubt on the projections of the most extreme rainfall intensities (O'Gorman, 2015). Hourly observations from Hong Kong suggest a super Clausius–Clapeyron scaling of 10%–14%/K increases in extreme precipitation with dew point temperature taken 4 hours prior to the event (Lenderink et al., 2011); this relationship does not extend above 23°C, which may link to microphysical constraints on precipitation extremes (Singh and O'Gorman, 2014), although the highest temperatures may be sampling unusual meteorological regimes (Hardwick Jones et al., 2010), and relative humidity changes linked to moisture availability may also play a role so further work is necessary to understand the relevance of scaling at the highest temperatures. Tantalizing evidence for changing characteristics of extreme storms at higher temperatures, including increased intensity at the expense of reduced spatial extent (Wasko et al., 2016), in addition to microphysical considerations, may explain super Clausius–Clapeyron responses. Yet cloud-resolving models are increasingly necessary to adequately simulate feedbacks and extreme rainfall responses to climate change while long standing biases in extreme rain rate simulated by models are difficult to evaluate consistently with satellite data (Wilcox and Donner, 2007). Thus, in evaluating and predicting extremes of tropical precipitation, a continued improvement in the capability to sample and represent precipitation intensity and spatial structure in models and observations is necessary in combination with an ongoing capacity to monitor changes in tropical climate.

ACKNOWLEDGMENTS

Thanks to Paul O'Gorman and an anonymous reviewer for providing comments on a draft of the chapter. Support was provided from the UK National Centre for Earth Observation (NCEO) and National Centre for Atmospheric Science (NCAS) and the Natural Environment Research Council SMURPHS (NE/N006054/1) projects. We acknowledge the World Climate Research Programme's Working Group on Coupled Modelling, which is responsible for CMIP, and we thank the climate modeling groups for producing and making available their model outputs; for CMIP, the US Department of Energy's PCMDI provided coordinating support and led development of software infrastructure in partnership with the Global Organization for Earth System Science Portals. CMIP5 and AMIP5 climate model data sets were extracted from the British Atmospheric Data Centre (http://badc.nerc.ac.uk/home) and the Program for Climate Model

Diagnosis and Intercomparison (pcmdi3.llnl.gov/esgcet). GPCP data were extracted from precip.gsfc.nasa.gov and GPCC precipitation observations from the Global Precipitation Climatology Centre (http://www.dwd.de/EN/ourservices/gpcc/gpcc.html). Merged radiation budget data are available from http://www.met.reading.ac.uk/~sgs02rpa/research/DEEP-C/GRL/ while SSM/I and SSMIS products were also obtained online (ftp.ssmi.com). HadCRUT4 data are available from http://www.metoffice.gov.uk/hadobs/hadcrut4/.

REFERENCES

Adler, R.F., Sapiano, M.R.P., Huffman, G.J., Wang, J.-J., Gu, G., Bolvin, D., Chiu, L., Schneider, U., Becker, A., Nelkin, E., Xie, P., Ferraro, R., Shin, D.-B., 2018. The Global Precipitation Climatology Project (GPCP) Monthly Analysis (New Version 2.3) and a Review of 2017 Global Precipitation. Atmosphere. 9, 138. https://doi.org/10.3390/atmos9040138.

Allan, R.P., 2012. The role of water vapour in Earth's energy flows. Surv. Geophys. 33, 557–564. https://doi.org/10.1007/s10712-011-9157-8.

Allan, R.P., Liu, C., Zahn, M., Lavers, D.A., Koukouvagias, E., Bodas-Salcedo, A., 2014a. Physically consistent responses of the global atmospheric hydrological cycle in models and observations. Surv. Geophys. 35, 533–552. https://doi.org/10.1007/s10712-012-9213-z.

Allan, R.P., Liu, C., Loeb, N.G., Palmer, M.D., Roberts, M., Smith, D., Vidale, P.-L., 2014b. Changes in global net radiative imbalance 1985-2012. Geophys. Res. Lett. 41, 5588–559710. https://doi.org/10.1002/2014GL060962.

Allen, M.R., Ingram, W.J., 2002. Constraints on future changes in climate and the hydrological cycle. Nature 419, 224–232. https://doi.org/10.1038/nature01092.

Allen, R.J., Sherwood, S.C., Norris, J.R., Zender, C.S., 2012. Recent Northern Hemisphere tropical expansion primarily driven by black carbon and tropospheric ozone. Nature 485, 350–355. https://doi.org/10.1038/nature11097.

Andrews, T., Forster, P.M., Boucher, O., Bellouin, N., Jones, A., 2010. Precipitation, radiative forcing and global temperature change. Geophys. Res. Lett. 37, L14701. https://doi.org/10.1029/2010GL043991.

Berg, A., Findell, K., Lintner, B., Giannini, A., Seneviratne, S.I., van den Hurk, B., Lorenz, R., Pitman, A., Hagemann, S., Meier, A., Cheruy, F., Ducharne, A., Malyshev, S., Milly, P.C.D., 2016. Land–atmosphere feedbacks amplify aridity increase over land under global warming. Nat. Clim. Change 6, 869–874. https://doi.org/10.1038/nclimate3029.

Bollasina, M., Ming, Y., Ramaswamy, V., 2011. Anthropogenic aerosols and the weakening of the South Asian summer monsoon. Science 334, 502–505. https://doi.org/10.1126/science.1204994.

Bony, S., Bellon, G., Klocke, D., Sherwood, S., Fermepinand, S., Denvil, S., 2013. Robust direct effect of carbon dioxide on tropical circulation and regional precipitation. Nat. Geosci. 6, 447–451. https://doi.org/10.1038/ngeo1799.

Boos, W.R., Storelvmo, T., 2016. Near-linear response of mean monsoon strength to a broad range of radiative forcings. Proc. Natl. Acad. Sci. U.S.A. 113 (6), 1510–1515. https://doi.org/10.1073/pnas.1517143113.

Byrne, M.P., O'Gorman, P.A., 2015. The response of precipitation minus evapotranspiration to climate warming: why the "wet-get-wetter, dry-get-drier" scaling does not hold over land. J. Clim. 28, 8078–8092. https://doi.org/10.1175/JCLI-D-15-0369.1.

Byrne, M.P., O'Gorman, P.A., 2016. Understanding decreases in land relative humidity with global warming: conceptual model and GCM simulations. J. Clim. 29, 9045–9061. https://doi.org/10.1175/JCLI-D-16-0351.1.

Byrne, M.P., Schneider, T., 2016. Narrowing of the ITCZ in a warming climate: physical mechanisms. Geophys. Res. Lett. 43, 11350–11357. https://doi.org/10.1002/2016GL070396.

Chadwick, R., Boutle, I., Martin, G., 2013. Spatial patterns of precipitation change in CMIP5: why the rich do not get richer in the tropics. J. Clim. 26, 3803–3822. https://doi.org/10.1175/JCLI-D-12-00543.1.

Chadwick, R., Good, P., Martin, G., Rowell, D.P., 2015. Large rainfall changes consistently projected over substantial areas of tropical land. Nat. Clim. Change 6, 177–181. https://doi.org/10.1038/nclimate2805.

Chou, C., Chiang, J.C.H., Lan, C.-W., Chung, C.-H., Liao, Y.-C., Lee, C.-J., 2013. Increase in the range between wet and dry season precipitation. Nat. Geosci. 6, 263–267. https://doi.org/10.1038/NGEO01744.

Dee, D.P., Uppala, S.M., Simmons, A.J., Berrisford, P., Poli, P., Kobayashi, S., Andrae, U., Balmaseda, M.A., Balsamo, G., Bauer, P., Bechtold, P., Beljaars, A.C.M., van de Berg, L., Bidlot, J., Bormann, N., Delsol, C., Dragani, R., Fuentes, M., Geer, A.J., Haimberger, L., Healy, S.B., Hersbach, H., Hólm, E.V., Isaksen, L., Kållberg, P., Köhler, M., Matricardi, M., McNally, A.P., Monge-Sanz, B.M., Morcrette, J.-J., Park, B.-K., Peubey, C., de Rosnay, P., Tavolato, C., Thépaut, J.-N., Vitart, F., 2011. The ERA-Interim reanalysis: configuration and performance of the data assimilation system. Q. J. R. Meteorol. Soc. 137, 553–597. https://doi.org/10.1002/qj.828.

Donat, M.G., Lowry, A.L., Alexander, L.V., O'Gorman, P.A., Maher, N., 2016. More extreme precipitation in the world's dry and wet regions. Nat. Clim. Change 6, 508–513. https://doi.org/10.1038/nclimate2941.

Dong, B., Sutton, R., 2015. Dominant role of greenhouse-gas forcing in the recovery of Sahel rainfall. Nat. Clim. Change 5, 757–760. https://doi.org/10.1038/nclimate2664.

Durack, P., Wijffels, S., Matear, R.J., 2012. Ocean salinities reveal strong global water cycle intensification during 1950–2000. Science 336, 455–458. https://doi.org/10.1126/science.1212222.

Emori, S., Brown, S.J., 2005. Dynamic and thermodynamic changes in mean and extreme precipitation under changed climate. Geophys. Res. Lett. 32, L17706. https://doi.org/10.1029/2005GL023272.

Feng, X., Porpotato, A., Rodriguez-Iturbe, I., 2013. Changes in rainfall seasonality in the tropics. Nat. Clim. Change 3, 811–815. https://doi.org/10.1038/NCLIMATE1907.

Frierson, D., Hwang, Y.-T., Fučkar, N.S., Seager, R., Kang, S.M., Donohoe, A., Maroon, E.A., Liu, X., Battisti, D.S., 2013. Contribution of ocean overturning circulation to tropical rainfall peak in the Northern Hemisphere. Nat. Geosci. 6, 940–944. https://doi.org/10.1038/ngeo1987.

Gregory, J.M., Andrews, T., 2016. Variation in climate sensitivity and feedback parameters during the historical period. Geophys. Res. Lett. 43, 3911–3920. https://doi.org/10.1002/2016GL068406.

Greve, P., Orlowsky, B., Mueller, B., Sheffield, J., Reichstein, M., Seneviratne, S.I., 2014. Global assessment of trends in wetting and drying over land. Nat. Geosci. 7, 716–721. https://doi.org/10.1038/ngeo2247.

Gu, G., Adler, R.F., Huffman, G.J., 2016. Long-term changes/trends in surface temperature and precipitation during the satellite era (1979–2012). Clim. Dyn. 46, 1091–1105. https://doi.org/10.1007/s00382-015-2634-x.

Hardwick Jones, R., Westra, S., Sharma, A., 2010. Observed relationships between extreme sub-daily precipitation, surface temperature, and relative humidity. Geophys. Res. Lett. 37, L22805. https://doi.org/10.1029/2010GL045081.

Hawcroft, M., Haywood, J.M., Collins, M., et al., 2016. Southern Ocean albedo, inter-hemispheric energy transports and the double ITCZ: global impacts of biases in a coupled model. Clim. Dyn. 48, 2279–2295. https://doi.org/10.1007/s00382-016-3205-5.

He, J., Soden, B.J., 2016. A re-examination of the projected subtropical precipitation decline. Nat. Clim. Change 7, 53–57. https://doi.org/10.1038/nclimate3157.

Hegerl, G.C., Black, E., Allan, R.P., Ingram, W.J., Polson, D., Trenberth, K.E., Chadwick, R.S., Arkin, P.A., Sarojini, B.B., Becker, A., Dai, A., Durack, P.J., Easterling, D., Fowler, H.J., Kendon, E.J., Huffman, G.J., Liu, C., Marsh, R., New, M., Osborn, T.J., Skliris, N., Stott, P.A., Vidale, P.-L., Wijffels, S.E., Wilcox, L.J., Willett, K.M., Zhang, X., 2015. Challenges in quantifying changes in the global water cycle. Bull. Am. Meteorol. Soc. 96, 1097–1115. https://doi.org/10.1175/BAMS-D-13-00212.1.

Held, I.M., Soden, B.J., 2006. Robust responses of the hydrological cycle to global warming. J. Clim. 19, 5686–5699. https://doi.org/10.1175/JCLI3990.1.

Hwang, Y.-T., Frierson, D.M.W., Kang, S.M., 2013. Anthropogenic sulfate aerosol and the southward shift of tropical precipitation in the 20th century. Geophys. Res. Lett. 40, 1–6. https://doi.org/10.1029/2012GL054022.

IPCC, 2013. Climate change 2013: the physical science basis. In: Stocker, T.F., Qin, D., Plattner, G.-K., Tignor, M., Allen, S.K., Boschung, J., Nauels, A., Xia, Y., Bex, V., Midgley, P.M. (Eds.), Contribution of Working Group I to the Fifth Assessment Report of the Intergovernmental Panel on Climate Change. Cambridge University Press, Cambridge, United Kingdom and New York, NY, USA. https://doi.org/10.1017/CBO9781107415324. 1535 p.

Kumar, S., Allan, R.P., Zwiers, F., Lawrence, D.M., Dirmeyer, P.A., 2015. Revisiting trends in wetness and dryness in the presence of internal climate variability and water limitations over land. Geophys. Res. Lett. 42, 10,867–10,875. https://doi.org/10.1002/2015GL066858.

Lenderink, G., Mok, H.Y., Lee, T.C., van Oldenborgh, G.J., 2011. Scaling of trends in hourly precipitation extremes in two different climate zones – Hong Kong and The Netherlands. Hydrol. Earth Syst. Sci. 15, 3033–3041. https://doi.org/10.5194/hess-15-3033-2011.

Levermann, A., Schewe, J., Petoukhov, V., Held, H., 2009. Basic mechanism for abrupt monsoon transitions. Proc. Nat. Acad. Sci. 106, 20572–20577. https://doi.org/10.1073/pnas.0901414106.

Lin, L., Wang, Z., Xu, Y., Fu, Q., 2016. Sensitivity of precipitation extremes to radiative forcing of greenhouse gases and aerosols. Geophys. Res. Lett. 43, 9860–9868. https://doi.org/10.1002/2016GL070869.

Lintner, B.R., Biasutti, M., Diffenbaugh, N.S., Lee, J.-E., Niznik, M.J., Findell, K.L., 2012. Amplification of wet and dry month occurrence over tropical land regions in response to global warming. J. Geophys. Res. 117, D11106. https://doi.org/10.1029/2012JD017499.

Liu, C., Allan, R.P., 2012. Multi-satellite observed responses of precipitation and its extremes to interannual climate variability. J. Geophys. Res. 117, D03101. https://doi.org/10.1029/2011JD016568.

Liu, C., Allan, R.P., 2013. Observed and simulated precipitation responses in wet and dry regions 1850-2100. Environ. Res. Lett. 8, 034002. https://doi.org/10.1088/1748-9326/8/3/034002.

Loeb, N.G., Wang, H., Cheng, A., Kato, S., Fasullo, J.T., Xu, K.-M., Allan, R.P., 2016. Observational constraints on atmospheric and oceanic cross-equatorial heat transports: revisiting the precipitation asymmetry problem in climate models. Clim. Dyn. 46, 3239–3257. https://doi.org/10.1007/s00382-015-2766-z.

Maidment, R.I., Allan, R.P., Black, E., 2015. Recent observed and simulated changes in precipitation over Africa. Geophys. Res. Lett. 42, 8155–8164. https://doi.org/10.1002/2015GL065765.

Marvel, K., Bonfils, C., 2013. Identifying external influences on global precipitation. Proc. Nat. Acad. Sci. 110, 19301–19306. https://doi.org/10.1073/pnas.1314382110.

Milly, P.C.D., Dunne, K.A., 2016. Potential evapotranspiration and continental drying. Nat. Clim. Change 6, 946–949. https://doi.org/10.1038/nclimate3046.

Ming, Y., Ramaswamy, V., Persad, G., 2010. Two opposing effects of absorbing aerosols on global-mean precipitation. Geophys. Res. Lett. 37, L13701. https://doi.org/10.1029/2010GL042895.

Mitchell, J., Wilson, C.A., Cunnington, W.M., 1987. On CO_2 climate sensitivity and model dependence of results. Q. J. R. Meteorol. Soc. 113, 293–322.

Muller, C.J., O'Gorman, P.A., 2011. An energetic perspective on the regional response of precipitation to climate change. Nat. Clim. Change 1, 266–271. https://doi.org/10.1038/nclimate1169.

Murray-Tortarolo, G., Friedlingstein, P., Sitch, S., Seneviratne, S.I., Fletcher, I., Mueller, B., Greve, P., Anav, A., Liu, Y., Ahlström, A., et al., 2016. The dry season intensity as a key driver of NPP trends. Geophys. Res. Lett. 43, 2632–2639. https://doi.org/10.1002/2016GL068240.

Myhre, G., Forster, P., Samset, B., Hodnebrog, Ø., Sillmann, J., Aalbergsjø, S., Andrews, T., Boucher, O., Faluvegi, G., Flaeschner, D., Iversen, T., Kasoar, M., Kharin, S., Kirkevåg, A., Lamarque, J., Olivié, D., Richardson, T., Shindell, D., Shine, K., Stjern, C., Takemura, T., Voulgarakis, A., Zwiers, F., 2017. PDRMIP: a precipitation driver and response model Intercomparison project, protocol and preliminary results. Bull. Am. Meteorol. Soc. 98, 1185–1198. https://doi.org/10.1175/BAMS-D-16-0019.1.

O'Gorman, P.A., 2012. Sensitivity of tropical precipitation extremes to climate change. Nat. Geosci. 5, 697–700. https://doi.org/10.1038/NGEO1568.

O'Gorman, P.A., Allan, R.P., Byrne, M.P., Previdi, M., 2012. Energetic constraints on precipitation under climate change. Surv. Geophys. 33, 585–608. https://doi.org/10.1007/s10712-011-9159-6.

O'Gorman, P.A., 2015. Precipitation extremes under climate change. Curr. Clim. Change Rep. 1, 49–59. https://doi.org/10.1007/s40641-015-0009-3.

Palmer, M.D., McNeall, D.J., 2014. Internal variability of Earth's energy budget simulated by CMIP5 climate models. Environ. Res. Lett. 9, 034016. https://doi.org/10.1088/1748-9326/9/3/034016.

Pendergrass, A.G., Hartmann, D.L., 2014. Two modes of change of the distribution of rain. J. Clim. 27, 8357–8370. https://doi.org/10.1175/JCLI-D-14-00182.1.

Pendergrass, A.G., Reed, K.A., Medeiros, B., 2016. The link between extreme precipitation and convective organization in a warming climate: global radiative-convective equilibrium simulations. Geophys. Res. Lett. 43, 11445–11452. https://doi.org/10.1002/2016GL071285.

Polson, D., Hegerl, G., 2016. Strengthening contrast between precipitation in tropical wet and dry regions. Geophys. Res. Lett. 43. https://doi.org/10.1002/2016GL071194.

Polson, D., Hegerl, G.C., Solomon, S., 2016. Precipitation sensitivity to warming estimated from long island records. Environ. Res. Lett. 11, 074024. https://doi.org/10.1088/1748-9326/11/7/074024.

Roca, R., Aublanc, J., Chambon, P., Fiolleau, T., Viltard, N., 2014. Robust observational quantification of the contribution of mesoscale convective systems to rainfall in the tropics. J. Clim. 27, 4952–4958. https://doi.org/10.1175/JCLI-D-13-00628.1.

Roderick, M.L., Sun, F., Lim, W.H., Farquhar, G.D., 2014. A general framework for understanding the response of the water cycle to global warming over land and ocean. Hydrol. Earth Syst. Sci. 18 (5), 1575–1589. https://doi.org/10.5194/hess-18-1575-2014.

Salzmann, M., 2016. Global warming without global mean precipitation increase? Sci. Adv. 2, e1501572. https://doi.org/10.1126/sciadv.1501572.

Scheff, J., Frierson, D.M., 2015. Terrestrial aridity and its response to greenhouse warming across CMIP5 climate models. J. Clim. 28, 5583–5600. https://doi.org/10.1175/JCLI-D-14-00480.1.

Seidel, D.J., Fu, Q., Randel, W.J., Reichler, T.J., 2008. Widening of the tropical belt in a changing climate. Nat. Geosci. 1, 21–24. https://doi.org/10.1038/ngeo.2007.38.

Singh, M.S., O'Gorman, P.A., 2014. Convective Precipitation Extremes Scaling with Temperature Limited by Droplet/ice Fall Speeds in Simulations. Geophys. Res. Lett. 41, 6037–6044. https://doi.org/10.1002/2014GL061222.

Simmons, A.J., Willett, K.M., Jones, P.D., Thorne, P.W., Dee, D.P., 2010. Low-frequency variations in surface atmospheric humidity, temperature, and precipitation: inferences from reanalyses and monthly gridded observational data sets. J. Geophys. Res. 115, D01110. https://doi.org/10.1029/2009JD012442.

Sippel, S., Zscheischler, J., Heimann, M., Lange, H., Mahecha, M.D., Van Oldenborgh, G.J., Otto, F.E.L., Reichstein, M., 2018. Have precipitation extremes and annual totals been increasing in the world's dry regions over the last 60 years? Hydrol. Earth Syst. Sci. Discuss. 21, 441–458. https://doi.org/10.5194/hess-21-441-2017.

Skliris, N., Zika, J.D., Nurser, G., Josey, S.A., Marsh, R., 2016. Global water cycle amplifying at less than the Clausius-Clapeyron rate. Sci. Rep. 6, 38752. https://doi.org/10.1038/srep38752.

Su, H., Neelin, J.D., 2003. The scatter in tropical average precipitation anomalies. J. Clim. 16, 3966–3977. https://doi.org/10.1175/1520-0442(2003)016, 3966:TSITAP.2.0.CO;2.

Sugiyama, M., Shiogama, H., Emori, S., 2010. Precipitation extreme changes exceeding moisture content increases in MIROC and IPCC climate models. Proc. Nat. Acad. Sci. 107, 571–575. https://doi.org/10.1073/pnas.0903186107.

Sukhatme, J., Venugopal, V., 2016. Waxing and waning of observed extreme annual tropical rainfall. Q. J. R. Meteorol. Soc. 142, 102–107. https://doi.org/10.1002/qj.2633.

Taylor, C.M., de Jeu, R.A.M., Guichard, F., Harris, P.P., Dorigo, W.A., 2012. Afternoon rain more likely over drier soils. Nature 489, 423–426. https://doi.org/10.1038/nature11377.

Taylor, C.M., Belušić, D., Guichard, F., Parker, D.J., Vischel, T., Bock, O., Harris, P.P., Janicot, S., Klein, C., Panthou, G., 2017. Frequency of extreme Sahelian storms tripled since 1982 in satellite observations. Nature 544, 475–478. https://doi.org/10.1038/nature22069.

Taylor, K.E., Stouffer, R.J., Meehl, G.A., 2011. An overview of CMIP5 and the experiment design. Bull. Am. Meteorol. Soc. 93, 485–498. https://doi.org/10.1175/BAMS-D-11-00094.1.

Trenberth, K.E., Dai, A., Rasmussen, R.M., Parsons, D.B., 2003. The changing character of precipitation. Bull. Am. Meteorol. Soc. 84, 1205–1217. https://doi.org/10.1175/BAMS-84-9-1205.

Trenberth, K.E., Dai, A., van der Schrier, G., Jones, P.D., Barichivich, J., Briffa, K.R., Sheffield, J., 2014. Global warming and changes in drought. Nat. Clim. Change 4, 17–22. https://doi.org/10.1038/nclimate2067.

Turner, A.G., Slingo, J.M., 2009. Uncertainties in future projections of extreme precipitation in the Indian monsoon region. Atmos. Sci. Lett. 10, 152–158. https://doi.org/10.1002/asl.223.

Vittal, H., Ghosh, S., Karmakar, S., Pathak, A., Murtugudde, R., 2016. Lack of dependence of indian summer monsoon rainfall extremes on temperature: an observational evidence. Sci. Rep. 6, 31039. https://doi.org/10.1038/srep31039.

Wasko, C., Sharma, A., Westra, S., 2016. Reduced spatial extent of extreme storms at higher temperatures. Geophys. Res. Lett. 43, 4026–4032. https://doi.org/10.1002/2016GL068509.

Westra, S., Fowler, H.J., Evans, J.P., Alexander, L.V., Berg, P., Johnson, F., Kendon, E.J., Lenderink, G., Roberts, N.M., 2014. Future changes to the intensity and frequency of short-duration extreme rainfall. Rev. Geophys. 52, 522–555. https://doi.org/10.1002/2014RG000464.

Wentz, F.J., 2013. SSM/I Version-7 Calibration Report, Report Number 011012. Remote Sensing Systems, Santa Rosa, CA. 46 p.

Wilcox, E.M., Donner, L.J., 2007. The frequency of extreme rain events in satellite rain-rate estimates and an atmospheric general circulation model. J. Clim. 20, 53–69. https://doi.org/10.1175/JCLI3987.1.
Xie, S.-P., Deser, C., Vecchi, G.A., Ma, J., Teng, H., Wittenberg, A.T., 2010. Global warming pattern formation: sea surface temperature and rainfall. J. Clim. 23, 966–986. https://doi.org/10.1175/2009JCLI3329.1.
Zahn, M., Allan, R.P., 2013a. Climate Warming related strengthening of the tropical hydrological cycle. J. Clim. 26, 562–574. https://doi.org/10.1175/JCLI-D-12-00222.1.
Zahn, M., Allan, R.P., 2013b. Quantifying present and projected future atmospheric moisture transports onto land. Water Resour. Res. 49, 7266–7277. https://doi.org/10.1002/2012WR013209.

CHAPTER 6

Extreme El Niño Events

Boris Dewitte[1,2,3,4], Ken Takahashi[5]
[1]Centro de Estudios Avanzado en Zonas Áridas (CEAZA), Coquimbo, Chile; [2]Universidad Católica del Norte, Coquimbo, Chile; [3]Millennium Nucleus for Ecology and Sustainable Management of Oceanic Islands (ESMOI), Coquimbo, Chile; [4]Laboratoire d'Etudes en Géophysique et Océanographie Spatiales, Toulouse, France; [5]Servicio Nacional de Meteorología e Hidrología, Lima, Peru

Contents

1. INTRODUCTION

El Niño–Southern Oscillation (ENSO) is the dominant climate mode of the tropical Pacific at interannual timescales impacting many regions around the globe through oceanic and atmospheric teleconnections. Owing to its signatures on many geophysical and biological phenomena, it has been studied extensively although it is only from the mid-80s that it has been observed comprehensively, thanks in particular to the development of the Tropical Atmosphere Ocean (TAO) array, under the Tropical Ocean/Global Atmosphere program (McPhaden et al., 1998) and the development of ENSO theories and models (Neelin et al., 1998).

El Niño was originally documented in 1891 as a warm ocean current that brought extreme flooding to northern coastal Peru (Carranza, 1891;

Tropical Extremes: Natural Variability and Trends
ISBN 978-0-12-809248-4
https://doi.org/10.1016/B978-0-12-809248-4.00006-6

Carrillo, 1893) and the reports of the very strong event in 1925 (Murphy, 1926; Takahashi and Martínez, 2017) first enticed the concern of the international scientific community to El Niño. Soon after it was recognized that El Niño was not restricted to the coast of Peru and that it was linked to the large-scale atmospheric Southern Oscillation (Cushman, 2004). It was, however, only in the mid-60s that it was proposed that ENSO involves a positive feedback between the warming in the eastern equatorial Pacific and the weakening of the easterly trade winds associated with the so-called atmospheric Walker circulation and that this ocean–atmosphere interaction allows an El Niño event to grow (Bjerknes, 1966). Later the role of ocean equatorial wave dynamics in stopping and reversing such growth was emphasized (Cane and Sarachik, 1977). The first coupled ocean–atmosphere physical ENSO models were proposed in the 1980s (Anderson and McCreary, 1985; Zebiak and Cane, 1987), and the first paradigms based on linear dynamics (Schopf and Suarez, 1988; Battisti and Hirst, 1989; Penland, 1996; Penland and Sardeshmukh, 1995; Jin, 1997) were successful in capturing the major features of ENSO, such as its amplitude and its oscillatory nature (i.e., transition from El Niño to La Niña). In particular, the so-called "recharge–discharge oscillator" (Jin, 1997) encapsulates in its principle, the previous theories, and has served thus far as the main theoretical background for most ENSO studies. Because of the fast adjustment time of the ocean wave dynamics compared with the adjustment time of sea surface temperature (SST) (i.e., "fast wave limit," Neelin, 1991), Jin (1997) proposed that the system evolution is controlled by the heat content (i.e., the averaged temperature within the thermocline) over the equatorial Pacific (so-called warm water volume, hereafter WWV; Meinen and McPhaden, 2000) that acts as a heat reservoir of the system. Along the equator, the trade winds during an El Niño event induce not only an eastward warm water transport but also a poleward transport because of the effect of the Coriolis force (i.e., the Sverdrup transport). During La Niña, this wind-driven transport replenishes the equatorial band with warm waters after the thermocline has risen (recharge process). This meridional transport is phase-shifted relative to the peak SST anomalies and therefore provides the memory required for the system to oscillate. It has been argued that when the recharge process is particularly strong (i.e., high value of the WWV), there is a possibility of the development of a strong to extreme El Niño event, such as the 1997/98 event, one of the largest El Niño's observed so far. However, recent observations and model simulations indicate that large heat content is not a necessary precondition for an extreme El Niño to develop (Menkes et al., 2014;

Takahashi and Dewitte, 2016), which points to both the stochastic and the nonlinear character of ENSO. In fact, one drawback of the linear theories is that they cannot explain the amplitude asymmetry of ENSO, i.e., the largest El Niño events are larger than those of the largest La Niña events (An and Jin, 2004). This is the fundamental ENSO property that underpins the notion of extreme El Niño events, which cannot be accounted for by the recharge–discharge paradigm in its simplest form, i.e., a linearly damped zero-dimensional recharge–discharge oscillator (Jin, 1997; Burger et al., 2005).

Later in this chapter (Section 3), we review the theories for the amplitude ENSO asymmetry that tend to separate into two broad classes: (1) those that consider that the source of the nonlinearity is embedded into the ocean dynamics and (2) those conferring to the atmospheric processes a central role in the nonlinear amplification of El Niño events. It is because ENSO amplitude is asymmetric that there are extreme events and vice versa. In the statistical sense, ENSO is positively skewed (i.e., its probability density is not symmetrical), which implies the occurrence of extreme El Niño events.

More recently, the concept of ENSO diversity has been proposed (see Capotondi et al., 2015 for a comprehensive review), which has been another approach to tackle the issue of extreme ENSO event dynamics. ENSO diversity is generally thought of in terms of the different spatial patterns of ENSO, in particular focusing on whether the El Niño events have peak SST anomalies in the eastern Pacific (hereafter EP El Niño) or in the central Pacific (hereafter CP El Niño) (Fig. 6.1). However, two classes of El Niños can also be identified in terms of intensity (i.e., moderate vs. extreme/strong) as part of the concept of ENSO diversity. These two types of classification are related, as "strong" El Niño events tend to be of the EP type (e.g., 1982/83, 1997/98), whereas CP El Niños are "moderate" (Takahashi and Dewitte, 2016). So, the mechanisms underlying amplitude diversity appear to be linked to those associated with the spatial diversity.

ENSO amplitude and spatial patterns are related to the relative importance of the different feedbacks throughout the equatorial Pacific. In the central Pacific, the oceanic zonal advection feedback dominates, whereas in the eastern Pacific the thermocline feedback dominates (An and Jin, 2001; Chen et al., 2015). The nonlinear response of atmospheric convection to SST provides a key feedback that can lead to extreme eastern Pacific warming, whereas nonlinearities in ocean advection allow this warming to continue for longer (Takahashi and Dewitte, 2016). The study of ENSO diversity thus consists in disentangling the processes that lead to the

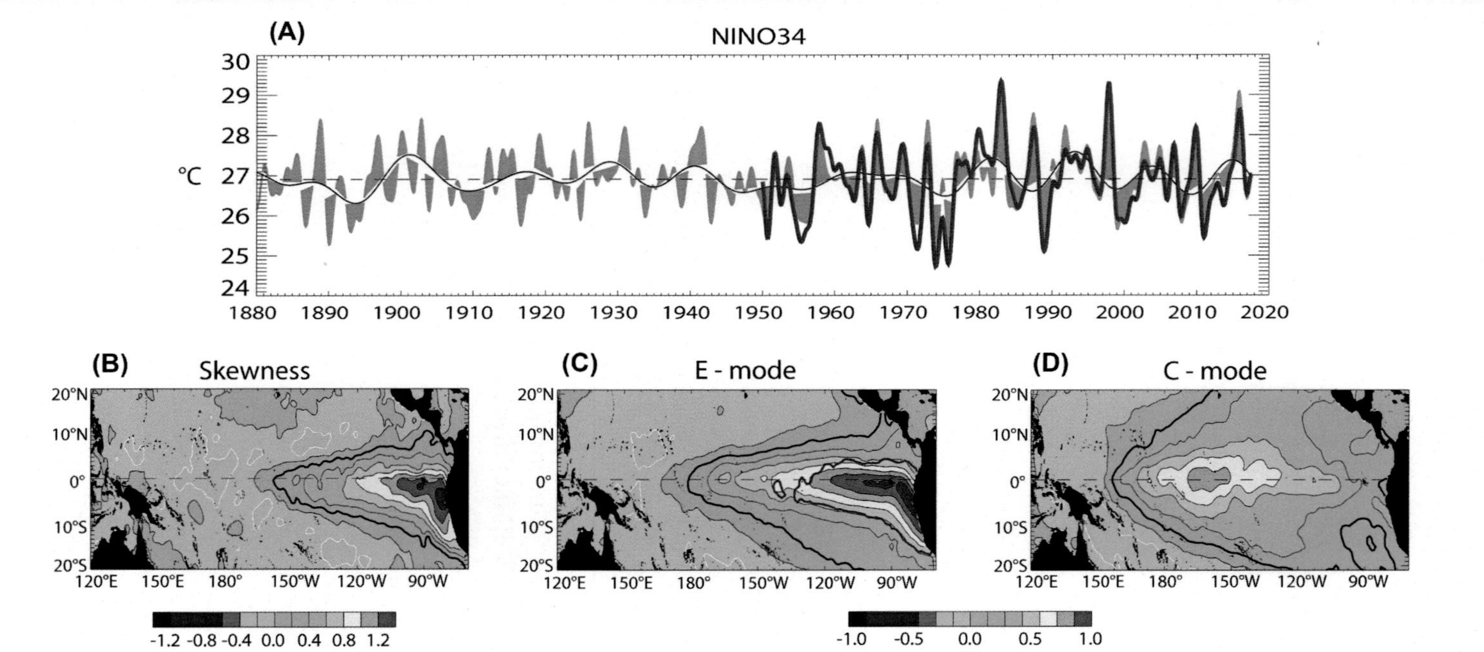

Figure 6.1 El Niño asymmetry and diversity: (A) NINO34 index over 1880–2017, (B) normalized skewness (in °C) of sea surface temperature (SST) anomalies over 1950–2017, which indicates the positive asymmetry of El Niño–Southern Oscillation (ENSO) in the far eastern Pacific, (C and D) the E and C modes calculated over 1950–2017. These two modes were proposed by Takahashi et al. (2011) and consists in the regression patterns of SST anomalies onto two indices defined as: $E = \frac{PC1 + PC2}{\sqrt{2}}$ and $C = \frac{PC1 - PC2}{\sqrt{2}}$ where the $PC1$ and $PC2$ are the PC time series of the first two empirical orthogonal function (EOF) modes of SST anomalies in the tropical Pacific (120°E–290°E; 21°S–21°N). Compared to the first two EOF mode patterns of SST anomalies, these two modes better capture the nonlinear evolution of El Niño grasping distinct ENSO regimes, with the E modes accounting for extreme EP events and the C mode accounting for both CP events, which includes Modoki events (Ashok et al., 2007) and Canonical El Nino events (Rasmuson and Carpenter, 1982), and La Niña events. Data have been linearly detrended over the period 1950–2017. The isocontour 0.5°C for skewness is overplotted in *red thick line* in (C) so as to indicate that extreme El Niño events are of EP variety. The *curve in thick brown line* corresponds to the approximated NINO34 obtained through bilinear regression onto the E and C indices over the period 1950–2017 (NINO34approx = 0.51E + 0.62C). *(Data are from the HadISST v1.1. data set Rayner, N.A., Parker, D.E., Horton, E.B., Folland, C.K., Alexander, L.V., Rowell, D.P., Kent, E.C., Kaplan, A., 2003. Global analyses of sea surface temperature, sea ice, and night marine air temperature since the late nineteenth century. J. Geophys. Res. https://doi.org/10.1029/2002JD002670.)*

amplification of one feedback through nonlinearities in the system that could have their roots either in the ocean or the atmosphere. We will see in Section 2 that the concept of ENSO diversity, by clarifying the definition of different types of El Niño events, has permitted to reinvigorate the interest in extreme ENSO dynamics.

The main difficulty in the investigation of extreme El Niño dynamics has been related to the fact that by definition these occur rarely and very few have been observed comprehensively. Over the satellite era, only three strong El Niño events (excluding the coastal type) have been observed (1982/83, 1997/98, and 2015/16) and with the 1972/73 El Niño event, four strong events have been observed since the 50s (Hong et al., 2014) (Fig. 6.2). Experimentation with models of various complexities has been a necessary surrogate approach to address this issue to explore the mechanisms at work during strong El Niño events, which led to significant advances thanks to the availability of a growing number of models and sensitivity experiments in the framework of the successive Coupled Model Intercomparison Projects (CMIP) of the World Climate Research Program. Additionally, an inherent difficulty is that the nonlinearity of extreme El Niño events limits the usefulness of classical techniques often used in geophysical analyses, such as principal component analysis (empirical orthogonal functions [EOFs] in climate sciences) and the assumption of Gaussianity.

Extreme El Niño events have two fundamental dimensions: their underlying physical dynamics (i.e., nonlinear) and their impacts. Extreme El Niño events impact the local climate and the global climate through oceanic and atmospheric teleconnections. However, those impacts have varied among the events, which in addition to their rarity over the contemporary record, have prevented drawing a clear picture of the extreme El Niño teleconnection patterns. In addition to the limitations imposed by the small sample (e.g., Deser et al., 2017), the nonlinear convective feedbacks also imply a different distribution of convective anomalies that can affect the teleconnection patterns (Johnson and Kosaka, 2016). Even in the tropical Pacific basin, the impacts of the 2015/16 El Niño on the rainfall season in northern Peru was much weaker than during the 1997/98 El Niño, despite comparable SST anomalies in the central-eastern Pacific (L'Heureux et al., 2017; Sanabria et al., 2017).

In this chapter, we propose a brief overview of the state of knowledge and of some current lines of research dedicated to extreme El Niño events. Section 2 provides a definition of extreme events based on physical principles and the recently developed concept of ENSO diversity. Section 3

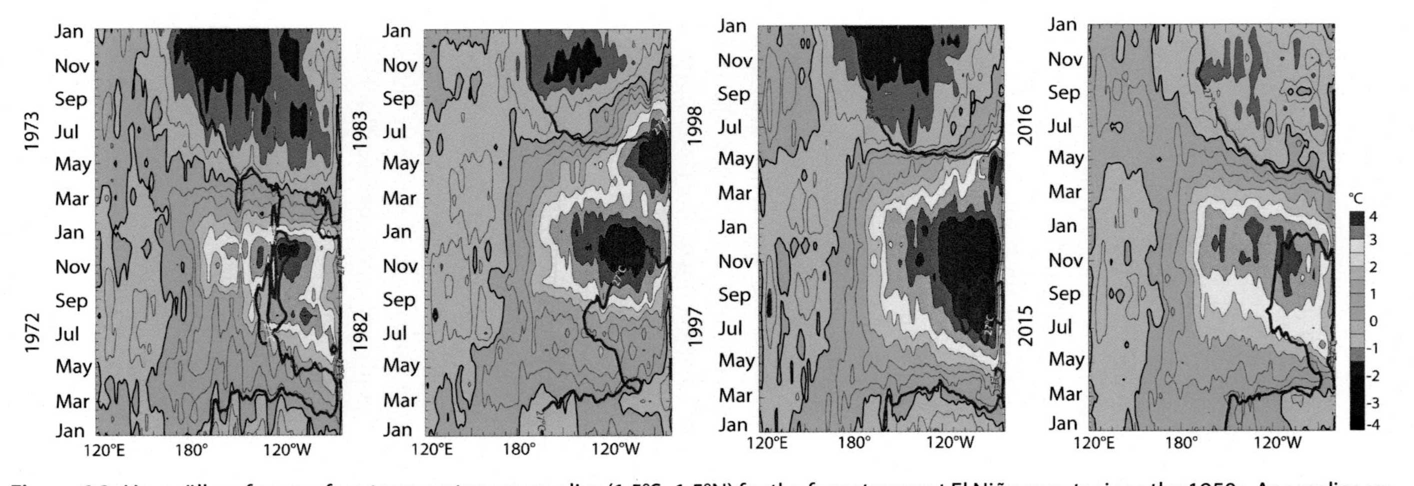

Figure 6.2 Hovmöller of sea surface temperature anomalies (1.5°S–1.5°N) for the four strongest El Niño events since the 1950s. Anomalies are relative to the seasonal cycle calculated over the period 1950–2014 and have been detrended. The isotherm 27°C is overplotted for the no-detrended data. *(Data are from the HadISST v1.1. data set Rayner, N.A., Parker, D.E., Horton, E.B., Folland, C.K., Alexander, L.V., Rowell, D.P., Kent, E.C., Kaplan, A., 2003. Global analyses of sea surface temperature, sea ice, and night marine air temperature since the late nineteenth century. J. Geophys. Res. https://doi.org/10.1029/2002JD002670.)*

reviews the current paradigms of extreme El Niño dynamics and the hierarchy of modeling approaches. Section 4 is devoted to the illustration of extreme El Niño impacts and difficulties in assessing their teleconnections. Section 5 addresses the state of knowledge on extreme El Niño events and their sensitivity to climate change, whereas Section 6 draws a nonexhaustive list of some challenges to the ENSO community for making progress in our understanding in extreme El Niño dynamics and teleconnections.

2. DEFINITION OF EXTREME EL NIÑO

According to the US National Oceanic and Atmospheric Administration (NOAA), strong El Niño events are those for which the NINO34 SST anomaly index (i.e., averaged over the region 170°W–120°W; 5°S–5°N) is above 1.5°C (approximately more than twice the standard deviation over the last six decades) over at least three consecutive overlapping 3-month periods. The merit of this definition is its simplicity, which facilitated ENSO-related research in the social and environmental sciences. However, because the probability density function of the NINO34 index is positively skewed, it is better accounted for by a non-Gaussian stable distribution (Boucharel et al., 2009). In this framework, the estimate of the statistical moments (mean, variance, skewness) is uncertain (i.e., sensitive to the length of the time series), so that the NOAA definition has inherent limitations. Thus, it is perhaps best to base the definition of extreme El Niño events on physical considerations.

In recent years, much effort has been dedicated to explain the so-called ENSO diversity, primarily the spatial diversity. There has been a debate on whether or not the two types of El Niño are part of a continuum in the position of the peak SST anomalies along the equator (Ray and Giese, 2012; Newman et al., 2011; Capotondi and Wittenberg, 2013; Johnson, 2013), or if the two types of event belong to two well-defined classes for which a bimodal distribution of some indices could be found (Capotondi et al., 2015; Takahashi and Dewitte, 2016). This debate cannot be closed from the analysis of the too short a data set at our disposal thus far (Ray and Giese, 2012), but it has provided evidence that the nonlinear evolution of ENSO (and so the existence of two types of events) could be accounted for by the first two EOF patterns of SST anomalies in the tropical Pacific (Takahashi et al., 2011; Newman et al., 2011; Takahashi and Dewitte, 2016; Karamperidou et al., 2016).

Fig. 6.3 presents the phase space of the two indices, E and C (see definition in caption), defined as in Takahashi et al. (2011) for the month of December. Five events (1972, 1982, 1997, 2009, and 2015) are highlighted and their evolution from July to December is shown to illustrate the diversity of the onset of El Niño events. The 2009/10 CP El Niño event (moderate) reached a high peak C value and the largest NINO4 anomaly to that date (Lee and McPhaden, 2010). The 1972/73 El Niño is usually considered as a strong event (e.g., Hong et al., 2014) because the value of the NINO34 index in December was comparable to the one for the 1982/83 and 1997/98. The uncertainty in the data so far precludes determining whether the 2015–16 El Niño was stronger by this metric than 1982/83 and 1997/98 but is surely in the top three (L'Heureux et al., 2017; see oblique lines in Fig. 6.3 that correspond to isocontours of the regressed NINO34 index onto the E and C indices). However, neither the 1972/73 nor the 2015/16 events had an EP warming as high as those in the 1982/83 and 1997/98 El Niño events (L'Heureux et al., 2017) and had E values less than half of those years, near the threshold value. Additionally, the 27°C isotherm at the edge of the western Pacific warm pool did not penetrate as far east in the equatorial Pacific in 2015 and 1972 as it did in 1982 and 1997 (Fig. 6.2). Consistent with this, the deep convection in the eastern Pacific was substantially weaker in the 2015/16 event than in 1982/83 and 1997/98 (L'Heureux et al., 2017).

The observations thus indicate that the extreme El Niño events are those that develop in the far eastern Pacific and for which the E index yields SST in excess of ~27.5°C, which can be interpreted as the convective threshold for the nonlinear amplification of the Bjerknes feedback (Takahashi and Dewitte, 2016). According to this definition, the 2015/16 and the 1972/73 El Niño events were not "strong" events although they were among the strongest El Niño events since 1950, with the NINO34 index values rivaling those of the 1997/98 and 1982/83 events. Additionally, the 2015/16 El Niño also presented the nonlinearly amplified convective feedback seen in 1982/83 and 1997/98 (Takahashi et al., 2018) but unlike those events, the 2015/16 El Niño had milder SST anomalies near the coast of South America, which was associated with a much weaker impact on precipitation along the coast of Ecuador and Northern Peru (L'Heureux et al., 2017; Sanabria et al., 2017). Still the 2015/16 El Niño was associated with extreme events at high latitudes (Stuecker et al., 2017).

From this, we see that even when the impacts are not considered and we focus on its physical characteristics only, the definition of an extreme El Niño can be ambiguous and will depend on the specific process or phenomena

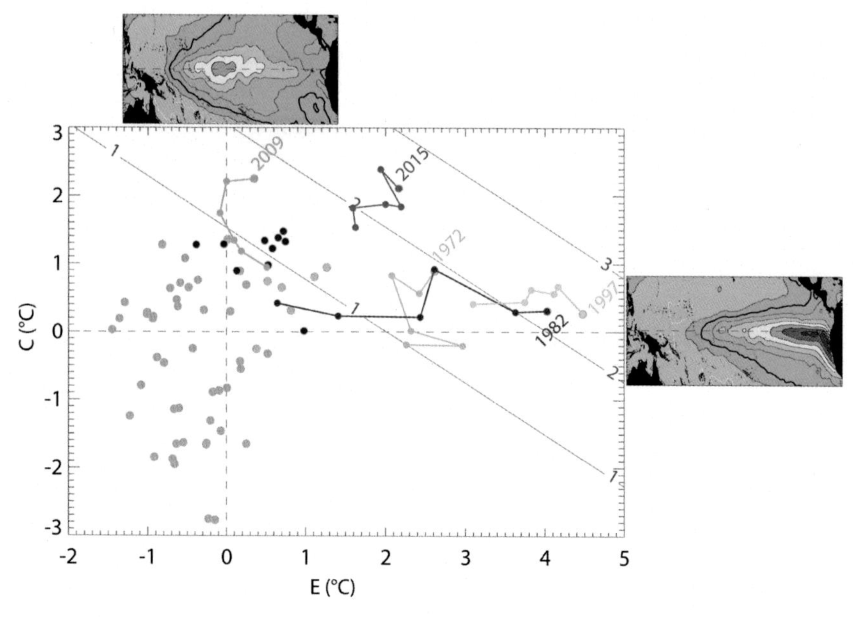

Figure 6.3 E versus C indices for the month of December. These indices accounts to a large extent for the variability of the NINO1 + 2 (90°W–80°W; 0°S–10°S) and NINO4 (160°E–150°W; 5°S–5°N) indices, respectively, since because their regression on the E and C indices yields approximated indices (NINO1 + 2approx and NINO4approx) that accounts for 90% and 93% of explained variance of the original indices over 1950–2016, with NINO1 + 2approx = 0.93 E+0.23 C and NINO4approx = 0.13 E+0.47 C. *Colored dots* correspond to El Niño years. Five El Niño events (1972, 1982, 1997, 2009, and 2015) are highlighted in different colors and their evolution from July to December are shown. The *dot* labeled with a year corresponds to the month of December of this particular year. The *light blue dot* corresponds to the El Niño events used for the Rasmuson and Carpenter (1982)'s "canonical" composite, i.e., 1951, 1953, 1957, 1963, 1965, 1969, and 1972. The oblique *gray lines* correspond to the iso-contours of the approximate NINO3.4 index (5°S–5°N; 170°W–120°W), that is the NINO3.4 index recomposed from its E and C contributions (NINO3.4approx = 0.47 E+0.67 C). The NINO3.4 index for the four El Niño events (1972, 1982, 1997, and 2015) is close to 2°C. The patterns associated to the E and C indices (see Fig. 6.1) is are recalled and displayed along the corresponding axis. *(Data are from the HadISST v1.1. data set Rayner, N.A., Parker, D.E., Horton, E.B., Folland, C.K., Alexander, L.V., Rowell, D.P., Kent, E.C., and Kaplan, A., 2003., Global analyses of sea surface temperature, sea ice, and night marine air temperature since the late nineteenth century., J. Geophys. Res. 2003, https://doi.org/10.1029/2002JD002670. Modified from Dewitte, B., and Takahashi, K., 2017. Diversity of moderate El Niño events evolution: role of air-sea interactions in the eastern tropical Pacific., Clim. Dyn. 2017, (revised). https://doi.org/10.1007/s00382-017-4051-9.)*

of interest. In general though, there is the notion that extreme El Niño events refer to a phenomenon that involves one or various nonlinear processes, whereas "regular" or moderate El Niño events can be understood in terms of a linear dynamics. However, because of the nonlinear character of the ENSO teleconnections (see Section 4), extreme El Niño events do not necessarily yield extreme impacts, whereas moderate events could be associated with extreme impacts in remote regions. Here we define extreme El Niño events based on the exceedance of the threshold value for the E index that indicates the activation of deep convection in the eastern equatorial Pacific.

Because of the nonlinear relationship between SST and convection and the observation that extreme El Niño events are associated with the activation of deep convection in the eastern equatorial Pacific, another definition of extreme El Niño event has been proposed recently, which is based on the precipitation anomalies averaged over the NINO3 (150°W–90°W; 5°S–5°N) region (Cai et al., 2014) (see Section 5). This definition, in its physical justification, is comparable to the previous definition based on the E index but can result in different estimate of the frequency of occurrence of extreme El Niño event in models due to decoupling between SST and precipitation in the high-convective regime associated with the mean state bias in the far eastern Pacific.

3. THEORY AND MODELING

3.1 Deterministic Processes

Extreme El Niño events involve nonlinear processes that allow for their rapid amplification. As a coupled system, the sources of the nonlinearity can be embedded either in the oceanic or the atmospheric components of the system, or in both. Pioneering theoretical studies on extreme El Niño events have focused on the oceanic nonlinear advection (or nonlinear dynamical heating, hereafter NDH) as a source of the so-called "bursting" behavior of ENSO. Timmermann et al. (2003) proposed the first model of ENSO bursting for which no external forcing is required. The low-order model is a two box model that accounts for SST variability in the eastern Pacific where the thermocline feedback (i.e., relationship between SST and thermocline anomalies) is strong because of the shallow mean thermocline and for SST variability in the western Pacific where zonal advection is considered as the dominant process. The models consider the rate of change of total temperature so that nonlinear advection contributes to SST variability. The model produces extreme El Niño events in the form of "bursts" as part of complex self-sustained oscillations, but these have a much larger amplitude than observations and sensitive to parameters values. In particular, the model can produce multiple equilibria that include a permanent El Niño state, which is questionable in theoretical model settings (Neelin and Dijkstra, 1995).

The role of NDH in explaining the large amplitude of warm events compared with cold events (i.e., positive asymmetry of ENSO) was further assessed by Jin et al. (2003) and An and Jin (2004) based on the analysis of reanalysis products. However, NDH appears more relevant for the maintenance rather than the growth of El Niño (Su et al., 2010; Takahashi and Dewitte, 2016), although uncertainties remain in the actual contribution of NDH to the heat budget due to sensitivity to resolution of the mixed-layer heat budget in models and/or uncertainties in oceanic reanalysis products. Additionally, it has been suggested that tropical instability waves that have a nonlinear character and are not uniformly accounted for in models with different resolutions may provide a weaker damping for El Niño during its development (An, 2008).

Although previous studies focused on oceanic nonlinear processes, a more recent body of literature suggests the key role of atmospheric nonlinear processes in the temporal and amplitude asymmetry between El Niño and La Niña events (Dommenget et al., 2012; Choi et al., 2013a; Lloyd et al., 2012), diversity in patterns (Xiang et al., 2013), and the El Niño amplitude (Takahashi and Dewitte, 2016). The underlying concept behind all these studies is that the Bjerknes feedback in the tropical Pacific has a nonlinear character associated with the activation of deep convection in the otherwise stable eastern equatorial Pacific, which can lead to an enhanced equatorial westerly wind response to SST, leading to a nonlinear Bjerknes feedback that can help El Niño events to become extreme. Takahashi et al. (2018) showed that the recharge–discharge model under stochastic (additive) forcing can produce a separation between the peak SST during moderate and strong El Niño events when the damping rate is set to zero when the SST exceeds 27.5°, which implies an anomaly exceeding ~1.5°C for the NINO3 region. This model provides so far the most parsimonious theory for El Niño regimes. Note that the latter does not consider processes associated to non-linear interactions of the annual cycle and ENSO, which can also produce ENSO amplitude modulation (e.g., Jin et al., 1994; Münnich et al., 1991; Neelin et al., 2000; Tziperman et al., 1994) and that can be understood in terms of the combination mode (Stuecker et al., 2013), causing rapid transition between ENSO phases by introducing higher frequencies (i.e., combination tones) than the slow interannual ENSO frequency.

Although, within the above conceptual framework, extreme El Niño is produced via a deterministic nonlinear process in the equatorial Pacific, there are other studies that give a central role to the characteristics of the external forcings and their nonlinear interactions with ENSO.

3.2 Role of External Forcing

Westerly wind bursts (WWBs) in the equatorial Pacific can lead to surface ocean warming and are thought of as random weather forcing ("noise"), but their frequency is modulated by SST anomalies associated with ENSO (Eisenman et al., 2005), leading to the concept of state-dependent noise. Jin et al. (2007) introduced this into the framework of the recharge–discharge ENSO model to explain ENSO irregularity and bursting behavior. Within this low-order modeling framework, extreme events emerge either because of the increased amplitude of WWEs or changes in their characteristics (e.g., frequency of occurrence) (Levine et al., 2016), which is consistent with previous works using more complex models (Lengaigne et al., 2004; Gebbie et al., 2007; Kapur and Zhang, 2012; Chen et al., 2015; Lopez et al., 2013).

The state-dependent component of the wind stress forcing has been estimated to increase during the extreme El Niño development relative to the additive component (Levine and Jin, 2010; Levine et al., 2016; Fig. 6.4). State-dependent noise in this modeling framework is enhanced with warmer SST, which encapsulates the response of convection to the eastward migration of the western Pacific warm pool during extreme El Niño development, which allows the intraseasonal tropical atmospheric variability from the western Pacific, including the Madden–Julian oscillation (MJO) and convective atmospheric waves, to penetrate further eastward (McPhaden et al., 2006; Gushchina and Dewitte, 2012; Puy et al., 2015). A simple model considering state-dependent noise, explicit wave dynamics, and the zonal contrast in SST across the equatorial Pacific (Chen and Majda, 2017), bridging the gap between the minimal model of Jin et al. (2006) and the intermediate complexity models similar to the Zebiak and Cane (1987)'s model, is able to capture the ENSO spatial diversity, simulating both CP El Niño and "super" El Niño events peaking in the eastern Pacific.

The relative realism of these simple models in simulating many aspects of the ENSO properties (i.e., asymmetry and diversity) has invigorated the interest in the external forcing for ENSO because the noise forcing in the form of WWBs during the early development of an event can either originate from outside the tropical Pacific or can be modulated by modes of variability independent of ENSO. For instance, Hong et al. (2014) showed

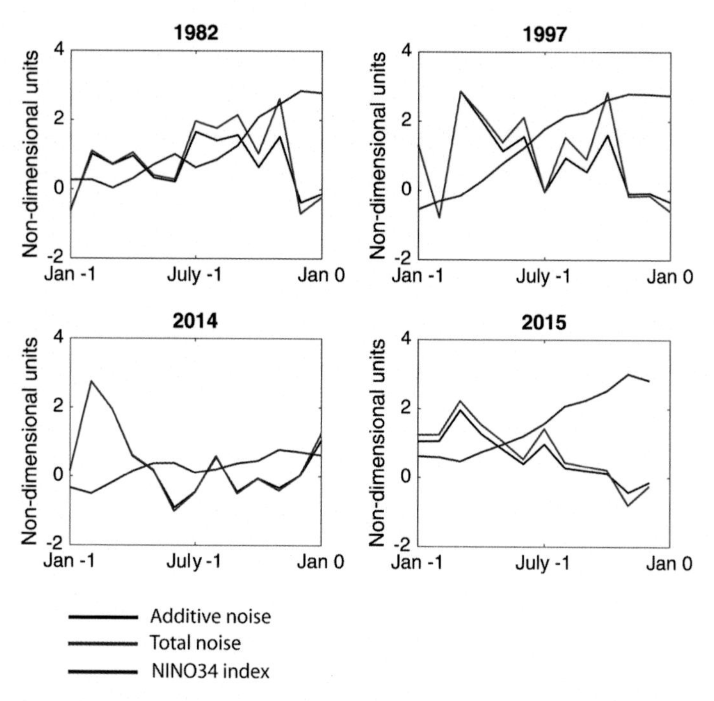

Figure 6.4 A comparison of the total noise forcing (red) and additive noise forcing (black) for the 1982/83, 1997/98, 2014/15, and 2015 (through December 2015) El Niño events. Data are from ERA-Interim. The NINO34 index is shown by the *blue lines*. Large initial additive forcing comprises the initial forcing in all cases, but the additional forcing in boreal summer and fall (July–November) exists mostly in 1982 and 1997. In these cases, the additive component no longer accounts for nearly all of the total forcing. State-dependent noise forcing acts to amplify the additive noise forcing for greater total noise forcing in these events. *(Data are from ERA-Interim. From Levine, A., Jin, F.F., McPhaden, M.J., 2016. Extreme noise-extreme El Niño: how state-dependent noise forcing creates El Niño-La Niña asymmetry. J. Clim. https://doi.org/10.1175/JCLI-D-16-0091.1.)*

that the strong El Niño event of 1982/83 had a precursor signal far away from the equatorial Pacific consisting of southerlies along the east coast of Australia. Consistently, Takahashi and Dewitte (2016) found that the zonal winds in the central equatorial Pacific in August prior to a strong ENSO peak are a good precursor and that it can be explained by the regression model using SST anomalies in the equatorial Pacific as a predictor only during the 1997/98 and 1972/73 El Niño events but not during the 1982/83 El Niño event, supporting the hypothesis of a weak convective coupling with SST in August 1982, so that external forcing at that time was key to trigger the 1982/83 El Niño event, perhaps contributed to by the radiative forcing from the eruption of the El Chichón volcano

(Khodri et al., 2017). This is to some extent consistent with the findings of Levine et al. (2016) that shows that the proportion of state-dependent noise during the Austral winter prior to the 1982 El Niño was weaker than that of the 1997/98 El Niño (Fig. 6.4). Within the simple model studies, Takahashi et al. (2018) also indicate that low-frequency forcing is needed for pushing the system into the strong El Niño regime in a recharge–discharge model with convective threshold nonlinearity. Dommenget and Yu (2017) showed, based on the experimentation with a zero-dimensional recharge–discharge model embedded into a slab ocean and coupled to an atmospheric general circulation model, that the coupling to the extratropical Pacific amplifies the ENSO spatial diversity, leading to a decoupling between the NINO3 and NINO4 indices. Their simulation also points to a more important role of the forcing from the southern Pacific.

Overall these recent simple model studies point to the important role of the low-frequency component of the external forcing for ENSO, which can favor the activation of the nonlinear ENSO regime and thus produce diversity. Retrospectively, this is consistent with the large body of literature showing the influence of remote regions from the tropical Pacific on ENSO. For instance, the Indian Ocean warming modulates atmospheric circulation over the western Pacific, which in turn influences ENSO evolution (Kug and Kang, 2006; Dommenget et al., 2006; Ohba and Ueda, 2007; Jansen et al., 2009; Izumo et al., 2010; Frauen and Dommenget, 2012; Santoso et al., 2012; Kajtar et al., 2015). Tropical Atlantic SST variability can also affect ENSO variability through the Walker circulation (Dommenget et al., 2006; Jansen et al., 2009; Rodriguez-Fonseca et al., 2009; Frauen and Dommenget, 2012; Ding et al., 2012; Ham et al., 2013; McGregor et al., 2014). A number of studies also suggest that the SST variability in the North Pacific influences El Niño events occurring in the central Pacific through a mode called the North Pacific meridional mode that operates through wind–evaporation–SST feedback (Vimont et al., 2001; Chiang and Vimont, 2004; Yu and Kim, 2011; Larson and Kirtman, 2013). The Southern Hemisphere also influences ENSO at different timescales (e.g., Matei et al., 2008; Terray, 2011; Okumura, 2013; Zhang et al., 2014; You and Furtado, 2017). Other forms of potentially important external forcing include volcanic eruptions (Khodri et al., 2017).

Observational evidence for the important role of external forcing of ENSO was also provided recently with the "aborted" 2014 El Niño. Although the ENSO community was expecting an extreme El Niño to occur following a strong WWBs in February–March 2014 with significant recharge of the heat content (Menkes et al., 2014), it has been argued that an easterly wind burst in June 2014 was able to shutdown the development of this suspected extreme El Niño event (Hu and Fedorov, 2016; McPhaden, 2015). On the other hand, Chiodi and Harrison (2017) argue that it was simply lack of a sequence of WWBs in the spring and summer (i.e., low-frequency forcing) that resulted in an interruption of the growth of El Niño in 2014.

3.3 Extreme El Niño Events in Models

The aforementioned theoretical studies offer paradigms for extreme El Niño dynamics useful for advancing our understanding of ENSO as a whole and interpreting more complex models that provide a higher dimensional view of the ENSO complexity and also allow for seasonal forecasting.

Although accounting for a wide range of processes involved in the ocean–atmosphere interaction during ENSO, the full-physics models still have difficulties in realistically simulating the ENSO diversity (Ham and Kug, 2012), which is associated with a bias in ENSO asymmetry (Zhang and Sun, 2014; Karamperidou et al., 2016). In addition, they also have a tendency to simulate too many extremes events compared to nature. For instance, the average number of extreme El Niño events of an ensemble of the Coupled Model Intercomparison Project Phase 5 (CMIP5) simulations (historical runs, period 1891–2004, see Table 6.1 for the list of models used) is 4.6 events per century, whereas the observational record suggests that it would be close to 2–3 events per century. DJF NINO34 SST above 1.75 times the standard deviation of the NINO34 SST index defines extreme events here. Also, in two models that realistically simulate the ENSO diversity (GFDL_CM2.1 (Delworth et al., 2006) and NCAR CESM (Kay et al., 2015)), the frequency of extreme events estimated from the definition based on a threshold value of the E index (see Section 2) reaches 7.4 and 5.2 events per century, respectively (Takahashi and Dewitte, 2016; Dewitte and Takahashi, 2017). Although the historical data set may not account for the "true" statistics of ENSO because of its too short span (Wittenberg, 2009; Takahashi et al., 2018), the realism of the sensitivity in the extreme events occurrence to mean state changes as inferred from these same models also remains debatable.

Levine et al. (2016) also showed that the current generation of state-of-the-art models (CMIP5) may not simulate realistically the amplitude of the state-dependent noise, which led to their failure in simulating correctly the

Table 6.1 List of models used for generating figure 6.8. The second and third columns indicate the models that simulate extreme El Niño events as defined in Cai et al. (2014), i.e., DJF NINO3 (150°W–90°W; 5°S–5°N) rainfall greater than 5 mm per day defines an extreme El Niño event. The numbers in parenthesis indicate the number of extreme El Niño events. Note that Cai et al. (2014) use a different baseline climatology for the historical (1891–1990) and rcp8.5 (1991–2090) runs and that data are quadratically detrended in their study instead of linearly detrended in the present paper

Model	Historical (1891–2005)	RCP8.5 (2006–2095)
ACCESS1-0	X (34)	X (30)
ACCESS1-3	X (17)	X (16)
CanESM2	X (11)	X (10)
CMCC-CESM	X (4)	X (12)
CMCC-CM		X (2)
CNRM-CM5	X (17)	X (20)
CMCC-CMS	X (2)	X (9)
GFDL-CM3	X (10)	X (16)
GFDL-ESM2G		
GFDL-ESM2M	X (13)	X (8)
HadGEM2-CC	X (6)	X (4)
HadGEM2-ES	X (1)	X (8)
INMCM4		
IPSL-CM5A-LR		
IPSL-CM5A-MR		
MIROC5	X (3)	X (7)
MIROC-ESM	X (1)	
MIROC-ESM-CHEM	X (1)	X (1)
MPI-ESM-LR		
MPI-ESM-MR		
MRI-CGCM3		X (2)
NorESM1-M	X (21)	X (7)
MRI-ESM1		X (2)
GISS-E2-H	X (3)	X (18)
GISS-E2-R	X (47)	X (60)
HadGEM2-AO	X (1)	X (6)
IPSL-CM5B-LR	X (4)	X (11)
GISS-E2-H-CC	X (13)	X (30)
GISS-E2-R-CC	X (48)	X (64)
NorESM1-ME	X (11)	X (9)
CESM	X (6)	X (12)

ENSO asymmetry (i.e., skewness) and, therefore, the diversity since state-dependent noise favor extreme events, which tend to be of EP variety. Note that state-dependent noise forcing has been also shown to preferentially excite the EP El Niño events in previous studies (Lopez et al., 2013; Lopez and Kirtman, 2014; Chen et al., 2015).

These deficiencies in coupled general circulation models (CGCMs) are consistent with current inability of prediction systems in forecasting extreme El Niño events with more than a 9-month lead time (Barnston et al., 2012; L'Heureux et al., 2017). Also models have a harder time making accurate predictions in the eastern Pacific, predicting not large enough SST anomalies in the far eastern Pacific during the extreme El Niño events (Takahashi et al., 2014).

The so-called intermediate complexity models can provide a useful tool for addressing extreme ENSO dynamics because they usually allow avoiding biases in mean states by prescribing climatologies derived from observations, and they have been as skillful as seasonal prediction systems based on full-physics models (Chen et al., 2004). However, they might miss processes important for ENSO such as those associated to mesoscale dynamics and, by construction, do not include the two-way interactions between ENSO and mean state. A current direction toward trying to improve ENSO dynamics in CGCMs has been to increase their resolution, which is reported to lead to important improvements in reducing mean state biases (Delworth et al., 2012).

4. IMPACT OF EXTREMES EL NIÑO EVENTS

This section is dedicated to illustrating some impacts of extreme El Niño events through teleconnections. Because of the smallness of the sample of extreme El Niño events, the inherent nonlinearity of the ENSO teleconnection (Frauen et al., 2014) and the likely impact of natural variability in modulating those, it is not possible to establish a statistically robust view of the impacts associated with extreme El Niño events away from the tropical Pacific where the signal-to-noise ratio becomes smaller (e.g., Kumar et al., 2000). However, even within the tropical Pacific, nonlinearity associated with the ENSO teleconnection is notable especially for precipitation. Rather than providing a list of likely impacts, we rather focus here on pointing out the main features of extreme ENSO events that would lead to differences in impacts over remote regions (i.e., seasonal evolution event and magnitude/pattern). We also illustrate possible impacts associated with the oceanic teleconnection resulting from changes in the statistics of the extreme El Niño events so as to point out the relevance of investigating extreme El Niño's for climate change studies and not only for weather extremes, which we will transition to in the next section.

Because of the eastward shift in the Walker circulation during extreme El Niño events, tropical teleconnections at the peak phase of ENSO can be inferred from basic principles (i.e., shifts/amplification of the Walker cells over the different tropical basins). Thus during a developing extreme El Niño event, one can expect (see also Fig. 6.5): dryer conditions over

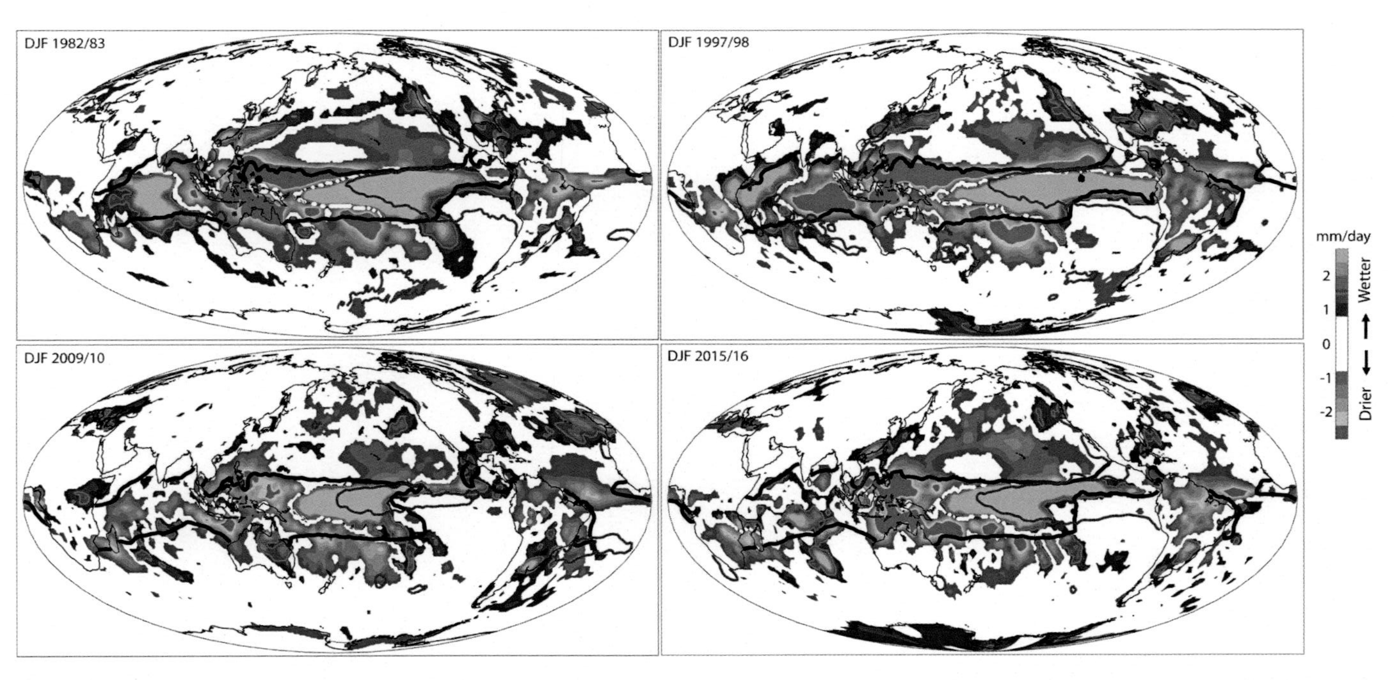

Figure 6.5 Precipitation anomalies in Dec–Jan–Feb (DJF) for the strongest El Niño events since 1980: 1982/83, 1997/98, and 2015/16. The 2009/10 CP El Niño is also presented for comparison. Anomalies are relative to the climatology calculated over 1980–2014 and data were linearly detrended over 1979–2017. The 28°C isotherm is shown in *black thick line*, whereas the isocontour 1°C of the corresponding sea surface temperature anomalies is displayed in *red thick line*. The *brown thick line* indicates the 1.5 mm per day anomalies. *(Precipitation data are from CPC Merged Analysis of Precipitation (Xie and Arkin, 1997). SST data from HadISST data set.)*

northeast Brazil (Rodrigues et al., 2011), anomalous atmospheric subsidence over the Indian Ocean leading to Indian Ocean warming through intensified solar radiation, and a weakening land–sea contrast in summer, which is associated with less Indian summer monsoon rainfall (Kumar et al., 2006), drier northeast Australia (Power et al., 2006; Cai et al., 2010), an intensified Pacific/North American–like atmospheric pattern (Horel and Wallace, 1981), wetter California, and Ecuador/northern Peru, the increased cyclone activity in the boreal summer following the ENSO peak over the northeastern Pacific associated to the significant heat discharge (Jin et al., 2014). However, the last two strongest El Niño events that have been observed comprehensively (i.e., 1997 and 2015) have shown that the severity of the impacts may vary from region to region, and the main precipitation anomalies may also change location. For instance, during the 1997 El Niño, monsoon rains were abundant and no drought occurred, whereas India witnessed below-normal monsoon in 2015 (Fig. 6.6). Similarly, the northern region of Peru experienced much weaker precipitations during the 2015 El Niño than during the 1997 El Niño event. Although decadal variability certainly comes into play to explain such differences, an important aspect to consider is the influence of the seasonal evolution of the El Niño that has been distinct across the strongest El Niño events of the last five decades and the details of the SST anomaly pattern in the warm waters region (Fig. 6.7).

As a result of the sensitivity of the amount of diabatic heating released to the upper atmosphere to the mean SST in the tropical Pacific, details in the patterns of extreme El Niño events are also likely to modulate the teleconnections at higher latitudes. Frauen et al. (2014) shows that strong regional differences in ENSO teleconnections can be associated with slight changes in the spatial patterns (decomposed into EP and CP modes) and amplitude, with the EP El Niño exhibiting larger nonlinearity in the teleconnection than CP El Niño. Hoerling et al. (2001) also showed that nonlinearities in the tropical precipitation response and in the extratropical atmospheric response mainly emerge for stronger events. Johnson and Kosaka (2016) showed that it is not the spatial pattern of SST itself but the associated anomalous presence of deep convection in the eastern Pacific that leads to different teleconnections in the Northern Hemisphere.

Besides its direct impact in remote regions through teleconnection, an extreme El Niño can also alter persistently the mean conditions over epochs when it occurs more frequently, which is due to its asymmetry (An, 2004). The repeated occurrence of extreme El Niño events (or on the contrary their "disappearance" over extended period) can therefore rectify the mean

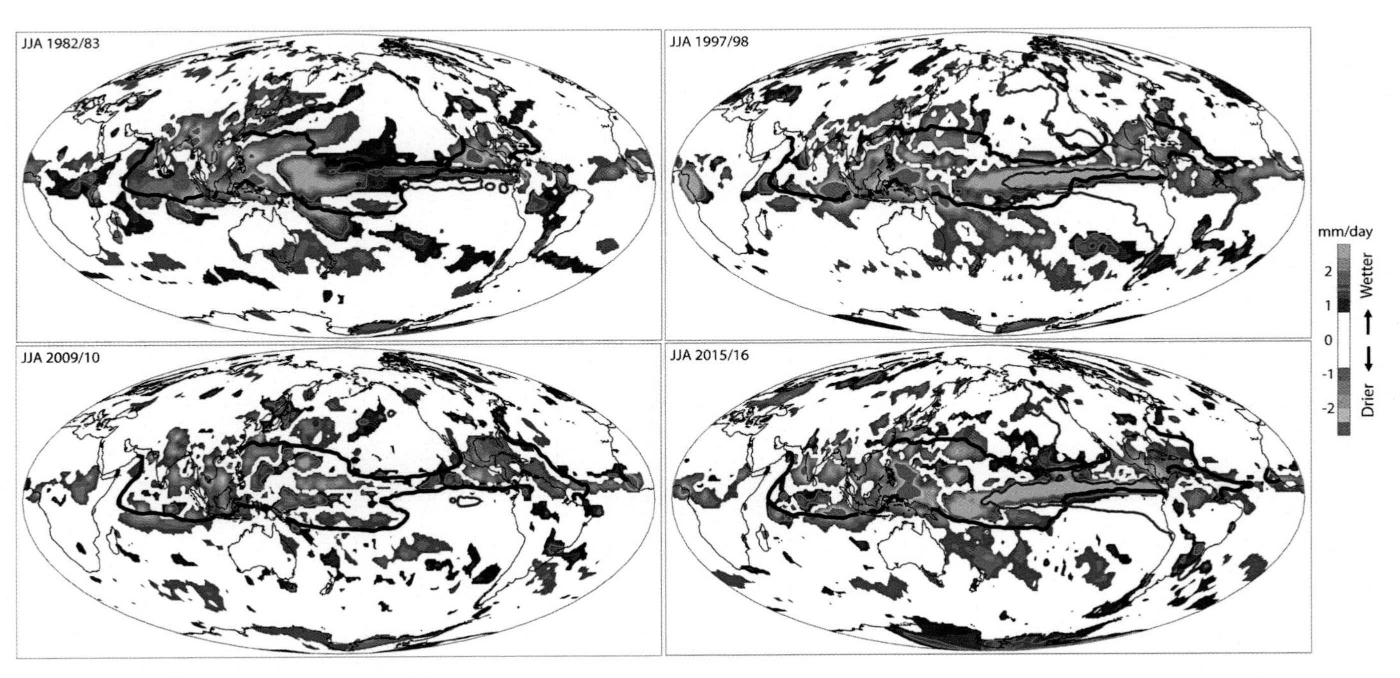

Figure 6.6 Precipitation anomalies in Jun–Jul–Aug (JJA) for the strongest El Niño events since 1980: 1982/83, 1997/98, and 2015/16. The 2009/10 CP El Niño is also presented for comparison. Anomalies are relative to the climatology calculated over 1980–2014 and data were linearly detrended over 1979–2017. The 28°C isotherm is shown in black thick line, whereas the isocontour 1°C of the corresponding sea surface temperature anomalies is displayed in red thick line. The brown thick line indicates the 1.5 mm per day anomalies. *(Precipitation data are from CPC Merged Analysis of Precipitation (Xie and Arkin, 1997). SST data from HadISST data set.)* .

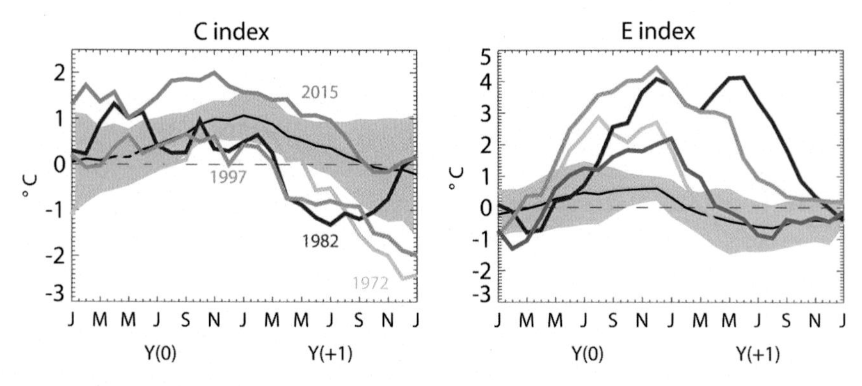

Figure 6.7 Evolution of the (left) C and (right) E indices of the four strong El Niño events (1972/73, 1982/83, 1997/98, and 2015/16) and the composite of moderate events (*black line*). The *gray shading* represents the dispersion (standard deviation) amongst the moderate El Niño events, i.e., all El Niño events since 1950 except the four strongest. Note the significant dispersion in the evolution of the E index for strong El Niño events. For instance, the 1997/98 El Niño event peaks in December 1997, while whereas the 1982/83 presents a double peak, one in December 1982 and the other one in June 1983. The 1972/73 El Niño peaked earlier in the year in August 1972 with still a secondary peak in December 1972, with also significantly lower values of the E index. The 2015/16 El Niño event is more comparable to the 1972/73 El Niño in terms of the magnitude of the E index. However, the 2015 El Niño event stands out in terms of the evolution of the C index, with values of the C index much larger than the other El Niño events that are hardly distinguishable from the moderate event composite. Note that the large value of the C index for the 2015/16 El Niño event is consistent with the larger contribution of the zonal advective feedback to the dynamics compared to with the 1997/98 El Niño event as shown by Paek et al. (2017). Although previous strong events were of eastern Pacific variety, the 2015/16 El Niño event is the strongest mixed type of El Niño ever recorded.

oceanographic and atmospheric regional conditions through teleconnections. One clear example is the upwelling system of Peru, which is connected to ENSO dynamics through the propagation of Kelvin waves along the coast. During extreme El Niño events, the coastal upwelling is switched off for several seasons. Conversely during a CP El Niño, near-normal to cool conditions prevail (Dewitte et al., 2012). The change in frequency of occurrence of the two types of El Niño as observed since the 1990s (i.e., more occurrence of CP El Niño events, cf. Lee and McPhaden, 2010) can therefore lead to amplified mean upwelling conditions at decadal timescales (Dewitte et al., 2012). Also, because extreme El Niño events tend to transfer energy toward the deep ocean through the vertical propagation of oceanic Rossby waves, extreme El Niño events also have the potential to modulate the intermediate to deep circulation in the eastern Pacific (Vergara et al., 2017), which can have implications for climate variability at long timescales.

5. CLIMATE CHANGE AND NATURAL VARIABILITY

Because of the important role of external forcing on ENSO, estimating its response to climate change represents a considerable challenge, which has been addressed since the 1980s with global model simulations performed for the Intergovernmental Panel on Climate Change (first IPCC Scientific Assessment in 1990). Changes in ENSO feedbacks and thus ENSO properties also depend on how the mean circulation in the equatorial Pacific will change. Although the mechanisms by which the mean equatorial state changes are still debated (Liu et al., 2005; Xie et al., 2013; Ying et al., 2016), there is now a certain consensus on the fact that the adjustment of the Pacific Walker circulation to anthropogenic forcing is through the Bjerknes feedback; i.e., the predicted El Niño-like warming in the Pacific is attributed to a weakening of the Walker circulation (Bayr et al., 2014). This is supported by the observations (Tokinaga et al., 2012) although decadal variability associated with natural variability has come into play since the 1990s with the so-called "hiatus" period when there was a pause in the global warming trend (Dai et al., 2015) and an increase in the trade winds across the Pacific (Luo et al., 2012; L'Heureux et al., 2013). A robust feature of the change in mean state in the equatorial Pacific predicted by models is also the increased stratification in the western-central Pacific (i.e., a shoaling thermocline) (Yeh et al., 2009; DiNezio et al., 2009), which results in an overall flattening of the thermocline mirroring the mean SST response.

How such changes will impact the ENSO properties (amplitude and frequency) remains uncertain, and no clear mechanism has emerged because of compensating effects among different possible processes. Based on ENSO linear dynamics (An and Jin, 2001), the following can be hypothesized: although the reduction of the zonal SST contrast is propitious to a reduction of the zonal advective feedback (and thus the ENSO variance in the central Pacific), the flattening thermocline could amplify ENSO through an increased transfer of momentum forcing into the surface circulation in the central western Pacific via wave dynamics (Dewitte et al., 2007; Thual et al., 2011) or on the contrary reduce ENSO amplitude by weakening the thermocline feedback in the eastern equatorial Pacific because of the deeper thermocline there. On the other hand, the increased SST is also favorable for an increase in vertical stratification (and hence the thermocline feedback) if the cool conditions below the mixed-layer are sustained by equatorial upwelling.

In fact, there is a considerable intermodel spread in the response of ENSO amplitude to anthropogenic forcing (Collins et al., 2010; Karamperidou et al., 2016, their Fig. 1), which is consistent with the nonlinear character of the ENSO dynamics and the fact that models have difficulties in simulating ENSO asymmetry (Zhang and Sun, 2014) and diversity (Ham and Kug, 2012).

A recent study (Cai et al., 2014) considers another measure of ENSO extreme than the one based on SST, noting that while there were uncertain changes in the ENSO amplitude, there were more robust changes in the pattern of the SST, which impacted ENSO-driven precipitation (Power et al., 2013). This measure is based on an index from precipitation that accounts for the meridional migration of the Intertropical Convergence Zone (ITCZ) in the eastern Pacific during ENSO. Based on this precipitation-based index, Cai et al. (2014) analyzed the CMIP3 and CMIP5 models and found a doubling in the occurrence of extreme El Niño events in the future in response to greenhouse warming. This increased frequency of extreme El Niño events is interpreted as resulting from the larger mean warming SST right along the equator compared with the region below the ITCZ, which reduces the meridional SST gradient and "eases" its migration southward to yield an extreme El Niño event (Fig. 6.8).

Although Cai et al. (2014) offer a vision of the future, the detailed mechanisms behind such an increase in extreme events are yet to be established and the too few extreme events in the observational record hamper an objective determination of the threshold values of the indices to select extreme events in the models. Model biases also limit the confidence in such projections, in particular the persistent double-ITCZ syndrome (Zhang et al., 2015), the warm SST biases in the far eastern Pacific (Richter, 2015; Takahashi et al., 2014), and the unrealistic ENSO asymmetry (Zhang and Sun, 2014) that influence ENSO feedbacks and nonlinearity (Karamperidou et al., 2016).

CMIP5 models also exhibit a wide range of behavior in terms of the relationship between change in mean state and the ENSO amplitude modulation without greenhouse gas forcing (Choi et al., 2012), which questions their ability to realistically simulate the natural variability that can interact with the forced signal (England et al., 2014).

Our ability to predict the fate of extreme El Niño events in the future is thus still tightly linked to our ability to understand patterns of climate change (in particular their regional features) and their interaction with natural variability. Active research is being carried out to provide a paradigm

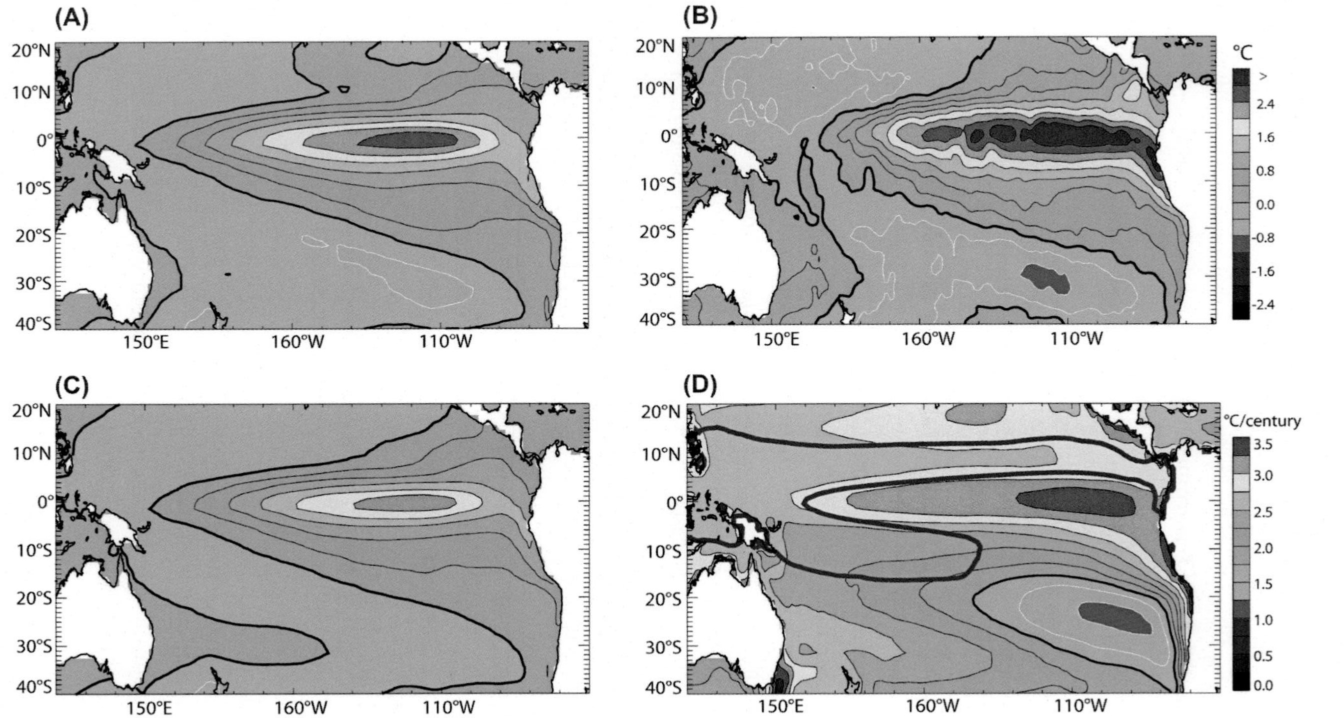

Figure 6.8 Composite of sea surface temperature (SST) anomalies for extreme events for (a) the historical simulations (1891–2004), (b) the observations (1972, 1982, 1997 and 2015 El Niño events), and (c) the RCP8.5 scenario simulations (2006–2095). DJF NINO3 (150°W–90°W; 5°S–5°N) rainfall greater than 5 mm per day defines an extreme El Niño event following Cai et al. (2014). Data have been linearly detrended before selecting events. 22 models out of 31 for the historical runs simulate 274 extreme El Niño events, whereas 24 models out of 31 for the rcp8.5 scenario simulations simulate 364 extreme El Niño events (see Table 6.1 for the list of models). Compared to Cai et al. (2014), we do not discriminate the models with the additional criteria that skewness of the NINO3 rainfall index be larger than 1°C. Note also that when using the DJF NINO34 SST anomalies for selecting extreme events (i.e., when DJF NINO34 SSTA > 1.75*var, where var is the variance of DJF NINO34 SSTA), 162 (122) extreme events for the historical (rcp8.5) runs are found, and all models simulate extreme events. Panel (d) shows the long-term trend (in °C/century) for the ensemble mean of the rcp8.5 scenario (24 models corresponding to those simulating extreme El Niño events). The iso-contour 5mm per day of the mean precipitation for the ensemble mean of the rcp8.5 scenario simulations is overplotted in thick blue line, which highlights that the mean warming in the near-equatorial region favors the southward intrusion of the ITCZ during El Niño. *(Data for observations are from HadISST data set.)*

for the mean climate change pattern in the tropics (Liu et al., 2005; Xie et al., 2013; Ying et al., 2016). The so-called "wet-get-wetter" hypothesis, by which increased moisture in a warmer atmosphere will firstly lead to increased evaporation and precipitation, supports the expectation that more ENSO-related extreme rainfall events will occur in the future. On the other hand, the greater diabatic heating released at upper levels in association with surface ocean warming will tend to stabilize the atmosphere, which should weaken the precipitation rate and therefore the tropical atmospheric general circulation (Held and Soden, 2006) (dry gets drier mechanism). Because a Bjerknes feedback (i.e., a dynamical adjustment) is expected for the climate change adjustment (Bayr et al., 2014) and that the ENSO nonlinearity, which shapes the regional pattern of mean changes (Karamperidou et al., 2016), is simulated differently in models (Zhang and Sun, 2014), much more work needs to be done to gain confidence in the projections of extremes El Niño events in a warmer world.

6. CHALLENGES

In this last section, we list, what we think are important issues to tackle to improve our understanding of extreme El Niño dynamics. Although certainly not exhaustive, they belong to the bulk of the concerns of the ENSO community, which has been challenged by the recent occurrence of the 2015/16 El Niño event almost 20 years after the emblematic 1997/98 El Niño event.

6.1 Natural Variability in Extreme El Niño Occurrence

Although much effort has been dedicated to the analysis of the sensitivity experiments to anthropogenic forcing of the CMIP models (Yeh et al., 2009; Power et al., 2013; Cai et al., 2014; Wang et al., 2017), which is spurred by societal concerns, many questions remain on the mechanisms behind natural variability (i.e., without external forcing) in extreme El Niño occurrences. Long integrations of coupled GCMs, without any changes in external forcings, can produce a low-frequency modulation of ENSO amplitude and diversity (Wittenberg, 2009; Kug et al., 2010), which implies they could be unpredictable (Wittenberg et al., 2014). On the other hand, the ENSO modulation can affect (rectify) the mean state at decadal timescales (Rodgers et al., 2004; Cibot et al., 2005; Schopf and Burgman, 2006; Choi et al., 2012; Ogata et al., 2013) to which the deterministic ENSO dynamics is tightly linked (An and Jin, 2001).

The existence of a two-way feedback for natural variability within the tropical Pacific is thus a possibility (Choi et al., 2009, 2013b). This issue is certainly key for gaining confidence in climate change projection of extreme El Niño events, all the more as natural variability has the potential to obscure the detection of climate change as evidenced by the recent debate on the "pause" or "hiatus" in global mean surface temperature changes (see Kosaka and Xie, 2013 among others). Some climate centers are heading to a promising experimental approach, beyond the "classical" CMIP exercise, by providing long-term integrations of a single model but with more diverse forcing conditions and many realizations (members) of the same scenario, which should allow exploring the critical processes behind the ENSO diversity–mean state interaction and its sensitivity to external forcing (Kay et al., 2015; Stevenson et al., 2017). Paleo ENSO studies are also necessary to test paradigms established from the models, although difficulties remain in capturing the ENSO spatial diversity based on the sparse network of proxy data (Karamperidou et al., 2015).

6.2 Nonlinearities and Extreme El Niño Events

The existence of extreme El Niño events implies the nonlinear dynamics of the tropical ocean–atmosphere coupled system (although it is likely that ENSO is at least weakly nonlinear, Takahashi et al., 2018) and the complex of processes operating in both components of the system leave room for a multitude of candidates for the source of the nonlinearity. Two main deterministic processes have been identified to produce ENSO asymmetry so far that consist of the oceanic nonlinear advection (so-called NDH) (Jin et al., 2003) and in a general sense, the nonlinear Bjerknes feedback that encapsulates a number of physical processes difficult to rank by importance. Although NDH is certainly important for explaining the amplitude of extreme ENSO, uncertainties remain on its role during the development of the event (Su et al., 2010; Takahashi and Dewitte, 2016). Further studies are certainly required to clarify its contribution to the mixed-layer heat budget during an extreme El Niño as more consistent and realistic oceanic reanalyses will become available. Mesoscale to submesoscale ocean dynamics might also be important for a better closure of the heat budget in the eastern tropical Pacific where the vertical component of NDH is important, which could be addressed from regional high-resolution ocean-atmosphere coupled modeling.

Regarding the Bjerknes feedback, the nature of the underlying atmospheric processes behind it has remained debatable because the low-level atmospheric circulation in the conceptual framework could be induced by several mechanisms, such as a Gill-type response (Gill, 1980) by which an SST anomaly under warm conditions leads to diabatic heating and a baroclinic response of the atmosphere or the Lindzen–Nigam process (Lindzen and Nigam, 1987; Battisti et al., 1999), in which an SST gradient produces a sea level pressure gradient that generates the low-level wind anomalies. Also, positive SST anomalies can enhance the mixing in the planetary boundary layer (PBL) and increase the downward transfer of momentum to the ocean surface, enhancing the wind speed and stress (e.g., Hayes et al., 1989; Wallace et al., 1989). Although this mechanism has been mostly invoked at the mesoscale (O'Neill et al., 2010; Koseki and Watanabe, 2010), it could be effective at larger scale around frontal zone and where the PBL is relatively shallow like the eastern tropical Pacific (Wallace et al., 1989). Nonlinearity associated with cloud feedback in the eastern equatorial Pacific has also been invoked (Frauen and Dommenget, 2010; Lloyd et al., 2012), which, along with the water vapor radiation feedback, remains the largest source of model uncertainty in climate change projections (Bony and Dufresne, 2005; Chen et al., 2013; Bellenger et al., 2014).

Therefore, disentangling the intricate scale interactions operating in the marine boundary layer during extreme ENSO development represents a keystone for further progress. Currently a well-popularized formalism has been proposed to estimate the Bjerknes feedback, which is based on linearized simplified mixed-layer heat budget of the equatorial Pacific (Jin et al., 2006). Although this formalism has been proven very useful for the analysis of full-physics model simulations (Karamperidou et al., 2016; Kim and Jin, 2010; Im et al., 2015; An et al., 2017), it may have structural limitation associated with not including contributions of the nonlinear terms of the heat budget (Graham et al., 2014) and not accounting realistically for the complex of atmospheric processes potentially important for the development of Bjerknes feedback. Regional air–sea interaction processes in the far eastern Pacific are also disregarded within this formalism although they play a role in the seasonality of the Bjerknes feedback (Dewitte and Takahashi, 2017).

6.3 External Forcing and Predictability

Although previous studies have investigated the role of mean state changes of the equatorial Pacific on the deterministic ENSO dynamics to explain changes in ENSO properties (An and Jin, 2001; Thual et al., 2013), there is more and more evidence that ENSO is to a large extent noise driven (Jin et al., 2006; Levine and Jin, 2010; Levine et al., 2016; Takahashi et al., 2018), which calls for investigating the multiple facets of external forcings of ENSO. Previous studies have focused on individual ENSO predictors in remote regions (e.g., Larson and Kirtman, 2014) or on those located in the near equatorial region (e.g., Capotondi and Sardeshmukh, 2015). However, it is likely that there is an interplay of various ENSO predictors originating from distinct regions that shape the conditions favorable for ENSO development (Boschat et al., 2013; Dommenget and Yu, 2017), which can also be modulated at decadal timescales. This complicates drastically the forecast of extreme El Niño events, limiting to some extent the traditional deterministic approach for seasonal forecasts. Because external forcing outside the tropical Pacific seems to be more important than previously thought for ENSO (and ENSO diversity) (Dommenget and Yu, 2017; Takahashi et al., 2018), it needs to be considered that the atmospheric teleconnections to the other ocean basins and their feedbacks to ENSO are important for the correct simulation of extreme El Niño events. This may also explain the limited skill in simulating ENSO diversity in state-of-the-art CGCMs because error accumulation is expected in the chain of processes.

Regarding the equatorial noise forcing, further studies are also important for better understanding the role WWBs and the different components of intraseasonal tropical variability (ITV) (i.e., MJO, tropical cyclones, convective waves) on ENSO development considering that their characteristics can influence not only the overall growth rate of ENSO (Levine et al., 2016) but also the evolution in space and time of the coupled ENSO instability (Thual et al., 2012). Because it is the low-frequency component of the forcing that matters for ENSO, ITV is relevant only if it produces a rectified effect, e.g., associated with a skewed distribution due to deep convection. But the relative importance of ITV and other sources of low-frequency phenomena in forcing ENSO are not really known.

Regarding predictability, another way to tackle this issue is to consider the large number of moderate events and investigate why these did not evolve into extreme El Niño events. This could be also valuable for improving the models considering that metrics for extreme El Niño events may have limited use because of the too few observed events of that type, whereas evaluating models on their ability to simulate moderate events would be statistically more robust (Dewitte and Takahashi, 2017).

At last, because extreme El Niño events are of EP variety, a denser observing system should be considered in that region considering the relative paucity of data east of 95°W where the easternmost TAO buoy is located along the equator. In particular because of the sloping thermocline from west to east (maximum gradient near 120°W), a number of processes have the potential to modify the characteristics of the equatorial Kelvin wave (flux, phase speed, amplitude, vertical structure), which is a main oceanic conduit by which WWBs (i.e., noise) impact ENSO dynamics. This region is also the siege of the mechanisms by which the thermocline fluctuations connect to the SST in the eastern Pacific (i.e., thermocline feedback), which remain largely parameterized in models, albeit fundamental for understanding extreme El Niño development. The recently initiated TPOS 2020 project (http://tpos2020.org/) aimed at designing the future of the tropical Pacific observing system is certainly a key opportunity for addressing such a need.

ACKNOWLEDGMENTS

B. Dewitte acknowledges supports from FONDECYT (projects 1171861 and 1151185) and from LEFE-GMMC (project STEPPE).

REFERENCES

An, S.-I., Heo, E.S., Kim, S.T., 2017. Feedback process responsible for intermodel diversity of ENSO variability. Geophys. Res. Lett. https://doi.org/10.1002/2017GL073203.

An, S.-I., Jin, F.-F., 2004. Why El Niño is stronger than La Niña. J. Clim. 17 (12), 2399–2412.

An, S.-I., 2004. Interdecadal changes in the El Nino-La Nina asymmetry. Geophys. Res. Lett. 31, L23210.

An, S.-I., Jin, F.-F., 2001. Collective role of zonal advective and thermocline feedbacks in ENSO mode. J. Clim. 14, 3421–3432.

An, S.-I., 2008. Interannual variations of the tropical ocean instability wave and ENSO. J. Clim. 21, 3680–3686.

Anderson, D.L.T., McCreary, J.P., 1985. Slowly propagating disturbances in a simple coupled ocean-atmosphere system. J. Atmos. Sci. 42, 615–629.

Ashok, K., Behera, S.K., Rao, S.A., Weng, H., Yamagata, T., 2007. El Niño Modoki and its possible teleconnection. J. Geophys. Res. 112, C11007. https://doi.org/10.1029/2006JC003798.

Bayr, T., Dommenget, D., Martin, T., Power, S., 2014. The eastward shift of the Walker circulation in response to global warming and its relationship to ENSO variability. Clim. Dyn. 43, 2747–2763.

Barnston, A., Tippett, M.K., L'Heureux, M.L., Li, S., DeWitt, D.G., 2012. Skill of real-time seasonal ENSO model predictions during 2002–2011: is our capability increasing? Bull. Am. Meteorol. Soc. 93, 631–651. https://doi.org/10.1175/BAMS-D-11-00111.1.

Battisti, D.S., Sarachik, E.S., Hirst, A.C., 1999. A consistent model for the large-scale steady surface atmospheric circulation in the tropics. J. Clim. 12, 2956–2964.

Battisti, D.S., Hirst, A.C., 1989. Interannual variability in a tropical atmosphere-ocean model: influence of the basic state, ocean geometry and nonlinearity. J. Atmos. Sci. 45, 1687–1712.

Bellenger, H., Guilyardi, E., Leloup, J., Lengaigne, M., Vialard, J., 2014. ENSO representation in climate models: from CMIP3 to CMIP5. Clim. Dyn. 42, 1999–2018.

Bjerknes, J., 1966. A possible response of the atmospheric Hadley circulation to equatorial anomalies of ocean temperature. Tellus 18, 820–829.
Bony, S., Dufresne, J.-L., 2005. Marine boundary layer clouds at the heart of tropical cloud feedback uncertainties in climate models. Geophys. Res. Lett. 32 (20), L20806.
Boschat, G., Terray, P., Masson, S., 2013. Extratropical forcing of ENSO. Geophys. Res. Lett. 40, 1605–1611.
Boucharel, J., Dewitte, B., Garel, B., duPenhoat, Y., 2009. ENSO's non-stationary and non-Gaussian character: the role of climate shifts. Nonlinear Process Geophys. 16, 453–473.
Burger, G., Jin, F.-F., van Oldenborgh, G.J., 2005. The simplest ENSO recharge oscillator. Geophys. Res. Lett. 32, L13706. https://doi.org/10.1029/2005GL022951.
Cai, W., van Rensch, P., Cowan, T., Sullivan, A., 2010. Asymmetry in ENSO teleconnection with regional rainfall, its multidecadal variability, and impact. J. Clim. 23, 4944–4955. https://doi.org/10.1175/2010JCLI3501.1.
Cai, W., Borlace, S., Lengaigne, M., van Rensch, P., Collins, M., Vecchi, G., Timmermann, A., Santoso, A., McPhaden, M., Wu, L., England, M.H., Wang, G., Guilyardi, E., Jin, F.-F., 2014. Increasing frequency of extreme El Niño events due to greenhouse warming. Nat. Clim. Change 4, 111–116. https://doi.org/10.1038/nclimate2100.
Cane, M.A., Sarachik, E.S., 1977. Forced baroclinic ocean motions part II: linear equatorial bounded case. J. Mar. Res. 35 (2), 395–432.
Capotondi, A., Wittenberg, A., Newman, M., Di Lorenzo, E., Yu, J.-Y., Braconnot, P., Cole, J., Dewitte, B., Giese, B., Guilyardi, E., Jin, F.-F., Karnauskas, K., Kirtman, B., Lee, T., Schneider, N., Xue, Y., Yeh, S.-W., 2015. Understanding ENSO diversity. Bull. Am. Meteorol. Soc. https://doi.org/10.1175/BAMS-D-13-00117.1.
Capotondi, A., Sardeshmukh, P.D., 2015. Optimal precursors of different types of ENSO events. Geophys. Res. Lett. https://doi.org/10.1002/2015GL066171.
Capotondi, A., Wittenberg, A.T., 2013. ENSO diversity in climate models. U.S. CLIVAR Var. 11 (2), 10–14.
Carranza, L., 1891. Contra-corriente maritima observada en Paita y Pacasmayo. Bol. Soc. Geogr. Lima 1 (9), 344–345.
Carrillo, C.N., 1893. Hidrografia oceanica. Bol. Soc. Geogr. Lima 72–110.
Chen, N., Majda, A.J., 2017. A simple stochastic dynamical model capturing the statistical diversity of El Nino southern oscillation. Proc. Natl. Acad. Sci. U.S.A. 114 (7), 1468–1473.
Chen, D., Cane, M.A., Kaplan, A., Zebiak, S.E., Huang, D.J., 2004. Predictability of El Niño over the past 148 years. Nature 428 (6984), 733–736.
Chen, D.K., Lian, T., Fu, C.B., Cane, M.A., Tang, Y.M., Murtugudde, R., Song, X.S., Wu, Q.Y., Zhou, L., 2015. Strong influence of westerly wind bursts on El Nino diversity. Nat. Geosci. 8 (5), 339–345. https://doi.org/10.1038/Ngeo2399.
Chen, L., Yu, Y., Sun, D.-Z., 2013. Cloud and water vapor feedbacks to the El Nino warming: are they still biased in CMIP5 models? J. Clim. https://doi.org/10.1175/JCLI-D-12-00575.1.
Choi, J., An, S.-I., Dewitte, B., Hsieh, W.W., 2009. Interactive feedback between the tropical Pacific decadal oscillation and ENSO in a coupled general circulation model. J. Clim. 22, 6597–6611.
Choi, J., An, S.-I., Yeh, S.-W., 2012. Decadal amplitude modulation of two types of ENSO and its relationship with the mean state. Clim. Dyn. 28, 2631–2644. https://doi.org/10.1007/s00382-011-1186-y.
Choi, J., An, S.-I., Yeh, S.-W., Yu, J.-Y., 2013b. ENSO-like and ENSO-induced tropical pacific decadal variability in CGCMs. J. Clim. 26, 1485–1501.
Choi, K., Vecchi, G.A., Wittenberg, A., 2013a. ENSO transition, duration, and amplitude asymmetries: role of the nonlinear wind stress coupling in a conceptual model. J. Clim. https://doi.org/10.1175/JCLI-D-13-00045.1.
Chiang, J.C.H., Vimont, D.J., 2004. Analogous Pacific and Atlantic meridional modes of tropical atmosphere–ocean variability. J. Clim. 17, 4143–4158.

Chiodi, A.M., Harrison, D.E., 2017. Observed El Niño SSTA development and the effects of easterly and westerly wind events in 2014–2015. J. Clim. 30 (4), 1505–1519. https://doi.org/10.1175/JCLI-D-16-0385.1.

Cibot, C., 138453 Maisonnave, E., Terray, L., Dewitte, B., 2005. Mechanisms of tropical Pacific interannual-to-decadal variability in the ARPEGE/ORCA global coupled model. Clim. Dyn. https://doi.org/10.1007/s00382-004-0513-y.

Collins, M., An, S.-I., Cai, W., Ganachaud, A., Guilyardi, E., Jin, F.-F., Jochum, M., Lengaigne, M., Power, S., Timmermann, A., Vecchi, G., Wittenberg, A., 2010. The impact of global warming on the tropical Pacific Ocean and El Niño. Nat. Geosci. 3, 391–397.

Cushman, G., 2004. Enclave vision: foreign networks in Peru and the internationalization of El Niño research during the 1920s. Hist. Meteorol. 1 (1), 65–74.

Dai, A., Fyfe, J.C., Xie, S.-P., Dai, X., 2015. Decadal modulation of global surface temperature by internal climate variability. Nat. Clim. Change 5, 555–559.

Delworth, T.L., Rosati, A., Anderson, W.G., Adcroft, A., Balaji, V., Benson, R., Dixon, K.W., Griffies, S.M., Lee, H.-C., Pacanowski, R.C., Vecchi, G.A., Wittenberg, A.T., Zeng, F., Zhang, R., 2012. Simulated climate and climate change in the GFDL CM2.5 high-resolution coupled climate model. J. Clim. 25 (8). https://doi.org/10.1175/JCLI-D-11-00316.1.

Delworth, T.L., et al., 2006. GFDL's CM2 global coupled climate models. Part I: formulation and simulation characteristics. J. Clim. 19, 634–674.

Deser, et al., 2017. The Northern Hemisphere Extratropical Atmospheric Circulation Response to ENSO: How Well Do We Know It and How Do We Evaluate Models Accordingly? . https://doi.org/10.1175/JCLI-D-16-0844.1.

Dewitte, B., Yeh, S.-W., Moon, B.-K., Cibot, C., Terray, L., 2007. Rectification of the ENSO variability by interdecadal changes in the equatorial background mean state in a CGCM simulation. J. Clim. 20 (10), 2002–2021.

Dewitte, B., Vazquez-Cuervo, J., Goubanova, K., Illig, S., Takahashi, K., Cambon, G., Purca, S., Correa, D., Gutierrez, D., Sifeddine, A., Ortlieb, L., 2012. Change in El Niño flavours over 1958-2008: implications for the long-term trend of the upwelling off Peru. Deep Sea Res. II. https://doi.org/10.1016/j.dsr2.2012.04.011.

Dewitte, B., Takahashi, K., 2017. Diversity of moderate El Niño events evolution: role of air-sea interactions in the eastern tropical Pacific. Clim. Dyn. (revised). https://doi.org/10.1007/s00382-017-4051-9.

DiNezio, P.N., Clement, A.C., Vecchi, G.A., Soden, B.J., Kirtman, B.P., Lee, S.K., 2009. Climate response of the equatorial Pacific to global warming. J. Clim. 22 (18), 4873–4892.

Ding, H., Keenlyside, N.S., Latif, M., 2012. Impact of the equatorial Atlantic on the El Nio southern oscillation. Clim. Dyn. 38, 1965–1972.

Dommenget, D., Bayr, T., Frauen, C., 2012. Analysis of the non-linearity in the pattern and time evolution of El Niño southern oscillation. Clim. Dyn. https://doi.org/10.1007/s00382-012-1475-0.

Dommenget, D., Yu, Y., 2017. The effects of remote SST forcings on ENSO dynamics, variability and diversity. Clim. Dyn. 49 (7–8), 2605–2624.

Dommenget, D., Semenov, V., Latif, M., 2006. Impacts of the tropical Indian and Atlantic oceans on ENSO. Geophys. Res. Lett. 33, L11701. https://doi.org/10.1029/2006GL025871.

Eisenman, I., Yu, L., Tziperman, E., 2005. Westerly wind bursts: ENSO's tail rather than the dog? J. Clim. 18, 5224–5238. https://doi.org/10.1175/JCLI3588.1.

England, M.H., McGregor, S., Spence, P., Meehl, G.A., Timmermann, A., Cai, W., Sen Gupta, A., McPhaden, M.J., Purich, A., Santoso, A., 2014. Recent intensification of wind-driven circulation in the Pacific and the ongoing warming hiatus. Nat. Clim. Change 4, 222–227. https://doi.org/10.1038/nclimate2106.

Frauen, C., Dommenget, D., 2010. El Niño and La Niña amplitude asymmetry caused by atmospheric feedbacks. Geophys. Res. Lett. 37, L18801. https://doi.org/10.1029/2010GL044444.

Frauen, C., Dommenget, D., Rezny, M., Wales, S., 2014. Analysis of the non-linearity of El Nino southern oscillation teleconnections. J. Clim. 27, 6225–6244.

Frauen, C., Dommenget, D., 2012. Inluences of the tropical Indian and Atlantic oceans on the predictability of ENSO. Geophys. Res. Lett. 39, L02706. https://doi.org/10.1029/2011GL050520.

Gebbie, G., Eisenman, I., Wittenberg, A., Tziperman, E., 2007. Modulation of westerly wind bursts by sea surface temperature: a semistochastic feedback of ENSO. J. Atmos. Sci. 64, 3281–3295. https://doi.org/10.1175/JAS4029.1.

Gill, A., 1980. Some simple solutions for heat-induced tropical circulation. Q. J. R. Meteorol. Soc. 106, 447–462.

Graham, F.S., Brown, J.N., Langlais, C., Marsland, S.J., Wittenberg, A.T., Holbrook, N.J., 2014. Effectiveness of the Bjerknes stability index in representing ocean dynamics. Clim. Dyn. 43, 2399–2414.

Gushchina, D., Dewitte, B., 2012. Intraseasonal tropical atmospheric variability associated to the two flavors of El Niño. Mon. Weather Rev. 140 (11), 3669–3681.

Ham, Y., Kug, J.S., 2012. How well do current climate models simulate two types of El Nino? Clim. Dyn. 39 (1–2), 383–398.

Ham, Y.G., Kug, J.S., Park, J.Y., Jin, F.F., 2013. Sea surface temperature in the north tropical Atlantic as a trigger for El Nino/southern oscillation events. Nat. Geosci. 6, 112–116.

Hayes, S.P., McPhaden, M.J., Wallace, J., 1989. The influence of sea surface temperature on surface wind in the Eastern Equatorial Pacific: weekly to monthly variability. J. Clim. 2, 1500–1506.

Held, I.M., Soden, B.J., 2006. Robust responses of the hydrological cycle to global warming. J. Clim. 19, 5686–5699.

Hoerling, M.P., Kumar, A., Xu, T.Y., 2001. Robustness of the nonlinear climate response to ENSO's extreme phases. J. Clim. 14, 1277–1293. https://doi.org/10.1175/1520-0442(2001)014, 1277:ROTNCR.2.0.CO;2.

Hong, L.-C., LinHo, Jin, F.-F., 2014. A southern hemisphere booster of super El Niño. Geophys. Res. Lett. 41, 2142–2149. https://doi.org/10.1002/2014GL059370.

Horel, J.D., Wallace, J.M., 1981. Planetary-scale atmospheric phenomena associated with the southern oscillation. Mon. Weather Rev. 109, 813–829.

Hu, S., Fedorov, A.V., 2016. An exceptional easterly wind burst stalling El Niño of 2014. Proc. Natl. Acad. Sci. U.S.A. https://doi.org/10.1073/pnas.1514182113.

Im, S.-H., An, S.-I., Kim, S.T., Jin, F.-F., 2015. Feedback processes responsible for El Nino-La Nina amplitude asymmetry. Geophys. Res. Lett. 42, 5556–5563.

Izumo, T., Vialard, J., Lengaigne, M., de Boyer, M., Behera, S.K., Luo, J.J., Cravatte, S., Masson, S., Yamagata, T., 2010. Influence of the state of the Indian ocean dipole on the following year's El Niño. Nat. Geosci. https://doi.org/10.1038/ngeo760.

Jansen, M.F., Dommenget, D., Keenlyside, N., 2009. Tropical atmosphere-ocean interactions in a conceptual framework. J. Clim. 22, 550–567.

Jin, F.-F., 1997. An equatorial recharge paradigm for ENSO: I. Conceptual model. J. Atmos. Sci. 54, 811–829.

Jin, F.-F., Neelin, J.D., Ghil, M., 1994. El Niño on the devil's staircase: annual subharmonic steps to chaos. Science. 264 (5155), 70–72. https://doi.org/10.1126/science.264.5155.70.

Jin, F.-F., An, S., Timmermann, A., Zhao, J., 2003. Strong El Niño events and nonlinear dynamical heating. Geophys. Res. Lett. 30 (3), 1120. https://doi.org/10.1029/2002GL016356.

Jin, F.-F., Kim, S.T., Bejarano, L., 2006. A coupled-stability index for ENSO. Geophys. Res. Lett. 33, L23708. https://doi.org/10.1029/2006GL027221.

Jin, F.-F., Lin, L., Timmermann, A., Zhao, J., 2007. Ensemble-mean dynamics of the ENSO recharge oscillator under state-dependent stochastic forcing. Geophys. Res. Lett. 34, L03807. https://doi.org/10.1029/2006GL027372.

Jin, F.F., Boucharel, J., Lin, I.I., 2014. Eastern Pacific tropical cyclones intensified by El Niño delivery of subsurface ocean heat. Nature. 516 (7529), 82–85.

Johnson, N.C., 2013. How many ENSO flavors can we distinguish? J. Clim. 26, 4816–4827.

Johnson, N.C., Kosaka, Y., 2016. The Impact of Eastern Equatorial Pacific Convection on the Diversity of Boreal Winter El Niño Teleconnection Patterns. Clim. Dyn. 47 (12), 3737–3765. https://doi.org/10.1007/s00382-016-3039-1.

Kajtar, J.B., Santoso, A., England, M.H., Cai, W.J., 2015. Indo–Pacific climate interactions in the absence of an Indonesian throughflow. J. Clim. 28, 5017–5029.

Kapur, A., Zhang, C., 2012. Multiplicative MJO forcing of ENSO. J. Clim. 25, 8132–8147.

Karamperidou, C., Jin, F.-F., Conroy, J., 2016. The importance of ENSO nonlinearities in tropical Pacific response to external forcing. Clim. Dyn. https://doi.org/10.1007/s00382-016-3475-y.

Karamperidou, C., DiNezio, P., Timmermann, A., Jin, F.F., Cobb, K.M., 2015. The response of ENSO flavors to mid-Holocene climate: implications for proxy interpretation. Paleoceanography. https://doi.org/10.1002/2014PA002742.

Karamperidou, C., Jin, F.F., Conroy, J.L., 2016. The importance of ENSO nonlinearities in tropical pacific response to external forcing. Clim. Dyn. https://doi.org/10.1007/s00382-016-3475-y.

Kay, J.E., et al., 2015. The community Earth system model (CESM) large ensemble project: a community resource for studying climate change in the presence of internal climate variability. Bull. Am. Meteorol. Soc. 96, 1333–1349. https://doi.org/10.1175/BAMS-D-13-00255.1.

Kim, S.-T., Jin, F.-F., 2010. An ENSO stability analysis. Part II: results from the twentieth and twenty-first century simulations of the CMIP3 models. Clim. Dyn. https://doi.org/10.1007/s00382-010-0872-5.

Kosaka, Y., Xie, S.-P., 2013. Recent global-warming hiatus tied to equatorial Pacific surface cooling. Nature. https://doi.org/10.1038/nature12534.

Khodri, M., Izumo, T., Vialard, J., Janicot, S., Cassou, C., Lengaigne, M., Mignot, J., Gastineau, G., Guilyardi, E., Lebas, N., Robock, A., McPhaden, M.J., 2017. Tropical explosive volcanic eruptions can trigger El Niño events by cooling tropical Africa. Nat. Commun. 8, 778,. https://doi.org/10.1038/s41467-017-00755-6.

Koseki, S., Watanabe, M., 2010. Atmospheric boundary layer response to mesoscale SST anomalies in the Kuroshio extension. J. Clim. 23, 2492–2507. https://doi.org/10.1175/2009JCLI2915.1.

Kug, J.-S., Choi, J., An, S.-I., Jin, F.-F., Wittenberg, A.-T., 2010. Warm pool and cold tongue El Nino events as simulated by the GFDL2.1 coupled GCM. J. Clim. 23, 1226–1239.

Kug, J.S., Kang, I.S., 2006. Interactive feedback between ENSO and the Indian Ocean. J. Clim. 19, 1784–1801.

Kumar, A., Barnston, A.G., Peng, P., Hoerling, M.P., Goddard, L., 2000. Changes in the spread of the variability of the seasonal mean atmospheric states associated with ENSO. J. Clim. 13 (17), 3139–3151.

Kumar, K.K., Rajagopalan, B., Hoerling, M., Bates, G., Cane, M., 2006. Unraveling the mystery of Indian monsoon failure during El Niño. Science 314, 115–119.

Larson, S.M., Kirtman, B.P., 2014. The pacific meridional mode as an ENSO precursor and predictor in the north American multi-model ensemble. J. Clim. 27, 7018–7032.

Larson, S., Kirtman, B., 2013. The Pacific meridional mode as a trigger for ENSO in a high-resolution coupled model. Geophys. Res. Lett. 40, 3189–3194.

Lee, T., McPhaden, M.J., 2010. Increasing intensity of El Niño in the central-equatorial pacific. Geophys. Res. Lett. 37, L14603. https://doi.org/10.1029/2010GL044007.

Lindzen, R.S., Nigam, S., 1987. On the role of sea surface temperature gradients in forcing low level winds and convergence in the tropics. J. Atmos. Sci. 44, 2418–2436.

Liu, Z.Y., Vavrus, S., He, F., Wen, N., Zhong, Y.F., 2005. Rethinking tropical ocean response to global warming: the enhanced equatorial warming. J. Clim. 18, 4684–4700.

L'Heureux, M., Takahashi, K., Watkins, A., Barnston, A., Becker, E., Di Liberto, T., Gamble, F., Gottschalck, J., Halpert, M., Huang, B., Mosquera-Vásquez, K., Wittenberg, A., 2017. Observing and predicting the 2015/16 El Niño. Am. Meteor. Soc. 98 (7), 1363–1382. https://doi.org/10.1175/BAMS-D-16-0009.1.

L'Heureux, M., Lee, S., Lyon, B., 2013. Recent multidecadal strengthening of the Walker circulation across the tropical Pacific. Nat. Clim. Change. 3 (6), 571–576.

Lengaigne, M., et al., 2004. Triggering of El Niño by westerly wind events in a coupled general circulation model. Clim. Dyn. 23, 601–620. https://doi.org/10.1007/s00382-004-0457-2.

Levine, A., Jin, F.F., McPhaden, M.J., 2016. Extreme noise-extreme El Niño: how state-dependent noise forcing creates El Niño-La Niña asymmetry. J. Clim. https://doi.org/10.1175/JCLI-D-16-0091.1.

Levine, A.F.Z., Jin, F.-F., 2010. Noise-induced instability in the ENSO recharge oscillator. J. Atmos. Sci. 67, 529–542. https://doi.org/10.1175/2009JAS3213.1.

Lloyd, J., Guilyardi, E., Weller, H., 2012. The role of atmosphere feedbacks during ENSO in the CMIP3 models. Part III: the shortwave flux feedback. J. Clim. 25, 4275–4293.

Lopez, H., Kirtman, B.P., Tziperman, E., Gebbie, G., 2013. Impact of interactive westerly wind bursts on CCSM3. Dyn. Atmos. Oceans 59, 24–51. https://doi.org/10.1016/j.dynatmoce.2012.11.001.

Lopez, H., Kirtman, B.P., 2014. WWBs, ENSO predictability, the spring barrier and extreme events. J. Geophys. Res. Atmos. 119. https://doi.org/10.1002/2014JD021908. pp. 10, 114–10, 138.

Luo, J.-J., Sakaki, W., Masumoto, Y., 2012. Indian Ocean warming modulates Pacific climate change. Proc. Natl. Acad. Sci. U.S.A. 109, 18701–18706.

Matei, D., Keenlyside, N., Latif, M., Jungclaus, J., 2008. Subtropical forcing of tropical Pacific climate and decadal ENSO modulation. J. Clim. 21, 4691–4709.

Meinen, C.S., McPhaden, M.J., 2000. Observations of warm water volume changes in the equatorial Pacific and their relationship to El Niño and La Niña. J. Clim. 13 (20), 3551–3559.

Menkes, C.E., Lengaigne, M., Vialard, J., Puy, M., Marchesiello, P., Cravatte, S., Cambon, G., 2014. About the role of westerly wind events in the possible development of an El Nino in 2014. Geophys. Res. Lett. 41 (18), 6476–6483.

McGregor, S., Timmermann, A., Stuecker, M.F., England, M.H., Merrifield, M., Jin, F.F., Chikamoto, Y., 2014. Recent Walker circulation strengthening and Pacific cooling amplified by Atlantic warming. Nat. Clim. Change 4, 888–892.

McPhaden, M.J., Busalacchi, A.J., Cheney, R., Donguy, J.-R., Gage, K.S., Halpern, D., Ji, M., Julian, P., Meyers, G., Mitchum, G.T., Niiler, P.P., Picaut, J., Reynolds, R.W., Smith, N., Takeuchi, K., 1998. The tropical ocean global atmosphere observing system: a decade of progress. J. Geophys. Res. 103 (C7), 14,169–14,240. https://doi.org/10.1029/97JC02906.

McPhaden, M.J., Zhang, X., Hendon, H.H., Wheeler, M.C., 2006. Large scale dynamics and MJO forcing of ENSO variability. Geophys. Res. Lett. 33, L16702. https://doi.org/10.1029/2006GL026786.

McPhaden, M.J., 2015. Playing hide and seek with El Niño. Nat. Clim. Change 5, 791–795. https://doi.org/10.1038/nclimate2775.

Münnich, M., Cane, M.A., Zebiak, S.E., 1991. A study of self-excited oscillations of the tropical ocean-atmosphere system. Part II: nonlinear cases. J. Atmos. Sci. 48 (10), 1238–1248. https://doi.org/10.1175/1520-0469(1991)048%3C1238:ASOSEO%3E2.0.CO;2.

Murphy, R.C., January 1926. Oceanic and climatic phenomena along the west coast of south America during 1925. Geogr. Rev. 16 (1), 26–54.

Neelin, J.D., 1991. The slow sea surface temperature mode and the fast-wave limit: analytic theory for tropical interannual oscillations and experiments in a hybrid coupled model. J. Atmos. Sci. 48, 584–606.
Neelin, J.D., Battisti, D.S., Hirst, A.C., Jin, F.-F., Wakata, Y., Yamagata, T., Zebiak, S., 1998. ENSO theory. J. Geophys. Res. 103 (C7), 14261–14290.
Neelin, J.D., Dijkstra, H.A., 1995. Ocean-atmosphere interaction and the tropical climatology. Part I: the angers of flux correction. J. Clim. 8, 1325–1342.
Neelin, J.D., Jin, F.-F., Syu, H.-H., 2000. Variations in ENSO phase locking. J. Clim. 13 (14), 2570–2590. https://doi.org/10.1175/1520-0442(2000)013%3C2570:VIEPL%3E2.0.CO;2.
Newman, M., Shin, S.-I., Alexander, M.A., 2011. Natural variation in ENSO flavors. Geophys. Res. Lett. https://doi.org/10.1029/2011GL047658.
Ogata, T., Xie, S.-P., Wittenberg, A., Sun, D.-Z., 2013. Interdecadal amplitude modulation of El Nino/Southern Oscillation and its impacts on tropical Pacific decadal variability. J. Clim. 26, 7280–7297.
Ohba, M., Ueda, H., 2007. An impact of SST anomalies in the Indian ocean in acceleration of the El Nino to La Nina transition. J. Meteorol. Soc. Jpn. 85, 335–348.
Okumura, Y.M., 2013. Origins of tropical Pacific decadal variability: role of stochastic atmospheric forcing from the south Pacific. J. Clim. 26, 9791–9796.
O'Neill, L.W., Esbensen, S., Thum, N., Samelson, R.M., Chelton, D.B., 2010. Dynamical analysis of the boundary layer and surface wind responses to mesoscale SST perturbations. J. Clim. 23, 559–581. https://doi.org/10.1175/2009JCLI2662.1.
Paek, H., Yu, J.-Y., Qian, C., 2017. Why were the 2015/16 and 1997/98 extreme El Niños different? Geophys. Res. Lett. 1–9. https://doi.org/10.1002/2016GL071515.
Penland, C., Sardeshmukh, P., 1995. The optimal growth of tropical sea surface temperature anomalies. J. Clim. 8, 1999–2024.
Penland, C., 1996. A stochastic model of IndoPacific sea surface temperature anomalies. Phys. Nonlinear Phenom. 98 (2–4), 534–558.
Power, S., Haylock, M., Colman, R., Wang, X., 2006. The predictability of interdecadal changes in ENSO activity and ENSO teleconnections. J. Clim. 19, 4755–4771.
Power, S., Delage, F., Chung, C., Kociuba, G., Keay, K., 2013. Robust twenty-first-century projections of El Niño and related precipitation variability. Nature 502, 541–545.
Puy, M., Vialard, J., Lengaigne, M., Guilyardi, E., 2015. Modulation of equatorial pacific westerly/easterly wind events by the Madden-Julian oscillation and convectively-coupled Rossby waves. Clim. Dyn. https://doi.org/10.1007/s00382-015-2695-x. (published online).
Rasmuson, E., Carpenter, T., 1982. Variations in tropical sea surface temperature and surface wind fields associated with the Southern Oscillation/El Niño. Mon. Weather Rev. 110, 354–384.
Ray, S., Giese, B.S., 2012. Changes in El Niño and La Niña characteristics in an ocean reanalysis and reconstructions from 1871-2008. J. Geophys. Res. 117, C11007. https://doi.org/10.1029/2012JC008031.
Rayner, N.A., Parker, D.E., Horton, E.B., Folland, C.K., Alexander, L.V., Rowell, D.P., Kent, E.C., Kaplan, A., 2003. Global analyses of sea surface temperature, sea ice, and night marine air temperature since the late nineteenth century. J. Geophys. Res. https://doi.org/10.1029/2002JD002670.
Richter, I., 2015. Climate model biases in the eastern tropical oceans: causes, impacts and ways forward. WIREs Clim. Change. https://doi.org/10.1002/wcc.338.
Rodgers, K.B., Friederichs, P., Latif, M., 2004. Tropical Pacific decadal variability and its relation to decadal modulations of ENSO. J. Clim. 17, 3761–3774.
Rodrigues, R.R., Haarsma, R.J., Campos, E.J.D., Ambrizzi, T., 2011. The impacts of inter-El Niño variability on the tropical Atlantic and Northeast Brazil climate. J. Clim. 24, 3402–3422.

Rodriguez-Fonseca, B., Polo, I., Garcia-Serrano, J., Losada, T., Mohino, E., Mechoso, C.R., Kucharski, F., 2009. Are Atlantic Niños enhancing Pacific ENSO events in recent decades? Geophys. Res. Lett. 36, L20705. https://doi.org/10.1029/2009GL040048.

Sanabria, J., Bourrel, L., Dewitte, B., Frappart, F., Rau, P., Olimpio, S., Labat, D., 2017. Rainfall along the coast of Peru during strong El Niño events. J. Int. Climatol. https://doi.org/10.1002/joc.5292.

Santoso, A., England, M.H., Cai, W., 2012. Impact of Indo–Pacific feedback interactions on ENSO dynamics diagnosed using ensemble climate simulations. J. Clim. 25, 7743–7763.

Schopf, P.S., Burgman, R.J., 2006. A simple mechanism for ENSO residuals and asymmetry. J. Clim. 19, 3167–3179.

Schopf, P.S., Suarez, M.J., 1988. Vacillations in a coupled ocean atmosphere model. J. Atmos. Sci. 45, 549–566.

Stevenson, S., Capotondi, A., Fasullo, J., Otto-Bliesner, B., 2017. Forced changes to 20th century ENSO diversity in a last millennium context. Clim. Dyn. https://doi.org/10.1007/s00382-017-3573-5.

Stuecker, M.F., Timmermann, A., Jin, F.F., McGregor, S., Ren, H.L., 2013. A combination mode of the annual cycle and the El Niño/southern oscillation. Nat. Geosci. 6 (7), 540–544. https://doi.org/10.1038/ngeo1826.

Stuecker, M.F., Bitz, C.M., Armour, K.C., 2017. Conditions leading to the unprecedented low Antarctic sea ice extent during the 2016 austral spring season. Geophys. Res. Lett. https://doi.org/10.1002/2017GL074691.

Su, J., Zhang, R., Li, T., Rong, X., Kug, J.S., Hong, C., 2010. Causes of the El Niño and La Niña amplitude asymmetry in the equatorial eastern pacific. J. Clim. 23 (3), 605–617. https://doi.org/10.1175/2009JCLI2894.1.

Suarez, M.J., Schopf, P.S., 1988. A delayed action oscillator for ENSO. J. Atmos. Sci. 45, 3283–3287.

Terray, P., 2011. Southern hemisphere extra-tropical forcing: a new paradigm for El Niño-Southern Oscillation. Clim. Dyn. 36, 2171–2199. https://doi.org/10.1007/s00382-010-0825-z.

Takahashi, K., Montecinos, A., Goubanova, K., Dewitte, B., 2011. ENSO regimes: reinterpreting the canonical and Modoki El Niño. Geophys. Res. Lett. 38, L10704. https://doi.org/10.1029/2011GL047364.

Takahashi, K., Dewitte, B., 2016. Strong and moderate nonlinear El Niño regimes. Clim. Dyn. 46 (5–6), 1627–1645.

Takahashi, K., Karamperidou, C., Dewitte, B., 2018. A theoretical model of strong and moderate El Niño regimes. Clim. Dyn. (revised). https://doi.org/10.1007/s00382-018-4100-z.

Takahashi, K., Martínez, R., Montecinos, A., Dewitte, B., Gutiérrez, D., Rodriguez-Rubio, E., 2014. Regional Applications of Observations in the Eastern Pacific: Western South America. Whitepaper for TPOS2020, 8a.

Takahashi, K., Martínez, A., 2017. The very strong coastal El Niño in 1925 in the far-eastern Pacific. Clim. Dyn. https://doi.org/10.1007/s00382-017-3702-1.

Thual, S., Dewitte, B., An, S.-I., Ayoub, N., 2011. Sensitivity of ENSO to stratification in a recharge-discharge conceptual model. J. Clim. 4, 4331–4348.

Thual, S., Dewitte, B., An, S.-I., Illig, S., Ayoub, N., 2013. Influence of recent stratification changes on ENSO stability in a conceptual model of the equatorial pacific. J. Clim. 26, 4790–4802.

Thual, S., Thual, O., Dewitte, B., 2012. Absolute and convective instability in the equatorial Pacific and implication for ENSO. Q. J. R. Meteorol. Soc. 139 (672), 600–606 ISSN: 0035-9009.

Timmermann, A., Jin, F.-F., Abshagen, J., 2003. A nonlinear theory for El Niño bursting. J. Atmos. Sci. 60, 152–165.

Tokinaga, H., Xie, S.-P., Deser, C., Kosaka, Y., Okumura, Y.M., 2012. Slowdown of the Walker circulation driven by tropical Indo-Pacific warming. Nature 491, 439–443. https://doi.org/10.1038/nature11576.

Tziperman, E., Stone, L., Cane, M.A., Jarosh, H., 1994. El Niño chaos: overlapping of resonances between the seasonal cycle and the Pacific Ocean-atmosphere oscillator. Science. 264 (5155), 72–74. https://doi.org/10.1126/science.264.5155.72.

Vergara, O., Dewitte, B., Ramos, M., Pizarro, O., 2017. Vertical energy flux at ENSO time scales in the subthermocline of the Southeastern Pacific. J. Geophys. Res. Oceans 122. https://doi.org/10.1002/2016JC012614.

Vimont, D.J., Battisti, D.S., Hirst, A.C., 2001. Footprinting: a seasonal connection between the tropics and mid-latitudes. Geophys. Res. Lett. 28, 3923–3926.

Wallace, J., Mitchell, T., Deser, C., 1989. The influence of sea-surface temperature on surface wind in the eastern equatorial Pacific: seasonal and interannual variability. J. Clim. 2, 1492–1499.

Wang, G., Cai, W., Gan, B., Wu, L., Santose, A., Lin, X., Chen, Z., J.McPhaden, M., 2017. Continued increase of extreme El Niño frequency long after 1.5C warming stabilization. Nat. Clim. Change. https://doi.org/10.1038/NCLIMATE3351.

Wittenberg, A.T., 2009. Are historical records sufficient to constrain ENSO simulations? Geophys. Res. Lett. 36, L12702. https://doi.org/10.1029/2009GL038710.

Wittenberg, A.T., Rosati, A., Delworth, T.L., Vecchi, G.A., Zeng, F., 2014. ENSO modulation: is it decadally predictable? J. Clim. 27, 2667–2681. https://doi.org/10.1175/JCLI-D-13-00577.1.

Xie, P., Arkin, P.A., 1997. Global precipitation: a 17-year monthly analysis based on gauge observations, satellite estimates, and numerical model outputs. Bull. Am. Meteorol. Soc. 78, 2539–2558.

Xie, S.-P., Lu, B., Xiang, B., September 1, 2013. Similar spatial patterns of climate responses to aerosol and greenhouse gas changes. Nat. Geosci. 6, 828–832. https://doi.org/10.1038/NGEO1931. (published online).

Xiang, B., Wang, B., Li, T., 2013. A new paradigm for the predominance of standing Central Pacific Warming after the late 1990s. Clim. Dyn. 41 (2), 327–340. https://doi.org/10.1007/s00382-012-1427-8.

Yeh, S.-W., Kug, J.-S., Dewitte, B., Kwon, M.-H., Kirtman, B.P., Jin, F.-F., 2009. El Niño in a changing climate. Nature 461, 511–514. https://doi.org/10.1038/nature08316.

Ying, J., Huang, P., Huang, R.H., 2016. Evaluating the formation mechanisms of the equatorial Pacific SST warming pattern in CMIP5 models. Adv. Atmos. Sci. 33 (4). https://doi.org/10.1007/s00376-015-5184-6.

You, Y., Furtado, J.C., 2017. The role of South Pacific atmospheric variability in the development of different types of ENSO. Geophys. Res. Lett. 1–9. https://doi.org/10.1002/2017GL073475.

Yu, J.Y., Kim, S.T., 2011. Relationships between extratropical sea level pressure variations and the Central Pacific and Eastern Pacific types of ENSO. J. Clim. 24, 708–720.

Zebiak, S.E., Cane, M.A., 1987. A model El-Niño southern oscillation. Mon. Weather Rev. 115 (10), 2262–2278.

Zhang, T., Sun, D.-Z., 2014. ENSO asymmetry in CMIP5 models. J. Clim. 27, 4070–4093. https://doi.org/10.1175/JCLI-D-13-00454.1.

Zhang, H.H., Clement, A., Di Nezio, P., 2014. The South Pacific meridional mode: a mechanism for ENSO-like variability. J. Clim. 27, 769–783.

Zhang, X., Liu, H., Zhang, M., 2015. Double ITCZ in coupled ocean-atmosphere models: from CMIP3 to CMIP5. Geophys. Res. Lett. https://doi.org/10.1002/2015GL065973.

CHAPTER 7

Hotspots of Relative Sea Level Rise in the Tropics

Mélanie Becker[1], Mikhail Karpytchev[1], Fabrice Papa[2,3]
[1]LIENSs/CNRS, UMR 7266, ULR/CNRS, La Rochelle, France; [2]LEGOS/IRD, UMR 5566, CNES/CNRS/IRD/UPS, Toulouse, France; [3]Indo-French Cell for Water Sciences, IRD-IISc-NIO-IITM, Indian Institute of Science, Bangalore, India

Contents

Tropical Extremes: Natural Variability and Trends
ISBN 978-0-12-809248-4
https://doi.org/10.1016/B978-0-12-809248-4.00007-8

1. INTRODUCTION

The pronounced impact of climate change on natural systems and human societies is a reality. Understanding the extent to which people, societies, ecosystems, and economy are exposed to risk under current and future climate is a challenging issue for modern science. One of the major consequences of the ongoing climate change is a rise in sea level (SL). The Intergovernmental Panel on Climate Change reported (IPCC AR5, 2013) that the global mean sea level (GMSL, 1.6–1.8 mm $year^{-1}$ rise over the 20th century (Church et al., 2013)) will continue rising in the 21st century and beyond, at probably a faster rate than observed today, even if the global temperature stabilized. Almost 90% of the coastlines worldwide will face challenges of rising SL (IPCC AR5, 2013), although to a different extent, as the rates of the SL rise can be several times larger in some regions than the rates of the GMSL rise (Church et al., 2013). Consequently, the part of coastal vulnerability reflecting a high and growing exposure and low adaptive capacity of the coastal populations to SL rise is not spatially uniform either (Nicholls et al., 2011). Certain regions throughout the world, especially in developing countries, are already recognized as particularly vulnerable to SL rise; for example, small islands in the Caribbean Sea, Maldives Archipelago in the Indian Ocean, Tuvalu Islands in the Pacific, or the West African coast from Morocco to Namibia, the south Asian coast from Pakistan to Burma, and the coasts in Southeast Asia from Thailand to Vietnam (Nicholls et al., 1999; Nicholls and Cazenave, 2010). Nicholls et al. (2011) defined these specific regions as areas where an efficient protection against SL rise will most likely fail, resulting in a significant portion of environmental refugees. It is worth mentioning here that those cases are related to relative sea level (RSL) changes, which are felt by coastal populations, i.e., the changes in SL relative to the land on which people live. Focusing on the analysis of the RSL variations is of obvious practical importance because it makes little difference to a person nearly submerged, whether the ocean is rising or land is subsiding (Milliman and Haq, 1996). Pronounced dispersion in the rates of RSL rise calls for detailed investigation of the processes responsible for SL changes not only at the global scale but also at the regional scale. The RSL changes are induced by a combination of various processes of a different nature and operating at different spatial and temporal scales, originated in the ocean, ice, atmosphere, sediment transport, and the solid Earth deformation inducing land subsidence or uplift (Stammer et al., 2013). Ocean temperature and salinity variations resulting from water heating, precipitation, or freshwater discharge from land can contribute to regional SL fluctuations by changing the seawater density. Additional freshwater fluxes from river discharge or land ice melting modify ocean currents, which in turn also have significant repercussions on regional SL variations (Stammer, 2008), with signals taking decades to propagate around the global ocean. Atmospheric pressure, at different scales, also plays a role in regional SL variations (Ponte, 1994; Wunsch and Stammer, 1997; Piecuch and Ponte, 2015). Concerning the vertical land movements, there exists a wide range of natural and anthropogenic processes, which can induce them. The water mass exchanges between land and ocean lead to changes in the Earth's surface and in the geoid that manifest themselves as part of observed RSL variations (Milne et al., 2009; Stammer et al., 2013). These result from different processes: (1) ice-water mass redistribution associated with ice-cap melting since the Last Glacial Maximum (called postglacial rebound or glacial isostatic adjustment/GIA (Peltier, 2004; Lambeck et al., 2010)), (2) ongoing land ice melting (Mitrovica et al., 2001; Tamisiea and Mitrovica, 2011), and (3) land water storage variation (Riva et al., 2010). GIA involves the viscoelastic response of the Earth's mantle to mass redistribution, whereas processes (2) and (3) involve the elastic response of the Earth's crust. GIA and present-day mass redistributions produce very different response of the solid Earth, and thus regional RSL variations (see, for example, Milne et al., 2009; Tamisiea, 2011; Tamisiea and Mitrovica, 2011). We now call these processes "static" effects (e.g., Stammer et al., 2013). The solid Earth also responds to sediment loading, referred to herein as sedimentary isostatic adjustment that often induces strong subsidence within the deltas (Blum and Roberts, 2009; Syvitski et al., 2009). Many other natural processes, such as tectonics and volcanism, can also generate land movements that are more local when compared with the "static" effects discussed above. Aside from most of these natural factors, an additional complex dimension to these changes is the nonnegligible impact of human activities; for instance, SL can be modified through building of dams and reservoirs, irrigation and hydrocarbon extraction, groundwater pumping, among many other processes (Fiedler and Conrad, 2010; Wada et al., 2012, 2016). These anthropogenic forcings affect directly the land water storage, and hence water mass exchange between land and ocean (Milly et al., 2010), and consequently can generate locally significant vertical land movement. Several Asian megacities subsided by several meters during the past few decades owing to groundwater withdrawal or hydrocarbon extraction (Syvitski, 2008).

In this chapter, we focus on the RSL changes within the Tropics, defined below as a region from 30°N to 30°S latitude. The Tropics are home to 40% of the world's population, and this proportion is projected to reach 50% by 2050 (Edelman et al., 2014). From today until 2050, the largest coastal population growth is expected to take place in Africa where the population will double (Edelman et al., 2014). Assessing the vulnerability of tropical coasts to future climate change and elaborating an efficient climate mitigation policy is one of the most important global issues of our time. Developing countries that make up a majority of tropical regions are the most vulnerable to RSL changes because they have limited resources to adapt themselves socially, technologically, and financially. Moreover, it is important to note that the Tropics host the largest deltas in the world. These low-lying delta plains are crucially affected by land subsidence that often makes the sea along the delta coasts to rise much faster than the GMSL rises. As the deltas are a home to tens of millions of people, the densely populated deltaic environments become a suitable site for springing up of megacities (greater than 5 million inhabitants) with the associated complex problems of their management.

One of the objectives of this chapter is to bring together SL observations to analyze similarities and differences in the RSL changes along the tropical coasts. It is crucial for all evaluations of coastal impacts, vulnerability, and adaptation, to account for the RSL rise, especially along the low-lying populated coasts where RSL is rising much faster than its global average rate. We call these sites hotspots of RSL rise (Sallenger et al., 2012). Our primary concern is to review the current knowledge about RSL in the tropical regions and to (1) comprehensively identify, document, and map the hotspots of RSL changes; (2) give an overview of available long-term SL records, and (3) update, where possible, previously published estimates of RSL trends over recent decades. Section 2 will review the different data sets currently available to study RSL. Then we will dedicate a specific section for each of the oceans in the tropical band, with a subsection dedicated to large oceanic subbasins. For each region, we will document both the societal and physical aspects of RSL. At the end of each section, we summarize the main features of the respective RSL hotspots.

2. DATA SETS

2.1 Tide Gauge Records

Tide gauge (TG) records are the main source of information available to assess coastal SL changes since the mid-19th century. The TGs were designed to measure RSL, namely, the water level relative to land on which they are installed (Pugh and Woodworth, 2014). Therefore, the TG measurements reflect not only absolute sea level (ASL, i.e., in respect to the center of the Earth) changes but also local vertical land movements along with changes in the geoid. The worldwide geographical distribution of TGs is particularly limited and irregular with an obvious lack of stations in the Southern Hemisphere, particularly in developing countries and island states. In our analyses, we use annually averaged SL series from the Permanent Service for Mean Sea Level (PSMSL) Revised Local Reference (RLR) database (Holgate et al., 2013). The PSMSL recommends using the RLR records, where the SL means were reduced to a common datum, for time series analysis. The PSMSL also provides the "Metric" data, without datum continuity checked and with, sometimes, large discontinuities. These metric records should only be used in studies pertaining to the seasonal cycle of mean SL (Holgate et al., 2013). The length of TG records and the number of missing values are of crucial importance for estimating long-term trends. Douglas (2001) has concluded that more than 50–60 years of continuous measurements are required for a long-term SL trend to be reliably estimated. In this study, we reduce this constraint by estimating trends at the stations with records longer than 30 years and with less than four consecutive years of missing data.

2.2 Satellite Altimetry

Since 1993, satellite altimetry has been used for measuring spatial and temporal variations of ASL rise. The ASL products, consisting of sea surface heights, are routinely processed and distributed by six groups: Archiving, Validation and Interpretation of Satellite Oceanographic data, Commonwealth Scientific and Industrial Research Organization, Colorado University, Goddard Space Flight Center, European Space Agency Climate Change Initiative (ESA-CCI), and Delft University of Technology (TUDelft-RADS). Here, we chose to use the newly reprocessed ESA-CCI Sea Level v1.1 gridded altimetry product (hereafter, called ESA) that is freely available at: http://www.esa-sealevel-cci.org (see details in Ablain et al., 2015). To remove the seasonal signal in the ASL time series, we used a 12-month running mean filter.

2.3 Reconstruction of Sea Level in the Past

Recently, a new approach was developed to reconstruct the ASL variations in the past. This method combines information from TG records with spatial patterns from altimetry and/or oceanic models (Church et al., 2004;

Llovel et al., 2009; Hamlington et al., 2011; Ray and Douglas, 2011; Meyssignac et al., 2012a). To get an overview of the regional ASL variation in the Tropics over a longer period, we employ an updated version of past SL reconstruction developed by Meyssignac et al. (2012a) for the period 1960–2014. This method is based on reduced optimal interpolation, combining long-term TG records with a time-varying linear combination of empirical orthogonal function–based spatial patterns derived from 2D SL grids based on oceanic model outputs.

2.4 Global Positioning System Stations

The Global Positioning System (GPS) is used for precisely positioning TG benchmarks with respect to the center of mass. These measurements, due to their relatively low cost and easy implementation, and maintenance, have become key components for SL studies as they provide accurate determination of coastal vertical land movements (Wöppelmann et al., 2007; Wöppelmann and Marcos, 2016). In this study, we use vertical velocities estimated by the University of La Rochelle from its latest GPS data reanalysis (called hereafter ULR6 (Santamaría-Gómez et al., 2017)). These estimates are made at the GPS stations that are directly collocated with TGs or situated not further than 15 km from them, provided that the GPS series have more than 3 years of data (Wöppelmann et al., 2007). The magnitudes of vertical velocity and their associated uncertainties are available (free of cost) at http://www.sonel.org; the GPS at TG data assembly center Système d'Observation du Niveau des Eaux Littorales (SONEL).

2.5 Urban Agglomerations and Low-Elevation Coastal Zones

We used the urban–rural Population and Land Area Estimates v2 data set, providing the number of people living on contiguous coastal elevations less than or equal to 10 m in 2010. This data set is from the low-elevation coastal zone collection (LECZ; McGranahan et al., 2007) and is freely downloadable from http://sedac.ciesin.columbia.edu/data/collection/lecz.

3. ATLANTIC OCEAN

3.1 Eastern South America

The tropical Atlantic Ocean is bordered in the west by the Brazilian coast extending through the Caribbean Sea to the Gulf of Mexico. The entire Brazilian coastline, extending from latitude 4°N to 34°S, has been experiencing erosion, although the erosion rates vary irregularly and are often enhanced within river outlets (Muehe, 2010). Since 1970s, rapid expansion of agglomerations and intensive construction of housing for residence and tourism bring more people to settle along the coast (Short and da Klein, 2016). At the end of the 1990s, already 20% of Brazilians live in coastal cities (Muehe and Neves, 1995) (Fig. 7.1). For example, the population density of the megacity of Rio de Janeiro has nearly doubled in four decades (27 inhab ha^{-1} in 1960 to 48 inhab ha^{-1} in 2000 (Saglio-Yatzimirsky, 2013)); presently, its population exceeds 12 million people. In the Northeast, Recife is a large metropolitan city with ~4 million inhabitants that ranks among the cities in Brazil, with the highest population density at the coast (Muehe and Neves, 1995; Neves and Muehe, 1995). This city is located at the mouth of two rivers, Beberibe and Capibaribe, within low-lying areas making it particularly vulnerable to RSL rise. All these large coastal cities, where the problems of urban drainage are nowadays permanent, have to deal with floods. In 2008, around 30% of more than 5.5 thousand municipalities in Brazil reported having inefficient drainage system and having suffered from floods in the past 5 years (Nali and Rigo, 2011). The consequences of drainage system deficiency in urban areas are important, ranging from impacts on human health, through groundwater contamination and proliferation of mosquitoes, to damage effects, inter alia, on housing, infrastructure, and psychological stress. These effects will become even more critical with a rise in RSL (Muehe, 2010).

In the north, the Brazil coastline of the Amazon delta extends from Cape Orange in the state of Amapa up to the French Guiana's border. Despite deforestation, dam construction, and land usage, the delta is in relatively good health (Syvitski et al., 2009). Mansur et al. (2016) estimated that over 1.2 million people are under the risk of flooding (fluvial and coastal) in this delta, and that 41% of urban sector inhabitants are exposed to potential flooding risks. The population of the Amazon delta is projected to grow by more than 60% over the 15-year period (Overeem and Syvitski, 2009), making this region particularly vulnerable to anthropogenic changes. The Orinoco delta in Venezuela is an area with small population and is less developed. However, it is estimated that by 2050, 21% of this delta population will be potentially inundated because of future RSL rise and 20% of the delta area could be lost (Ericson et al., 2006).

Over the last few decades, the observed retreat of mangrove vegetation along the delta coastline seems to be compatible with a long-term relative SL rise trend (Cohen and Lara, 2003; França et al., 2012). Gratiot et al. (2008) has shown that the mangrove retreat of the 1500 km-long flat muddy coast

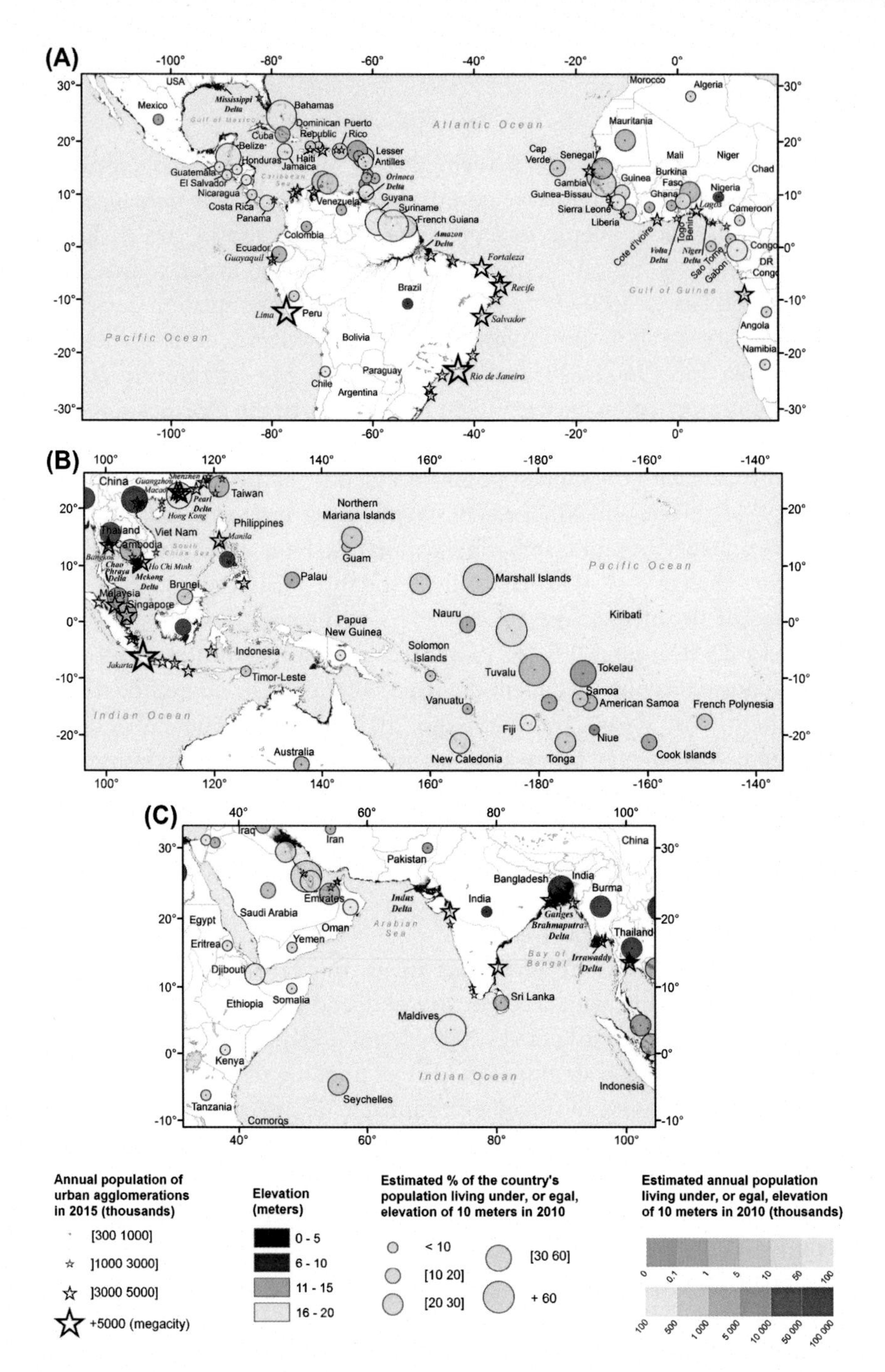

Figure 7.1 Map of the sub regions between 30°N to 30°S included in this global literature review. The stars correspond to large urban agglomerations with 300,000 inhabitants or more in 2015 (from United Nations World Urbanization Prospects: The 2014 Revision, https://esa.un.org/unpd/wup/cd-Rom/). The circle sizes represent the percentage, and their colors the number (in thousands), of the country's population living in contiguous coastal elevations less than or equal to 10 meters in 2010. The global digital elevation model GTOPO30 (https://lta.cr.usgs.gov/GTOPO30) is used to map elevation less than 20 meters.

from the Amazon to the Orinoco (Venezuela) rivers over the last 20 years has been governed primarily by the lunar 18.6-year low-frequency tide constituent. These findings highlight an extreme sensitivity of this region to global environmental changes in general, and, in particular, to SL changes.

Populations of the northern Brazil neighboring countries will also face SL rise adaptation problems: thus, in terms of population impacted by a 1 m SL rise, at least 6% of people living in Guyana, Suriname, and French Guiana population would be displaced (Dasgupta et al., 2009). These countries are among the top 10 countries/territories worldwide affected by climate-induced massive population relocation.

From the PSMSL RLR data set, 11 TG records are available along the Brazilian coast and one recent record from French Guiana (Ile Royale, 10 years, 2006–15). Of these SL records, only two from southeastern Brazil cover recent years and are long enough to allow long-term trend estimates: Cananeia (53 years, 1954–2006) and Rio de Janeiro (Ilha Fiscal station: 51 years, 1963–2013). The length of the other records is less than 21 years. Emery and Aubrey (1991) reviewed all records from Brazil available at PSMSL and noted a coherent RSL rise of about 2–4 mm $year^{-1}$ between 1950 and 1970; this was interpreted as land subsidence except for RSL observations at Recife, Belem, and Imbituba, where the trends are close to zero. The lower trends were suggested to result from land movement produced at Recife by the Pernambuco fault, and to sediment-induced subsidence at Imbituba and Belem. More recent work has revisited these long-term trends and estimated RSL trend in the range 3–5 mm $year^{-1}$ over the past 50 years (Neves and Muehe, 1995; Mesquita, 2003; Muehe, 2006; de Mesquita et al., 2013). We searched for new records in PSMSL to update the Emery and Aubrey results but have found only two recent series: One at Cananeia and another at Ilha Fiscal (Table 7.1). The Ilha Fiscal record exhibits no significant RSL trend over 1967–2013, the signal being dominated by strong multidecadal fluctuations. The presence of the multidecadal SL signal explains the low statistical confidence of the trend estimate at Ilha Fiscal noticed by Emery and Aubrey (1991). Our estimate of the RSL trend at Cananeia over a 50-year span (Table 7.1) is of 4.1 mm $year^{-1}$ (Table 7.1) that is surprisingly coherent with 4.2 mm $year^{-1}$ obtained by Emery and Aubrey over the first 30 years of apparently the same SL record. A different trend was found, however, by Ducarme et al. (2007) who estimated a larger RSL trend of 5.6 ± 0.07 mm $year^{-1}$ after having identified and corrected two periodicities of 24.2 and 10.7 year dominating the very low-frequency spectrum of SL at Cananeia. This rate, over the last 50 years, largely exceeds the observed GMSL trend from satellite altimetry over the last 22 years

Table 7.1 Atlantic—Eastern South America

Tide gauge from PSMSL

Country	ID	Name	LAT	LON	Date	Length (year)	RSL trend (mm year^{-1})	Error (mm year^{-1})
Brazil	726	Cananeia	−25.02	−47.93	1955–2004	50	4.1	0.7
Brazil	1032	Ilha Fiscal	−22.90	−43.17	1971–2013	43	⋆	⋆

GPS—ULR6 from SONEL

Country	ID	LAT	LON	Date	Length (year)	Tide gauge	Distance	Vertical velocity (mm year^{-1})	Error (mm year^{-1})
French Guiana	CAYN	4.95	−52.30	2005–2013	9	Cayenne	11 km	1.0	1.0
Brazil	MAPA	0.05	−51.10	2003–2013	11	Santana	14 km	1.0	0.4
Brazil	RECF	−8.05	−34.95	1999–2013	15	Recife	9 km	−2.4	0.3
Brazil	SAVO	−12.93	−38.43	2007–2013	7	Salvador	10 km	0.4	0.4
Brazil	SSA1	−12.98	−38.52	2007–2013	7	Salvador	150 m	−0.5	0.4
Brazil	SALV	−13.00	−38.51	1999–2008	10	Salvador	4 km	0.2	0.4
Brazil	NEIA	−25.02	−47.92	2002–2013	12	Cananeia	10 m	0.0	0.3
Brazil	IMBT	−28.23	−48.66	2007–2013	7	Imbituba	700 m	−1.1	0.4

Locations, time spans, and trends of RLR PSMSL tide gauges and SONEL GPS stations. Error corresponds to 95% margin of error for the linear trend. The symbol ⋆ corresponds to nonsignificant trend (P-value > 0.1). *GPS*, Global Positioning System; *PSMSL*, Permanent Service for Mean Sea Level; *RSL*, relative sea level; *SONEL*, Système d'Observation du Niveau des Eaux Littorales.

(3.1–3.3 mm year^{-1} (Cazenave and Le Cozannet, 2013)), and Cananeia should be classed as a strong positive anomaly, *a hotspot*, in the global SL rise pattern (de Mesquita et al., 2013). The reasons of the increased rate of RSL rise at Cananeia have not been completely explained yet, but they are unlikely due to land subsidence alone. (It is worth noting here that Aubrey et al., 1988; Muehe and Neves, 1995; de Mesquita et al., 2013 previously presented evidence that the Brazilian coast may be sinking.) An apparent contradiction comes from the NEIA GPS station collocated (Table 7.1) with the Cananeia TG (10 m distance from the TG and 15 years in operation). Indeed, no significant trend in vertical movement was detected by the NEIA GPS at the Cananeia TG over the last 15 years (Table 7.1). Yet, it does not exclude an increased RSL rise at Cananeia, as it can also be due to oceanic processes. Notice, however, that the ASL trends near Cananeia vary between 1.8 and 3 mm year^{-1} over 1993–2014 similarly to the SL trends of about 2.5 mm year^{-1} reconstructed near the Brazilian coast over 1960–2014 (Fig. 7.4).

The other two GPS stations on the Brazilian coast are collocated with Recife and Imbituba TG, the two sites likely influenced by local land movement (Emery and Aubrey, 1991). Land subsidence of 2.4 mm year^{-1} is observed at RECF GPS, at 9 km from the Recife TG, and slower subsidence of 1.1 mm year^{-1} at IMBT GPS, 700 m from Imbituba TG. Unfortunately, the lack of modern TG records at Recife and Imbituba does not allow separating the contribution of land movement from oceanic component in the observed RSL. The lack of data and insufficient density of the TG network is a major obstacle for accurate evaluation of regional SL changes in this region. Since 2007, efforts are being made to implement a Permanent Brazilian Sea Level Monitoring Network called Global Sea Level Observing System (GLOSS) Brazil Network. Under this program, 12 new TG stations have been installed and are now fully operational (data available on http://www.goosbrasil.org/gloss). This gives hope for obtaining more precise and accurate long-term SL measurements along the coast of Brazil (Lemos and Ghisolfi, 2011).

3.2 Caribbean Sea

The Caribbean Sea is bounded in the west by Central America and, in the south, by Venezuela and Colombia. It is connected to the Gulf of Mexico through the Yucatan Straits in the north. Cuba, the Greater Antilles and the Lesser Antilles, separate the Caribbean Sea from the Atlantic Ocean to the north and northeast. The Caribbean Sea includes more than 7000 islands that are particularly vulnerable to SL rise because of high population density. Indeed, about half of the island population lives within 1.5 km from the sea (Mimura et al., 2007) because of its dependence on coastal and sea resources (Nicholls and Cazenave, 2010). Dasgupta et al. (2009) identified, among 84 coastal developing countries, the Bahamas as one of the five most impacted countries of a 1-m SL rise. In terms of potential land loss, Belize, Puerto Rico, Cuba, and Jamaica are ranked in the top 10 in the SL vulnerability classification (from 1% to 2% of loss, Dasgupta et al., 2009). Similarly, Jamaica and Belize are among the top 5 in the classification of the largest wetland loss triggered by SL rise (~30% of loss, Dasgupta et al., 2009). Moreover, the unique biodiversity of the Caribbean Sea islands (Mittermeier et al., 2011) appears to be particularly threatened by the projected SL rise. With a 1 m of SL increase, ~9% of the islands (i.e., 63 islands among the 723 identified as biodiversity hotspot by Bellard et al., 2014) are expected to be entirely submerged, and the worst-case scenario of a 6-m increase would lead to a loss of half of the islands (i.e., 356 islands).

Updating Palanisamy's et al. (2012) work, over the 1960–2014 reconstruction period, we observed strong positive ASL trends in the Caribbean of about 2.5–3 mm year^{-1} (Fig. 7.4), except for Cuba, the Lesser and Greater Antilles where the ASL trends are lower, at around 1.8–2.5 mm year^{-1} (Fig. 7.4). The RLR TG records from the PSMSL data set corroborate these findings. Only seven SL records span more than 30 years (Table 7.2). Two stations are located on the continent: Cartagena (1949–92, 44 year) in Colombia shows an RSL trend of 5.2 mm year^{-1}, and Cristobal (1909–79, 71 year) in Panama shows an RSL trend of 1.5 mm year^{-1}. The former trend is the fastest of the long-term Caribbean SL observations (Fig. 7.2) that places Cartagena among the cities directly threatened by rising SL. The RSL measurements along the Antilles chain reveal trends of (1) about 3 mm year^{-1} in the Virgin Islands; and (2) about 2 mm year^{-1} in Puerto Rico and Cuba. Over the 1993–2014 altimetry period, we observe strong positive ASL trends from Nicaragua, southward through Venezuela to the Lesser Antilles, in the range of 3–5 mm year^{-1} (Fig. 7.3), which is greater than the GMSL trend over the same period. In the eastern part of the Caribbean Sea, the ASL trends are smaller; they range from 1.8 to 3 mm year^{-1} (Fig. 7.3) along the Greater Antilles islands, in particular along the coasts of Cuba, Jamaica, Haiti, and Puerto Rico. It is worth noting here that the seismically active Lesser Antilles subduction zone is a potential source of tsunami-induced flooding all along the Caribbean coasts (McCann, 2006).

Table 7.2 Atlantic—Caribbean Sea

			Tide gauge from PSMSL					
Country	**ID**	**Name**	**LAT**	**LON**	**Date**	**Length (year)**	**RSL trend (mm year^{-1})**	**Error (mm year^{-1})**
Colombia	572	Cartagena	10.40	−75.55	1949–1992	44	5.2	0.5
Panama	169	Cristobal	9.35	−79.92	1909–1979	71	1.4	0.3
Virgin Is. US	1447	Lime Tree Bay	17.69	−64.75	1984–2015	32	3.1	1.0
Virgin Is. US	1393	Charlotte Amalie	18.34	−64.92	1985–2015	31	3.3	1.2
Puerto Rico	1001	San Juan	18.46	−66.12	1963–2015	53	2.1	0.5
Puerto Rico	759	Magueyes Island	17.97	−67.05	1955–2015	61	1.7	0.4
Cuba	418	Guantanamo Bay	19.91	−75.15	1938–1971	34	1.8	0.8
Cuba	563	Gibara	21.11	−76.13	1976–2014	39	2.0	1.0

Continued

Table 7.2 Atlantic—Caribbean Sea—cont'd

GPS—ULR6 from SONEL

Country	ID	LAT	LON	Date	Length (year)	Tide gauge	Distance	Vertical velocity (mm year^{-1})	Error (mm year^{-1})
Mexico	UNPM	20.87	−86.87	2007–2013	7	Puerto Morelos	146 m	−1.9	0.4
Colombia	CART	10.39	−75.53	2000–2008	9	Cartagena	2 km	−2.2	0.5
Cayman Is.	GCGT	19.29	−81.38	2005–2011	7	South Sound	7 km	−1.4	0.2
Puerto Rico	BYSP	18.40	−66.16	2008–2013	6	San Juan	9 km	−1.3	0.8
Puerto Rico	PRMI	17.97	−67.05	2006–2013	8	Magueyes	500 m	−0.4	0.2
Puerto Rico	MAYZ	18.22	−67.16	2010–2013	4	Mayaguez	2 m	⋆	⋆
Puerto Rico	ZSU1	18.43	−65.99	2003–2011	9	San Juan	10 km	−1.1	0.3
Virgin Is. US	STVI	18.34	−64.97	2008–2013	6	Charlotte	6 km	−1.6	0.5
Virgin Is. US	VITH	18.33	−64.92	2006–2013	8	Charlotte	5 km	−1.3	0.3
Virgin Is. US	VIKH	17.71	−64.80	2006–2013	8	Lime tree bay	4 km	−2.9	0.3
Virgin Is. US	CR01	17.76	−64.58	1994–2013	20	Christiansted	13 km	−1.1	0.4
French West Indies	ABMF	16.26	−61.52	2008–2013	6	Pointe-à-Pitre	4 km	⋆	⋆
French West Indies	LMMF	14.59	−60.99	2008–2013	6	Fort de France	7 km	−3.6	0.5
Barbados	BDOS	13.09	−59.61	2004–2013	10	Bridgetown	3 km	0.4	0.5

Locations, time spans, and trends of RLR PSMSL tide gauges and SONEL GPS stations. Error corresponds to 95% margin of error for the linear trend. The symbol ⋆ corresponds to nonsignificant trend (P-value > 0.1). *GPS*, Global Positioning System; *PSMSL*, Permanent Service for Mean Sea Level; *RLR*, Revised Local Reference; *RSL*, relative sea level; *SONEL*, Système d'Observation du Niveau des Eaux Littorales.

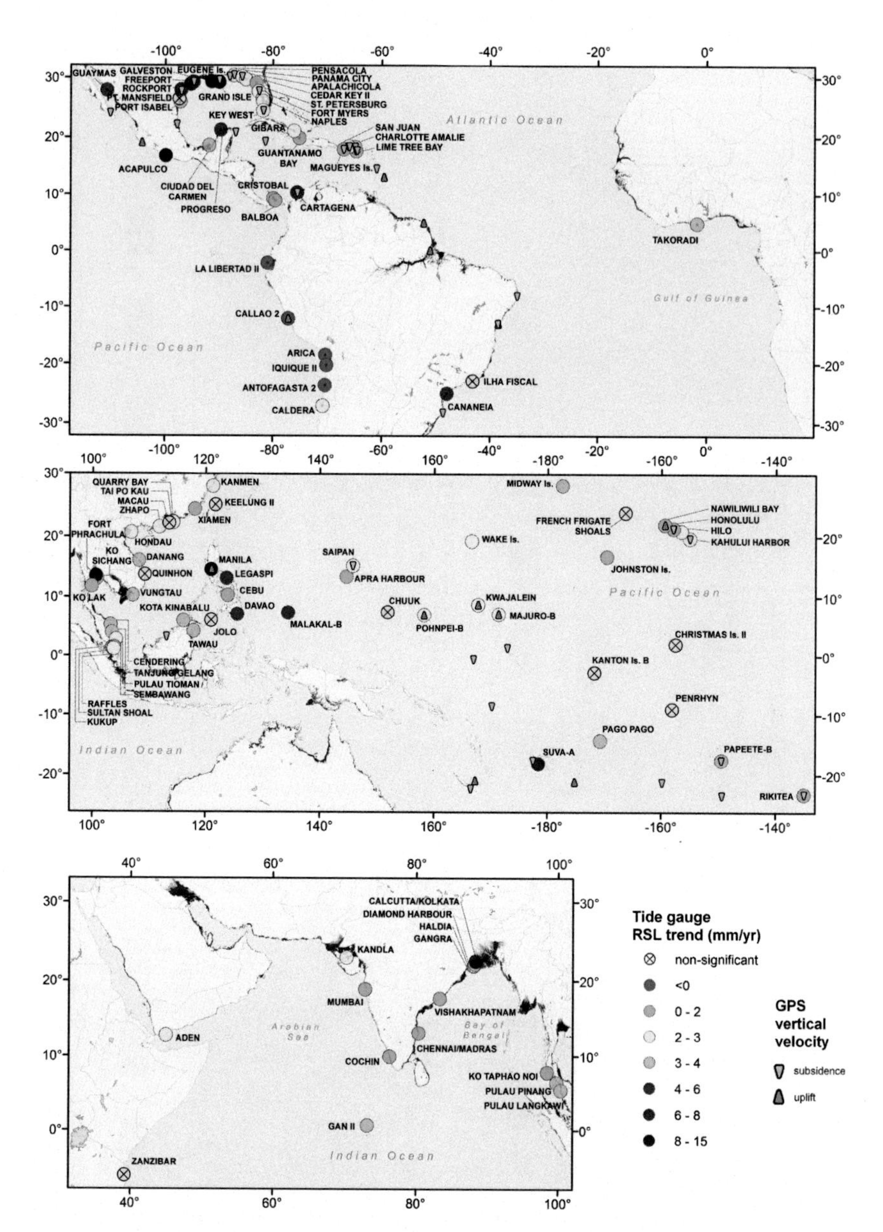

Figure 7.2 Geographic distribution of the tide gauge records, and their linear trends (mm year^{-1}), available from the RLR PSMSL data set and GPS stations from ULR6 SONEL database. *PSMSL*, Permanent Service for Mean Sea Level; *RLR*, Revised Local Reference; *SONEL*, Système d'Observation du Niveau des Eaux Littorales.

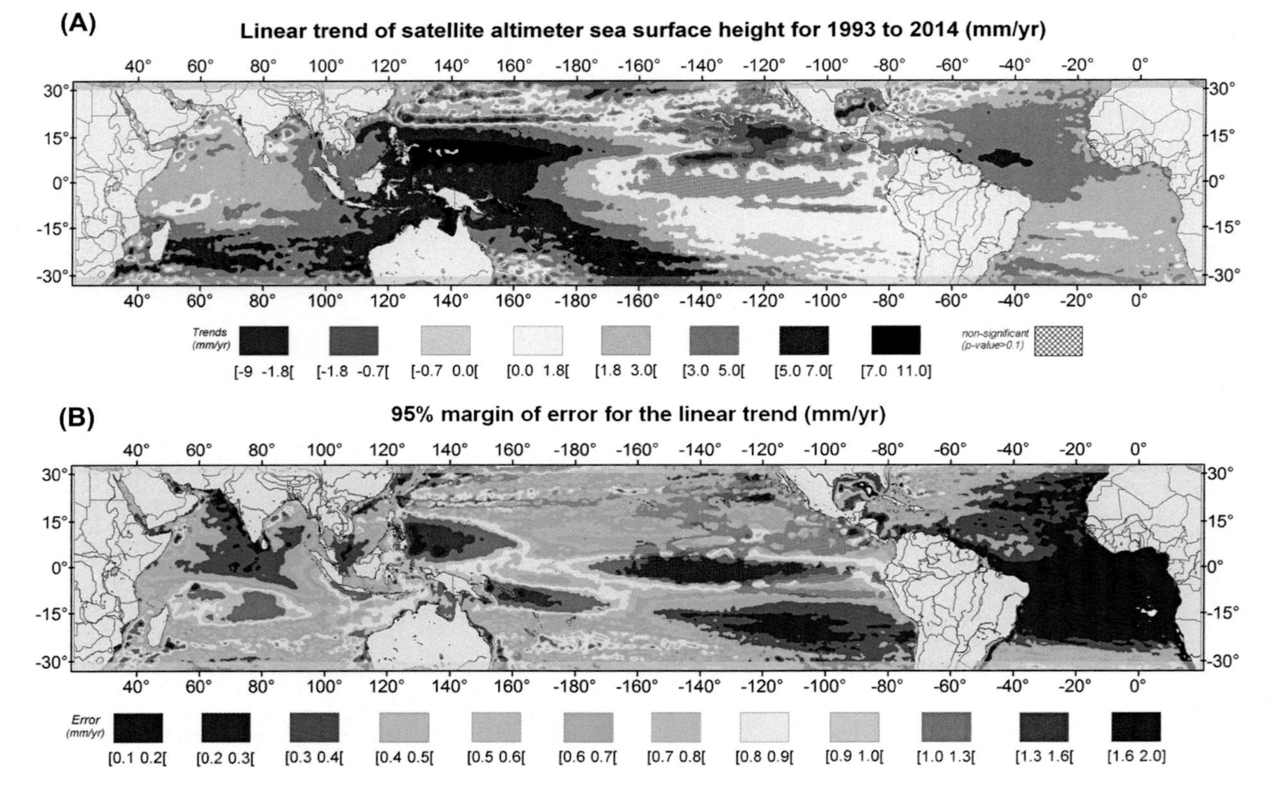

Figure 7.3 (A) Geographic distribution of sea surface height linear trends (mm year^{-1}) for 1993–2014 based on satellite altimetry. *Shaded area* represents nonsignificant trends (*P*-value > 0.1). (B) 95% margin of error for the linear regression equation (mm year^{-1}).

GPS stations (Table 7.2) are concentrated in the US Virgin Islands with the only station available on the continental coast, at Cartagena. All GPS records span about 9 years, except the station in Lime Tree Bay where the record is for 21 years. Analysis of the GPS data at Cartagena reveals a trend of 1.7 mm $year^{-1}$ that is most likely due to land movement along the fault (Emery and Aubrey, 1991). In this case, the ocean contribution to the 5.2 mm $year^{-1}$ RSL trend at Cartagena would be about 3.5 mm $year^{-1}$. It is difficult to answer whether the 9-year long GPS series are long enough for estimating land movement in these regions. To get an insight into this issue, we compared the trends derived from two GPS stations in Lime Tree Bay. The 21-year long CR01 GPS station shows a trend of 2.9 mm $year^{-1}$, whereas another one, which has a 9-year long record, has a much smaller trend of 1.1 mm $year^{-1}$. This points out to a possible nonstationary character or significant spatial variation in vertical land movements in Lime Tree Bay.

Along the Bahamas Islands, the ASL trends range from 0 to 3 mm $year^{-1}$, but some trends in this region are statistically insignificant, and some have high uncertainty (Fig. 7.3) because of pronounced interannual SL variability in the Caribbean region. Torres and Tsimplis (2013) show that the interannual fluctuations in this region can be partly explained by the influence of El Niño–Southern Oscillation (ENSO) at different time and space scales; however, they found no significant link with the North Atlantic Oscillation.

3.3 Gulf of Mexico

Along the US Gulf Coast, the population grew up 150% and housing construction by 246% from 1960 to 2008 (Wilson and Fischetti, 2010). In August 2005, hurricane Katrina (followed by hurricane Rita a few days later), resulted in the largest natural disaster in US history and devastated human and economic landscape along the US Gulf Coast. This disaster brought to the forefront a problem, recognized for decades, of adaptation to the Mississippi Delta sinking, which results in extensive wetland loss and increases the exposure of population, economic activities, and infrastructure to hurricane-induced storm surges (Syvitski et al., 2009). Dai et al. (2009), has shown that during the 20th century ~25% of the Mississippi wetlands were lost to the ocean. The largest factor contributing to the wetlands loss is the construction of artificial levees, reducing the number of sediment pathways into adjacent flood plain basins (Kesel, 2003). This land loss problem is exacerbated by trapping of 50% of the total sediment load by upstream dams, and there is not enough supply to keep pace with subsidence and accelerated SL rise (Blum and Roberts, 2012). Assuming an acceleration of

SL rise from 3 to 4 mm $year^{-1}$ and a subsidence rate from 1 to 1.7 mm $year^{-1}$, coupled with the absence of sediment input, Blum and Roberts (2009) projected a potential submergence of 25%–30% of the delta (~10,000–13,500 km^2) by the year 2100. Blum and Roberts (2012) concluded that significant drowning is inevitable, even if sediment loads are restored, because the SL is now rising at least three times faster than during the period of the delta plain formation. Moreover, anthropogenic effects, including locally accelerated subsidence can exacerbate this problem. Becker et al. (2014) estimated that 68% (~4 mm $year^{-1}$) of the SL rise recorded at Galveston over the last century is too large to be due to natural SL variability and, by consequence, should be dominated by land subsidence probably induced by extraction of subsurface fluids, hydrocarbons, and groundwater withdrawal (Morton et al., 2006; Kolker et al., 2011).

Kolker et al. (2011) used the Grand Isle, Galveston, and Pensacola TG records from the RLR PSMSL data set to investigate the subsidence rate in the northern part of US Gulf over 1947–2006. They assumed that the Pensacola record, located on a stable carbonate platform, experiences a linear land movement and therefore subtracted the Pensacola from the others records to remove interannual variability. In doing so, they underlined three distinct significant subsidence phases (1) 1947–58: 3.1 and 2.6 mm $year^{-1}$; (2) 1959–91: 9.8 and 6.2 mm $year^{-1}$; and (3) 1992–96: 1 and −2 mm $year^{-1}$ at Grand Isle and Galveston, respectively. They argued that the recent subsidence rates are lower than predictions of the subsidence scenario suggested by Blum and Roberts (2009) and, perhaps, future land losses linked to the subsidence will be limited. However, in updating Kolker et al. (2011) work we obtained a subsidence rate of ~3 mm $year^{-1}$ at Grand Isle over 1992–2015 (and ~0.8 mm $year^{-1}$ at Galveston, Table 7.3). This result is closer to the estimation of Morton et al. (2006) who reported a subsidence rate of ~4 mm $year^{-1}$ over 1993–2006. Moreover, our estimates agree with recent work by Letetrel et al. (2015) who combined satellite altimetry data and the long-term Grand Isle TG record and estimated a subsidence rate of ~5 mm $year^{-1}$ over 1992–2008 and ~7 mm $year^{-1}$ over 1947–2011. These values are close to the GPS-derived vertical velocity of ~−6.5 mm $year^{-1}$ over 2005–16, estimated from GRIS GPS station, 100 m from the Grand Isle TG.

The observed subsidence results from a combination of different processes such as tectonics, sedimentation, glacial isotactic adjustment, and anthropogenic fluid withdrawal (Douglas, 2001). Various studies estimated present-day subsidence rates in the range of 2–10 mm $year^{-1}$, as a response to the delta sedimentary load (Jurkowski et al., 1984; Ivins et al., 2007;

Table 7.3 Atlantic—Gulf of Mexico

Tide gauge from PSMSL								
Country	**ID**	**Name**	**LAT**	**LON**	**Date**	**Length (year)**	**RSL trend (mm year^{-1})**	**Error (mm year^{-1})**
Mexico	690	Progreso	21.30	−89.67	1952–1984	33	5.2	1.0
Mexico	796	Ciudad del Carmen	18.63	−91.85	1957–1987	31	3.6	1.2
USA	497	Port Isabel	26.06	−97.22	1945–2015	71	3.9	0.4
USA	1038	Port Mansfield	26.55	−97.42	1964–1995	32	★	★
USA	538	Rockport	28.02	−97.05	1964–2014	51	6.1	0.8
USA	725	Freeport	28.95	−95.31	1955–2007	53	8.8	1.2
USA	828	Galveston I	29.29	−94.79	1958–2010	53	6.7	0.8
USA	161	Galveston II	29.31	−94.79	1909–2015	107	6.4	0.3
USA	440	Eugene Island	29.37	−91.39	1940–1974	35	9.7	1.5
USA	526	Grand Isle	29.26	−89.96	1947–2015	69	9.0	0.5
USA	246	Pensacola	30.40	−87.21	1924–2015	92	2.2	0.3
USA	1641	Panama City	30.15	−85.67	1985–2015	31	3.0	1.1
USA	1193	Apalachicola	29.73	−84.98	1968–2015	48	2.1	0.8
USA	428	Cedar Key II	29.14	−83.03	1939–2015	77	1.8	0.3
USA	520	St. Petersburg	27.76	−82.63	1947–2015	69	2.7	0.3
USA	1106	Fort Myers	26.65	−81.87	1966–2015	50	2.9	0.6
USA	1107	Naples	26.13	−81.81	1966–2015	50	2.5	0.6
USA	188	Key West	24.56	−81.81	1913–2015	103	2.4	0.2

Continued

Table 7.3 Atlantic—Gulf of Mexico—cont'd

GPS—ULR6 from SONEL

Country	ID	LAT	LON	Date	Length (year)	Tide gauge	Distance	Vertical velocity (mm year^{-1})	Error (mm year^{-1})
Mexico	TAMP	22.23	−97.86	2007–2013	7	Ciudad Madero	7 km	−0.8	0.4
USA	ARP3	27.83	−97.06	1995–2006	12	Port Aransas	1 km	−1.3	0.4
USA	TXGA	29.33	−94.77	2005–2013	9	Galveston	3 km	−3.4	0.8
USA	GAL1	29.33	−94.74	1995–2003	9	Galveston	6 km	−4.6	0.8
USA	GRIS	29.62	−89.96	2004–2013	10	Grand Isle	100 m	−6.5	0.5
USA	MOB1	30.23	−88.02	1996–2009	14	Dauphin Is.	6 km	−3.1	0.4
USA	PCLA	30.47	−87.19	2004–2013	10	Pensacola	8 km	−0.4	0.4
USA	PNCY	30.20	−85.68	2001–2010	10	Panama City	6 km	−0.2	0.4
USA	MCD5	27.85	−82.53	2007–2013	7	St Petersburg	14 km	−1.6	0.4
USA	MCD1	27.85	−82.53	2001–2007	7	St Petersburg	13 km	−0.3	0.7
USA	KWST	24.55	−81.75	2002–2013	12	Key West	5 km	−1.1	0.4
USA	CHIN	24.55	−81.81	2007–2013	7	Key West	400 m	⋆	⋆

Locations, time spans and trends of RLR PSMSL tide gauges and SONEL GPS stations. Error corresponds to 95% margin of error for the linear trend. The symbol ⋆ corresponds to nonsignificant trend (P-value > 0.1). *GPS*, Global Positioning System; *PSMSL*, Permanent Service for Mean Sea Level; *RSL*, relative sea level; *SONEL*, Système d'Observation du Niveau des Eaux Littorales.

Syvitski, 2008; Törnqvist et al., 2008). However, Wolstencroft et al. (2014) argued that the viscoelastic deformation due to sediment loading alone is unlikely to exceed ~0.5 mm $year^{-1}$. Thus, the current high rates of observed subsidence are likely to be linked to sediment compaction and fluid extraction.

In the RLR PSMSL data set, we found 20 TG stations, with time spans of more than 30 years, distributed along the coast of the Gulf of Mexico. Seventeen stations are located in the United States, two in Mexico, and one in Cuba. The long-term RSL trends are gathered in three clusters. The first one represents the western coast from Progreso to Rockport (4 TGs) where the SL rises at 3–5 mm $year^{-1}$. The fastest RSL rise is observed at 6 TG situated along the northern coast from Freeport to Grand Isle in the Mississippi Delta. Conjugation of land subsidence with rising ASL results in an RSL of 6–10 mm $year^{-1}$. The third cluster contains moderate RSL trends of 2–4 mm $year^{-1}$ observed at 10 TGs in the eastern Gulf, from Dauphin Island to Key West. As in other regions, the main driver of the enhanced SL rise in the Gulf of Mexico is in the deltaic region and is due to land subsidence.

3.4 Atlantic Eastern Border: Gulf of Guinea

The Gulf of Guinea, located in the eastern equatorial Atlantic, is constituted of 18 coastal States from Senegal to Angola. Its 12,000 km-long coastline is characterized by typical low-lying topography, coastal lagoons, and by two large deltas: the Niger delta and the Volta River delta (Fig. 7.1).

This coastline hosts 12 townships, each with a population of over 1 million, which is highly vulnerable to the impacts of climate change (UN-HABITAT, 2014). Moriconi-Ebrard et al. (2016) highlight the formation of an urban band of high population density by 2020, in the coastal area of the Gulf of Guinea. Yet, this region is already extremely vulnerable to projected SL rise impacts (erosion, submersion, saline intrusion into coastal aquifers and agricultural areas, fisheries, mangrove degradation) (Nicholls and Mimura, 1998).

Jallow et al. (1999) estimated, by modeling the effects of coastal erosion and a rise in SL, that Banjul, the capital of the Gambia, can disappear by 2050. Dasgupta et al. (2009) ranked Benin in the top ten of 84 developing coastal countries worldwide, which would be most impacted by a 1-m SL rise in terms of population to be displaced (4%) and wetland area loss (14%) and Gambia in terms of land area loss (1%). According to Brown et al. (2011), Cameroon ranks in the top ten African countries likely to be impacted by flooding and forced migration by 2100. Hinkel et al. (2012) concluded that Nigeria is one of the most vulnerable African countries both in terms of the people-based SL impacts and in terms of economic costs. Some 25 million people are estimated to live currently within its coastal zones, with about 8.5 million beneath the 2-m inundation contour (French et al., 1995). The largest city, Lagos, is expanding rapidly across the land standing below a meter above SL. As much as 70% of the city's population live in slums characterized by extremely poor environmental conditions, including regular flooding of homes that lasts several hours and that sweeps raw sewage (Adelekan, 2009). In the Niger delta region, even in absence of acceleration in ASL rise, the land loss through edge erosion alone can cause shoreline recession of 3 km by the year 2100 (French et al., 1995). Moreover, the Niger delta is sinking much faster than global SL is rising (Syvitski et al., 2009). The high subsidence rate (25–125 mm year^{-1} (Abam, 2001)), due to oil and gas extraction, combined with a reduction in sediment deposition plus accelerated compaction of sediment, makes this delta along with the Nile River delta the most threatened of the African deltas (Syvitski et al., 2009).

Relatively little research on long-term SL change has been undertaken previously over the African continent because the existing African data set is shorter than that in other parts of the world (Emery and Aubrey, 1991; Woodworth et al., 2007). The lack of historical data on SL rise in Africa makes it difficult to assess coastal impacts and vulnerability with accuracy.

Woodworth et al. (2007) reviewed the African SL changes by using the PSMSL data set. In the Gulf of Guinea, some records exist but with less than 20 years of data available and no recent data. In the RLR PSMSL data bank, only two TG records from this region have relatively recent data but with substantial missing or inconsistent data: Dakar 2 (1992–2014, 73% of completeness, Senegal) and Takoradi (1929–2012, 79% of completeness, Ghana). In conclusion, along of the Gulf of Guinea coastline, only the Takoradi TG record, with reliable datum continuity, can be used to estimate a long-term RSL trend over 36 years (1930–65), which is ~3 mm year^{-1} (Woodworth et al., 2007).

In this context of lack of data, Wöppelmann et al. (2008) have initiated investigations at Dakar (Senegal) to find and rescue past SL records. Several decades of SL observations at Dakar have been found, the earliest dating back to 1889. The secular RSL trend estimated from this long reconstructed TG is 1.6 ± 0.2 mm year^{-1} from 1900 to 2011. Using satellite synthetic aperture radar interferometry (SAR), Le Cozannet et al. (2015) showed that despite a complex geology, a rapid population growth and development in Dakar, the historical TG does not seem to be affected by local vertical coastal land motion, and therefore can be a good candidate for SL studies in the Gulf of Guinea and for past SL reconstruction. The rate of ASL rise along the coast of the Gulf of Guinea is in the range 1.8–3 mm year^{-1} (Figs. 7.3 and 7.4) during the shorter (1993–2014) and longer periods (1960–2014).

Because of the lack of TGs, it is difficult to assess all the causes of SL variations along the West African coast. Melet et al. (2016) determined the processes responsible for coastal SL variability in the Gulf of Guinea over the 1993–2012 period. They showed that in Cotonou (Benin), the SL trend is largely dominated by the same ocean signal as observed in the altimetric data and, to a lesser extent, by interannual variability of the wave run-up height.

In the late 1990s, the Ocean Data and Information Network for Africa (www.odinafrica.org) project was initiated to develop an African SL observing network as part of the GLOSS Core Network and rescue historical SL data. Today, this project brings together more than 40 marine-related institutions from 25 African countries to address the challenges of accessing data and information for coastal management.

3.5 Tropical Atlantic Relative Sea Level Hotspots: Summary

- ***Guyana, Suriname, and French Guiana*** are in the world top 10 countries mostly impacted by a 1-m SL rise (Dasgupta et al., 2009). About 6% of people in these regions would be displaced, leading to high probability of climate-induced massive population displacements.
- ***Brazilian coast:*** The enhanced SL rise at Cananeia makes it a SL hotspot. Is the RSL trend at Cananeia a local anomaly or should it be seen as a typical value along the Brazilian coast? It is difficult to answer this question now, as the number of long-term TGs is insufficient to resolve RSL trends variations along the western South America coast.
- ***Cartagena in Colombia:*** With an RSL trend faster than 5 mm year^{-1} and 1 million inhabitants, this is a site of great concern. The problem is complicated by the fact that contribution of land movement to the observed RSL is not yet reliably established. Consequently, any projections of future RSL changes should be assessed with due care.
- ***Northeastern coast of the Gulf of Mexico:*** The region within and around Mississippi Delta is experiencing the fastest RSL rise measured by TGs in the tropical Atlantic. Land movement, due to sedimentation processes and water/oil/gas withdrawals, drives the long-term RSL changes in this region.

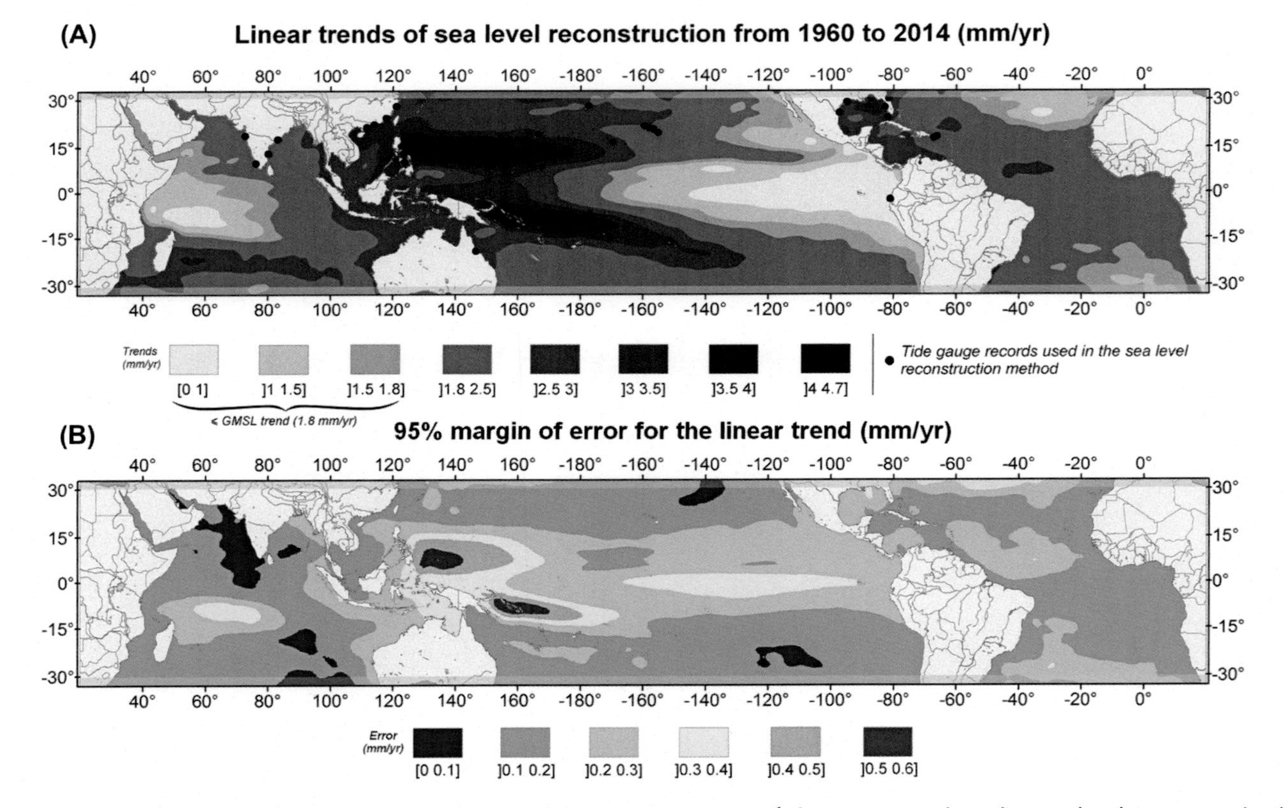

Figure 7.4 (A) Geographic distribution of sea surface height linear trends (mm year^{-1}) for 1960– 2014 based on sea level reconstruction in the past. *Shaded area* represents nonsignificant trends (*P*-value > 0.1). *Black dots* represent the tide gauge records used in the reconstruction method. (B) 95% margin of error for the linear regression equation (mm year^{-1}).

- ***Niger delta*** is sinking much faster than GMSL is rising (Syvitski et al., 2009). The high subsidence rate (25–125 mm year^{-1}), due to oil and gas extraction, combined with reduction in aggradation plus accelerated compaction of sediment, makes this delta, along with the Nile River delta, the most threatened among the African deltas (Syvitski et al., 2009).

4. PACIFIC OCEAN

4.1 Central America and South America

The Pacific coast of South America is a tectonically active zone driven by subduction of the Pacific plate. Little information about the long-term RSL trends along this coast is available, except from the earlier analysis by Aubrey et al. (1988) and Emery and Aubrey (1991) who have reported highly variable SL trends with changing signs all along the coast of Chile and Peru. These trend variations were attributed to nonuniform tectonism, faulting and segmentation of subducting lithosphere. Inspecting the updated PSMSL RLR data set, we found nine TG records spanning more than 30 years from Mexico to Chile. An interesting result is that, along the west coast of South America, five of six long-term stations (Fig. 7.2 and Table 7.4) reveal a decreasing RSL with a trend of about −1 mm year^{-1}. This value indicates a coastal uplift at a rate of ~2 mm year^{-1} provided that we take ~1 mm year^{-1} as a trend of ASL rise along the coast of Chile–Peru from altimetry (Fig. 7.3). Obviously, this evaluation should be taken with care because trend uncertainties are quite large (Table 7.4 and Fig. 7.3). Notice, nevertheless, that a 2 mm year^{-1} land emergence was detected by GPS at Callao, although this value was estimated from 5-year long measurements (Table 7.4). As to the southern most tropical Chilean TG Caldera, it manifests a positive RSL trend of about 2.8 mm year^{-1}, which is larger than 1.7 mm year^{-1} estimated by Emery and Aubrey (1991) from a shorter period. The noticeable difference between our estimates and those of Emery and Aubrey (1991) might result from significantly longer time series used in our analysis. The long-term series are necessary to separate the trend from interannual and, especially, decadal SL fluctuations that are particularly strong in this region. These low-frequency SL variations are driven by El Niño and have been extensively investigated since 1960s (Roden, 1963; Wyrtki, 1973, 1975; Mitchum and Wyrtki, 1988; Enfield, 1989; Clarke, 2014). Recently, Losada et al. (2013) estimated that ENSO explains more than 65% of the mean SL variance along the Peruvian coast. According to Reguero et al. (2015), the number of inhabitants affected by El Niño events, in addition to future SL rise,

Table 7.4 Pacific—Central America and South America

Tide gauge from PSMSL

Country	ID	Name	LAT	LON	Date	Length (year)	RSL trend (mm year^{-1})	Error (mm year^{-1})
Mexico	693	Guaymas	27.92	−110.90	1952–1989	38	4.4	1.4
Mexico	686	Acapulco	16.83	−99.92	1967–2000	34	8.4	3.0
Panama	163	Balboa	8.97	−79.57	1908–2015	108	1.5	0.2
Ecuador	544	La Libertad II	−2.20	−80.92	1950–2002	53	−1.3	1.0
Peru	1274	Callao 2	−12.05	−77.15	1970–2014	45	−0.3	1.2
Chile	618	Arica	−18.47	−70.33	1952–1991	40	−0.7	1.5
Chile	2261	Iquique II	−20.20	−70.15	1986–2015	30	−1.1	1.8
Chile	510	Antofagasta 2	−23.65	−70.40	1946–2015	70	−0.8	0.5
Chile	619	Caldera	−27.07	−70.83	1951–1991	41	2.8	0.9

GPS—ULR6 from SONEL

Country	ID	LAT	LON	Date	Length (year)	Tide gauge	Distance	Vertical velocity (mm year^{-1})	Error (mm year^{-1})
Mexico	SLCR	16.17	−95.20	2008–2012	5	Salina Cruz	1 m	⋆	⋆
Mexico	LPAZ	24.14	−110.32	2006–2012	7	La Paz	3 km	−1.1	0.3
Mexico	UCOM	19.12	−104.40	2007–2012	6	Manzanillo	12 km	0.7	0.6
Mexico	ACYA	16.84	−99.90	2004–2012	9	Acapulco	1 m	⋆	⋆
Peru	CALL	−12.06	−77.15	2009–2013	5	Callao	1 km	2.0	0.6

Locations, time spans, and trends of RLR PSMSL tide gauges and SONEL GPS stations. Error corresponds to 95% margin of error for the linear trend. The symbol ⋆ corresponds to nonsignificant trend (P-value > 0.1). *GPS*, Global Positioning System; *PSMSL*, Permanent Service for Mean Sea Level; *RSL*, relative sea level; *SONEL*, Système d'Observation du Niveau des Eaux Littorales.

will be substantial not only in Peru and Ecuador but also in Panama, El Salvador, Costa Rica, and Guatemala impacting more than 30% of population in these countries. Hallegatte et al. (2011), in a global study of losses due to future floods in coastal cities, identified Guayaquil, the largest and the most populated city in Ecuador, to be at particularly high risk.

Farther northward, in Central America, the century-scale Balboa record, the longest on the American tropical coast, shows an RSL trend of about 1.5 mm $year^{-1}$ that is comparable to the ASL trend measured by altimetry (Fig. 7.3). The two available 30-year long Mexican TGs have large, statistically significant, RSL trends. Acapulco, with more than 700,000 inhabitants, faces SL rising at a rate of 8.4 mm $year^{-1}$ that places this city as an RSL hotspot: the RSL is rising here at the rate among the fastest measured worldwide. A smaller (4.4 mm $year^{-1}$) but yet appreciable RSL trend was estimated at Guaymas, a low-lying city in northwestern Mexico. Along the Mexican South Pacific coast, the altimetry data set has nonsignificant ASL trend over the last 22 years. Buenfil-López et al. (2012) showed that the RSL in this region is affected by seismic activity that can generate instantaneous fall in SL. The GPS stations in Mexico have not yet provided reliable long-term estimates and we cannot reliably evaluate the land movement contribution to the observed RSL rise at Acapulco and Guaymas.

4.2 Southeast Asia

Approximately 20% (~134 million) of the world's population living in a contiguous area along the coast, within less than 10 m above SL, can be found in seven Southeast Asian countries: Vietnam, Cambodia, Thailand, Indonesia, Philippines, Malaysia, and Singapore (LECZ database, Fig. 7.1). The first four are among the top 10 countries in the world with the highest number of people living within less than 10 m above SL (McGranahan et al., 2007). Most of the megacities in this region are located either in coastal areas or within a large delta, with rich alluvial soils used for agriculture and aquaculture. A series of rapidly developing megacities is located in large deltas, such as Bangkok (~6 million inhabitants), the capital of Thailand in the Chao Phraya River delta (Fig. 7.1) and Ho Chi Minh city (Vietnam), of ~8 million inhabitants situated in the Mekong River delta (Fig. 7.1). The natural resources in this region will also be profoundly impacted by RSL. Thus, concerning the mangrove forest persistence in Indo-Pacific region, Lovelock et al. (2015) projected that some sites subject to SL rise, with low tidal range and low sediment supply, could be submerged by 2070s. This is the case in Chao Phraya and Mekong deltas, where vulnerability to SL rise is exacerbated by anthropogenic activities, as groundwater extraction and dam construction (Lovelock et al., 2015). In southern China, the Pearl River Delta, one of the most populated areas in the Chinese mainland (Wolanski, 2006), is home to several megacities (~8 million inhabitants each) as Shenzhen, Guangzhou, and Hong Kong. Hanson et al. (2011) evaluated the exposure of the population of the world's large cities to coastal flooding hazard by 2070s and concluded that only 12 countries contain 90% of the total of 148 million people exposed. (China [21%], Vietnam [9%], Thailand [3%], and Indonesia [2%] are among the top 10 countries.) They also pointed that the exposure in 2070s varies disproportionately in deltas among the top 10 cities: Guangzhou (~10 million people exposed), Ho Chi Minh City (~9 million), Bangkok (~5 million), and Hai Phòng (~5 million, Vietnam).

We updated analyses of TG records from eastern Asia previously performed by Emery and Aubrey (1986, 1991) and Yanagi and Akaki (1994). The SL records selected from the PSMSL database are the RLR series spanning at least 30 years, except for two stations Kota Kinabalu and Tawau (28 years), which are the only available data from Borneo Island (Table 7.5). We investigated for significant RSL trends in this region, from Vietnam to South China. In Vietnam, we found three significant RSL trends: in the north at Hondau, 2 mm year^{-1}, and in the south at Danang and Vungtau, on an average, 3.4 mm year^{-1}. There are no TGs from Cambodia available at PSMSL; the same is the case with the Mekong delta, though more than 20% of the national population lives in this area, which is also a vital agricultural zone. Fujihara et al. (2015) analyzed water level trends from 24 river gauge stations (over 1987–2006) managed by the Mekong River Commission. These stations located in the delta, and influenced by both inflow from upstream and tidal action from the South China Sea and the Gulf of Thailand, can also deliver relevant information about the RSL. Fujihara et al. (2015) estimated an RSL trend of ~7.4 mm year^{-1} over 1987–2006 in the Mekong delta, attributing 20% of this trend to ASL rise and 80% to land subsidence. Erban et al. (2014), using interferometric synthetic aperture radar (InSAR), estimated a rate of land subsidence, mainly due to groundwater pumping, throughout the Mekong delta in the range 10–40 mm year^{-1} during 2006–10. Their projection is that, if pumping continues at this rate, a land subsidence of ~0.9 m (0.35–1.4 m) is to be expected by 2050.

There are three long-term TG records available from Thailand. In the cities of Ko Sichang and Ko Lak, we estimated an RSL rate of 0.8 ± 0.5 mm year^{-1}. The Fort Phrachula TG is located at the coast of the Chao Phraya delta, just south of Bangkok, and it has an RSL trend of ~15 mm year^{-1}. This very fast

Table 7.5 Pacific—Southeast Asia

Tide gauge from PSMSL								
Country	ID	Name	LAT	LON	Date	Length (year)	RSL trend (mm year^{-1})	Error (mm year^{-1})
Vietnam	841	Hondau	20.67	106.80	1957–2013	57	2.1	0.6
Vietnam	1449	Quinhon	13.77	109.25	1977–2013	37	⋆	⋆
Vietnam	1475	Danang	16.10	108.22	1978–2013	36	3.2	1.0
Vietnam	1495	Vungtau	10.33	107.07	1979–2013	35	3.6	1.4
Thailand	449	Ko Sichang	13.15	100.82	1940–2002	63	0.8	0.5
Thailand	444	Fort Phrachula	13.55	100.58	1940–2015	76	14.7	0.9
Thailand	174	Ko Lak	11.80	99.82	1940–2015	76	0.8	0.5
Malaysia	1592	Cendering	5.27	103.19	1985–2014	30	3.3	1.1
Malaysia	1589	Tanjung Gelang	3.98	103.43	1984–2015	32	3.3	0.9
Malaysia	1678	Pulau Tioman	2.81	104.14	1986–2015	30	2.8	1.2
Malaysia	1677	Kukup	1.33	103.44	1986–2015	30	3.6	1.3
Singapore	724	Sembawang	1.47	103.83	1972–2015	44	1.8	0.7
Singapore	1248	Sultan Shoal	1.23	103.65	1972–2015	44	2.9	0.9
Singapore	1351	Raffles	1.17	103.75	1980–2015	36	2.7	1.1
Malaysia	1733	Kota Kinabalu	5.98	116.07	1988–2015	28	3.9	1.9
Malaysia	1734	Tawau	4.23	117.88	1988–2015	28	4.0	2.8
Philippines	260	Jolo. Sulu	6.07	121.00	1948–1994	47	⋆	⋆
Philippines	145	Manila	14.58	120.97	1948–2015	68	13.8	0.7
Philippines	394	Cebu	10.30	123.92	1948–2015	68	0.9	0.7
Philippines	522	Legaspi	13.15	123.75	1949–2009	61	5.5	0.7
Philippines	537	Davao	7.08	125.63	1949–1992	44	5.3	1.2
Taiwan	545	Keelung II	25.13	121.73	1956–1994	39	⋆	⋆
China	934	Kanmen	28.08	121.28	1959–2015	57	2.2	0.4
China	727	Xiamen	24.45	118.07	1954–2003	50	1.1	0.8
China	933	Zhapo	21.58	111.82	1959–2015	57	2.2	0.5
Hong Kong	1034	Tai Po Kau	22.44	114.18	1963–2015	53	3.0	0.8
Hong Kong	333	North Point	22.30	114.20	1950–1985	36	⋆	⋆
Hong Kong	1674	Quarrybay	22.29	114.21	1986–2015	30	2.8	1.7
Macau	269	Macau	22.20	113.55	1925–1982	58	⋆	⋆

Continued

Table 7.5 Pacific—Southeast Asia—cont'd

GPS—ULR6 from SONEL

Country	ID	LAT	LON	Date	Length (year)	Tide gauge	Distance	Vertical velocity (mm year^{-1})	Error (mm year^{-1})
Malaysia	GETI	6.22	102.11	1998–2002	5	Geting	5 m	⋆	⋆
Malaysia	KUAL	5.32	103.14	2007–2013	7	Cendering	8 km	⋆	⋆
Malaysia	UMSS	6.04	116.11	2007–2013	7	Kota Kinabalu	8 km	⋆	⋆
Malaysia	JUML	2.21	102.26	2004–2013	10	Tanjung Keling	11 km	⋆	⋆
Malaysia	BIN1	3.24	113.09	2007–2011	5	Bintulu	4 km	−3.2	0.5
Singapore	NTUS	1.35	103.68	1997–2013	17	Jurong	7 km	⋆	⋆
Philippines	PIMO	14.64	121.08	1998–2010	13	Manila	13 km	2.7	0.6

Locations, time spans, and trends of RLR PSMSL tide gauges and SONEL GPS stations. Error corresponds to 95% margin of error for the linear trend. The symbol ⋆corresponds to nonsignificant trend (P-value > 0.1). *GPS*, Global Positioning System; *PSMSL*, Permanent Service for Mean Sea Level; *RSL*, relative sea level; *SONEL*, Système d'Observation du Niveau des Eaux Littorales.

RSL rise is due to land subsidence induced partly by natural compaction of deltaic sediments and amplified by overpumping of groundwater, changing nonlinearly with time since 1955 (Emery and Aubrey, 1991; Phien-wej et al., 2006). Over the past 35 years, the land subsidence rate reached 120 mm $year^{-1}$ and nowadays ranges from 20 to 30 mm $year^{-1}$ (Phien-wej et al., 2006). The work of Phien-wej et al. (2006) suggests that for each 1 m^3 of groundwater pumped out in the Bangkok Plain, it is ~0.10 m^3 of ground that is lost at surface.

In Peninsular Malaysia, the average RSL trend, estimated from four TGs, in operation since 1980s to 2015, is about ~3 mm $year^{-1}$. At the southern tip of the Malaysian Peninsula, in Singapore, the RSL trend is about 2–3 mm $year^{-1}$ since 1970s. These estimates are consistent with the results of Tkalich et al. (2013) who reported an RSL trend of ~2.3 mm $year^{-1}$. The Singapore mainland is subsiding at a rate of 1.5–7 mm $year^{-1}$ (Catalao et al., 2013).

Along the Indonesian Pacific coast, there are no RLR TGs available at PSMSL. However, Fenoglio-Marc et al. (2012) used two TGs located on the Pacific coast of Java province from the Metric PSMSL database: Jakarta (1993–2011) and Surabaya (1993–2009). In Surabaya, they estimated an RSL trend of 8.8 mm $year^{-1}$ and −21.3 mm $year^{-1}$ at Jakarta, compared with an ASL trend of 3.8 mm $year^{-1}$ from altimetry at both locations. Combining these two techniques, they detected a high land subsidence rate at Jakarta of −19.7 mm $year^{-1}$ and of −5.3 mm $year^{-1}$ at Surabaya. The megacity of Jakarta (~10 million inhabitants) is located in a lowland area in the northern coast of West Java and is subject to land subsidence mainly induced by excessive groundwater extraction (Abidin et al., 2010). From leveling surveys, GPS observations, and InSAR analysis, Abidin et al. (2015) estimated the rate of land subsidence in Jakarta in the range 30–100 mm $year^{-1}$ during 1974–2010. Chaussard et al. (2013) performed a global survey of Sumatra and Java, using a method of differential SAR interferometry (D-InSAR) and identified land subsidence in five major coastal cities, mainly due to groundwater extraction, in the range 20–240 mm $year^{-1}$ during 2006–09. Moreover, at Jakarta, Hanson et al. (2011) estimated that more than 2 million people will be exposed to coastal flooding by 2070s. Considering the Coral Triangle countries, including Indonesia, Malaysia, Philippines, East Timor, Papua New Guinea, and the Solomon Islands, Mcleod et al. (2010) demonstrated that the SL rise (scenario: SL rises up to 0.4 m by 2100 and without adaptation) will significantly affect coastal population and habitats, and Indonesia will be a country, which is likely to be most affected by coastal flooding, with ~6 million people impacted annually by 2100.

In East Malaysia, located on the island of Borneo, only two long TGs are available and both they manifest a strong RSL trend of ~4 mm $year^{-1}$ since 1990s, consistent with the ASL trend from altimetry (Fig. 7.3) over the same period.

In Philippines, five acceptable TGs (Fig. 7.2) are available, and they span 44–68 years. The three TGs located on the eastern side of the archipelago have high RSL trends of: ~5.5 mm $year^{-1}$ at Davao and Legaspi; and ~14 mm $year^{-1}$ in Manila. In the west, the Cebu TG has an RSL trend of 0.9 mm $year^{-1}$. These differences in trend can be explained by land subsidence, which is larger on the eastern side of the subduction zone, and there is a marginal land uplift on the opposite side (Emery and Aubrey, 1991). In Manila, Rodolfo and Siringan (2006) showed that a much higher rate of RSL is induced by land subsidence, linked with the increase in groundwater pumping and consistent with the population growth curve over the same period. The GPS station (PIMO 2.7 mm $year^{-1}$, Table 7.5), located 13 km northeast from the TG, and the DORIS station (3.2 mm $year^{-1}$), located 10 km southeast from the TG, shows, in agreement, an uplift rather than subsidence (Santamaría-Gómez et al., 2017). This indicates a significant spatial variation in vertical displacements around the TG. Raucoules et al. (2013) demonstrated, from D-InSAR, that Manila was locally affected by vertical ground motions of about 15 cm $year^{-1}$ from 1993 to 2010. Therefore, the impact related to human-induced subsidence is already evident in Manila city. In this context, the results of Perez's et al. (1999) on vulnerability analysis suggest that most areas along the coast of Manila Bay (including Manila city) could succumb, from both physical and socioeconomic standpoints, to a 1 m SL rise by 2100.

In South China, we found three TGs with a time span of at least 50 years. We estimated RSL trends of ~2 mm $year^{-1}$ at Kanmen and Zhapo, and ~1 mm $year^{-1}$ at Xiamen. Tseng et al. (2010) estimated from TGs, around Taiwan, an RSL trend of 2.4 mm $year^{-1}$ from 1961 to 2003 and 5.7 mm $year^{-1}$ during the period 1993–2003. Ding et al. (2001) estimated an RSL trend at Hong Kong around ~2 mm $year^{-1}$ over 1954–99. Two TG records in Hong Kong from the RLR PSMSL data set have similar RSL trends of ~3 mm $year^{-1}$ (Table 7.5). Guo et al. (2015) estimated vertical land movement along the South China coast from TGs and satellite altimetry and found subsidence rates varying from 6 to 17 mm $year^{-1}$. At Shenzhen, land subsidence at a rate of 25 mm $year^{-1}$ was detected over 2007–10, by the method of Small Baseline Subset InSAR (SBAS-InSAR, Xu et al., 2016). In the Pearl River Delta, the RSL changes seem to be essentially controlled by vertical movements of active faults (Mei-e, 1993). He et al. (2014) reconstructed the regional SL, by combining TGs and altimetry, over the period 1959–2011 and estimated that SL has risen at a rate of 4 mm year^{-1} in the Pearl River Delta. They determined different spatial patterns of variability in the river mouth and along the coastline. In this region, there is no clear consensus on the causes of long-term RSL changes. Many more studies are urgently needed to understand the causes of observed RSL changesin order to mitigate potential disasters associated with future SL rise.

In light of the results mentioned above, the major concern of this region is that the rates of RSL rise are one to two times higher (and much more at Bangkok and Manila) than the GMSL trend over the 20th century. These results are confirmed by estimates from the SL reconstruction (Fig. 7.4) that vary around ~3 mm year^{-1} since 1960 and 3–5 mm year^{-1} since 1993 (Fig. 7.3).

Recent analysis of the regional SL variability in the Gulf of Thailand, including GPS-derived rate of vertical land movements, provides a rate of ASL rise of about 5 mm year^{-1} since 1940s and 3–6 mm year^{-1} over the altimetry era (Trisirisatayawong et al., 2011). In this region, the impact of the postseismic motion due to the 2004 Sumatra–Andaman earthquake on the RSL rate is of the order of -10 mm year^{-1} (Trisirisatayawong et al., 2011). Furthermore, there are indications that RSL rates increased significantly at all locations (20–30 mm year^{-1} almost everywhere (Saramul and Ezer, 2014)) after this earthquake.

Many studies have shown that in South China Sea, the interannual SL variations are linked to ENSO (Rong et al., 2007; Han and Huang, 2009; Peng et al., 2013) and to the Pacific Decadal Oscillation (PDO; Deng et al., 2013; Wu et al., 2014; Strassburg et al., 2015). The Indian Ocean Dipole (IOD) influences interannual SL variations in the southwestern (Malaysia Peninsula and Singapore Strait) and southeastern (Borneo Island) coastal regions (Soumya et al., 2015). The SL trends are greatly masked by a low-frequency variability associated with the PDO (Strassburg et al., 2015; Cheng et al., 2016). Since 1990s, there has been a major phase shift of PDO; this phase shift is associated with an intensification of the trade winds at the equator, storing warm water and increasing SL in the western tropical Pacific (WTP), and reducing it along the west coast of the Americas (Merrifield et al., 2012). Hence, the accelerated SL rise seems to be a part of global adjustment to this PDO phase shift (Cheng et al., 2016). Thus, it is important to take into account this natural decadal variability in the future SL trend estimates in the South China Sea, where SL rise expected to be much more intense.

4.3 Western Tropical Pacific Islands

Over the past several decades, there is a large scientific consensus on the threat hanging over small islands, and, particularly, on the WTP islands, due to rising SLs associated with global warming (Nurse et al., 2014). The future stability, and survival, of the nations of these small islands is a major international concern. A large number of studies, using TG data, altimetry observations, past SL reconstruction and global models, have revealed patterns of a recent enhanced SL trend in the WTP (among others Church et al., 2004; Merrifield, 2011; Becker et al., 2012; Merrifield et al., 2012; Meyssignac et al., 2012b; Zhang and Church, 2012). Inspecting the updated PSMSL RLR data set, we found 22 TG records spanning more than 30 years in the WTP region. Interestingly, 12 out of 22 long-term stations (Table 7.6) reveal an increasing RSL, with a trend greater than the 20th century GMSL.

Merrifield (2011) highlighted an abrupt SL rise in WTP since the early 1990s, compared to the last 40 years. Becker et al. (2012) showed that the RSL rate at Funafuti Island (Tuvalu) is ~5 mm year^{-1} over 1950–2009, which is about three times larger than the GMSL rise over the same period. These results are confirmed by our estimates from the SL reconstruction that estimates the trends in the range 4–5 mm year^{-1} since 1960 (Fig. 7.4) and 5–11 mm year^{-1} since 1993 (Fig. 7.3). In the WTP region, superimposed on these trends are transient interannual and decadal SL variations of the order of ±20–30 cm (Becker et al., 2012). This interannual and decadal SL variability is attributed to low-frequency Pacific trade wind fluctuations, associated with low-frequency modulations of ENSO and PDO (Merrifield, 2011; Zhang and Church, 2012; Moon et al., 2015; Palanisamy et al., 2015). However, the processes operating over longer timescales, and especially the influence of the Indian Ocean, are still under debate (Han et al., 2014; Moon et al., 2015; Mochizuki et al., 2016). Han et al. (2014) argued that the intensified decadal and multidecadal SL variability results from a phase shift in sea surface temperature between the Indian Ocean and tropical Pacific. In addition, at many islands in this region, the RSL can be affected by crustal deformation due to volcanic and tectonic activities. For example, Ballu et al. (2011) reported large earthquake-related land subsidence at the Torres Islands (Vanuatu) between 1997 and 2009, which added to the ASL, generating RSL rise of ~20 mm year^{-1}.

The increased island sensitivity to changes in human settlement patterns, and in socioeconomic and environmental conditions, makes it far more difficult to detect and attribute climate change effects. This also

Table 7.6 Pacific—Western Tropical Pacific Islands

Tide gauge from PSMSL								
Country	ID	Name	LAT	LON	Date	Length (year)	RSL trend (mm year^{-1})	Error (mm year^{-1})
Palau	1252	Malakal-B	7.33	134.47	1976–2014	39	4.1	2.5
Guam	540	Apra Harbour	13.44	144.65	1948–2015	68	1.8	0.9
Northern Mariana Is.	1474	Saipan	15.23	145.75	1979–2014	36	2.8	2.1
Micronesia	528	Chuuk	7.45	151.85	1953–1986	34	★	★
Micronesia	1370	Pohnpei-B	6.98	158.23	1976–2014	39	2.6	1.7
Marshall Is.	595	Wake Island	19.29	166.62	1951–2015	65	2	0.5
Marshall Is.	513	Kwajalein	8.73	167.74	1947–2015	69	2.2	0.7
Marshall Is.	1217	Majuro-B	7.10	171.37	1969–2001	33	3	1.8
Fiji	1327	Suva-A	−18.14	178.42	1988–2015	28	6.7	1.9
French Polynesia	1253	Rikitea	−23.12	−134.97	1970–2014	45	1.7	0.6
French Polynesia	1397	Papeete-B	−17.53	−149.57	1970–2014	45	3.3	0.8
USA	300	Hilo	19.73	−155.06	1947–2015	69	2.9	0.5
USA	521	Kahului Harbor	20.90	−156.48	1951–2015	65	1.9	0.5
USA	155	Honolulu	21.31	−157.87	1905–2015	111	1.4	0.2
USA	756	Nawiliwili Bay	21.95	−159.36	1955–2015	61	1.5	0.5
USA	1372	French Frigate Shoals	23.87	−166.28	1975–2005	31	★	★
USA	598	Johnston Island	16.74	−169.53	1950–2002	53	0.8	0.7
USA	523	Midway Island	28.21	−177.36	1982–2015	34	3.8	1.2
Cook islands	1450	Penrhyn	−9.02	−158.07	1978–2014	37	★	★
American Samoa	539	Pago Pago	−14.28	−170.69	1949–2015	67	3.2	0.7
Kiribati	1329	Kanton Island-B	−2.82	−171.72	1973–2011	39	★	★
Kiribati	1371	Christmas Island II	1.98	−157.48	1981–2014	34	★	★

Continued

Table 7.6 Pacific—Western Tropical Pacific Islands—cont'd

GPS—ULR6 from SONEL									
Country	**ID**	**LAT**	**LON**	**Date**	**Length (year)**	**Tide gauge**	**Distance**	**Vertical velocity (mm year^{-1})**	**Error (mm year^{-1})**
Cook Is.	CKIS	−21.20	−159.80	2002–2013	12	Rarotonga B	3 km	−0.5	0.4
Fiji	LAUT	−17.61	177.45	2002–2013	12	Lautoka	1 km	−1.2	0.3
French Loyalty Is.	LPIL	−20.92	167.26	1997–2013	17	Lifou	2 km	0.2	0.5
French New Caledonia	NRMD	−22.23	166.48	2006–2013	8	Noumea	9 km	−1.9	0.2
French Austral Is.	TBTG	−23.34	−149.48	2008–2013	6	Tubuai	1 m	−0.33	0.5
French Polynesia	GAMB	−23.13	−134.96	2000–03	4	Rikitea	900 m	−1	0.4
French Polynesia	PAPE	−17.53	−149.57	2003–2013	11	Papeete	1 m	−1.9	0.2
French Polynesia	FAA1, TAH2	−17.55	−149.61	2007–2011	5	Papeete	6 km	−1.8	0.5
French Polynesia	TAH1, THTI	−17.58	−149.61	2000–2013	14	Papeete	6 km	−1	0.3
Kiribati	KIRI	1.35	172.92	2002–2013	12	Tarawa C	2 km	−0.2	0.2
Marshall Is.	KWJ1	8.72	167.73	1996–2002	7	Kwajalein	1 km	0.5	0.4
Marshall Is.	MAJU	7.12	171.36	2007–2013	7	Majuro	2 km	0.8	0.4
Micronesia	POHN	6.96	158.21	2003–2013	11	Pohnpei	3 km	0.8	0.4
Palau	PALA	7.34	134.48	1996–2001	6	Malakal	3 km	⋆	⋆
Rep. of Nauru	NAUR	−0.55	166.93	2003–2013	11	Nauru	3 km	−1	0.3
Samoa	SAMO	−13.85	−171.74	2001–2013	13	Apia B	4 km	⋆	⋆
Solomon Is.	SOLO	−9.43	159.95	2010–13	4	Honiara B	1 km	⋆	⋆
Tonga	TONG	−21.14	−175.18	2002–2013	12	Nuku'Alofa B	800 m	3	0.4
USA	CNMR	15.23	145.74	2003–2013	11	Saipan	600 m	−1.2	0.2
USA	HNLC	21.30	−157.86	1997–2013	17	Honolulu	1 m	−0.2	0.2
USA	HILO	19.72	−155.05	1999–2009	11	Hilo	1 km	−1.1	0.2
USA	ASPA	−14.32	−170.72	2001–09	9	Pago Pago	7 km	⋆	⋆
USA	LHUE	21.98	−159.34	1999–2004	6	Nawiliwili	4 km	0.5	0.6
USA	ZHN1	21.31	−157.92	2002–2013	12	Honolulu	6 km	−0.6	0.3
Vanuatu	VANU	−17.74	168.32	2002–2012	11	Port Vila B	1 km	⋆	⋆
Tuvalu	TUVA	−8.53	170.20	2002–2013	12	Funafuti B	3 km	−1.7	0.2

Locations, time spans and trends of RLR PSMSL tide gauges and SONEL GPS stations. Error corresponds to 95% margin of error for the linear trend. The symbol ⋆ corresponds to nonsignificant trend (P-value > 0.1). *GPS*, Global Positioning System; *PSMSL*, Permanent Service for Mean Sea Level; *RSL*, relative sea level; *SONEL*, Système d'Observation du Niveau des Eaux Littorales.

remains a source of debate in the scientific community (Nurse et al., 2014). Over the past few decades, from a limited number of studies, no clear linkage between WTP island shoreline recession and recent SL rise was found (Webb and Kench, 2010; Le Cozannet et al., 2013; Kench et al., 2015; McLean and Kench, 2015; Duvat and Pillet, 2017), but net changes in shoreline position have been observed. However, Kench et al. (2015) question the islands' capacity to continue maintaining their current dynamic adjustment to higher rates of SL change, as those expected by 2100. A recent study by Albert et al. (2016) highlights that the rates of some Solomon Islands shoreline recession are substantially higher in areas exposed to high wave energy, indicating a synergistic interaction between SL rise and waves. Therefore, shoreline changes and floods seem to result from extreme events and from maladaptive trajectories exacerbated by the SL rise (Duvat et al., 2013).

4.4 Tropical Pacific Relative Sea Level Hotspots: Summary

- ***Acapulco (Mexican South Pacific coast)*** with more than 700,000 inhabitants faces a SL rise at a rate of 8 mm year^{-1}, one of the fastest rates along the Pacific coast of America.
- ***Mekong delta*** is a hotspot with an RSL rise of 7 mm year^{-1} over 1987–2006. Fujihara et al. (2015) estimated that 80% of this rate is due to land subsidence. The delta is likely to subside even faster, at a rate of 10–40 mm year^{-1}, as revealed by InSAR analysis over 2000–2010.
- ***Chao Phraya delta*** *(Bangkok)* faces an RSL rise of 15 mm year^{-1}, but the current subsidence is probably larger, being about 20–30 mm year^{-1} (Phien-wej et al., 2006) with a milder ASL trend of 3–5 mm year^{-1}.
- ***Jakarta megacity*** *(Indonesia)* is one of the world's cities most threatened by rising RSL with a high population density, fast land subsidence of 20 mm year^{-1} or larger (InSAR 20–240 mm year^{-1}), and an enhanced ASL rate of 5–7 mm year^{-1}.
- ***Manila megacity*** *(Philippines)* is an indisputable RSL hotspot due to land movement induced by a variety of processes in this region. The contribution of ASL rise (5–7 mm year^{-1}) and the interannual variations due to ENSO are not negligible either.
- ***Almost all the WTP Islands*** are subject to pronounced ASL rise. In combination with land subsidence induced by tectonic faults and the Pacific subduction zone (e.g., Vanuatu), some of the WTP islands can face rapid coastal submergence in the future.

5. INDIAN OCEAN

5.1 Bay of Bengal

The Bay of Bengal (BoB) located in the northern Indian Ocean is surrounded to the east by Bangladesh and Myanmar and to the west by India. The BoB is the largest bay in the world and is unique in many ways. Today, a quarter of the world's population lives in its vicinity (~1.5 billion people from World Development Indicators, Mundial, 2014) and more than 170 million people live below 10 m of coastal elevation (from LECZ, India 7%, Bangladesh 40%, Myanmar 25%, and Sri Lanka 13% in percentage of the respective national population, Fig. 7.1). The population is being concentrated in megacities such as Kolkata (India, ~15 million inhabitants), Chennai (India, ~10 millions), and Dhaka (Bangladesh, ~18 millions), and in large urban agglomerations (~4.5 millions) such as Chittagong in southeastern Bangladesh and Yangon in Myanmar (Fig. 7.1). Additionally, Dhaka and Kolkata are megacities located in the low-lying Ganges–Brahmaputra–Meghna (GBM) delta and Yangon in the Irrawaddy River delta. Other major deltas along the India's east coast are the Krishna, Godavari, and Mahanadi. Syvitski et al. (2009) revealed that all these deltas are already threatened by rising RSL. They classified the deltas as subject to (1) high risk for the Krishna delta, because of virtually no deposition of sediment and accelerating compaction, (2) greater risk: GBM and Irrawaddy deltas because of compaction of the soil exacerbating the low rate of sediment deposition; and (3) significant risk: the Mahanadi and Godavari deltas because of lower sediment deposition rates than that of ASL rise. The geographic and socioeconomic situation of the BoB coast places it among the most vulnerable to climate change and to RSL rise not only in Southeast Asia but also in the world. Rao et al. (2008) demonstrated that, over the four past decades, pronounced coastal erosion along the Krishna and Godavari deltas is apparently due to sediment retention at dams. This result was confirmed by Gupta et al. (2012), who showed that increasing number of mega dams and reservoirs between 1978 and 2003 on the Krishna River (9 mega-dams), Godavari (9 mega-dams), and Mahanadi (2 mega-dams) could be an obvious reason for the observed decrease (>70%) in sediment supply. Concerning the GBM delta, Sarwar and Woodroffe (2013), using 20 years of Landsat satellite images, noticed that the entire delta coast changed little and erosion and accretion are relatively balanced. However, Wilson and Goodbred (2015) highlighted three regions where sediment supply is insufficient to offset subsidence or erosion: in the northeast (Sylhet Basin), along the Indian tidal delta plain and the fluvio-tidal transition in the western and central parts of the delta. Shearman et al. (2013) documented, from 20 years of Landsat satellite images, a net contraction of delta mangrove area, including the Sundarbans region.

The RSL changes along the eastern coast of India from West Bengal to Sri Lanka have been previously estimated from two long-term (>60 years) TGs at Vishakhapatnam and Chennai (RLR PSMSL) analyzed by: Emery and Aubrey (1989), Unnikrishnan and Shankar (2007), and Palanisamy et al. (2014). These studies found consistent RSL trends equal to 0.6 and 0.8 mm $year^{-1}$ at Chennai and Vishakhapatnam (Table 7.7), respectively. Both values are significantly smaller than the 20th century GMSL trend (still valid if GIA correction of ~−0.4 mm $year^{-1}$ is applied). Both stations are located at the border of the tectonically stable Precambrian shield and their lower RSL trends were interpreted by Emery and Aubrey (1991) as consequence of land submergence. However, the altimetry-derived and reconstructed ASL trends (Figs. 7.3 and 7.4) near the eastern coast of India are about 1 mm $year^{-1}$ or larger than the RSL trends at Chennai and Vishakhapatnam. Thus, the subsidence of the eastern coast of India does not seem to be supported by these long-term RSL measurements.

In the northwest BoB, along the Hooghly River in West Bengal, Emery and Aubrey (1989) found erratic RSL rates between −7 and 6 mm $year^{-1}$ at Saugor (1937–82, 45 year, 4 mm $year^{-1}$), Diamond Harbour (1948–82, 35 year, −7 mm $year^{-1}$), Kidderpore (1881–1931, 24 year, 6 mm $year^{-1}$), and Kolkata (1932–82, 50 year, −7 mm $year^{-1}$). They finally omitted all these records because of a great influence of cyclonic storm surges, floods, sediment compaction, and datum shifts. Nandy and Bandopadhyay (2011) estimated RSL trends based on three TGs (>30 year) from the RLR PSMSL data set (Table 7.7): 1.2 mm $year^{-1}$ at Gangra (31 km from the sea coast), 2.8 mm $year^{-1}$ at Haldia (43 km from the sea), and 4 mm $year^{-1}$ at Diamond Harbour (70 km from the sea). They argued that this trend variability appears to originate from the morphology of the landward-narrowing estuary, with some contribution from sediment compaction. Brammer (2014) detected a shift in 1975 in the Diamond Harbour TG, coinciding with the construction of the Farakka barrage across the Ganges. This construction, and probably other upstream engineering works, may have altered the RSL at Diamond Harbour, increasing the dry season volume of freshwater discharge, extending toward the freshwater zone to the mouth of the estuary and impacting the tidal regime (Sinha et al., 1997).

Table 7.7 Indian Ocean—Bay of Bengal

Tide gauge from PSMSL								
Country	**ID**	**Name**	**LAT**	**LON**	**Date**	**Length (year)**	**RSL trend (mm year^{-1})**	**Error (mm year^{-1})**
India	205	Chennai/Madras	13.10	80.30	1953–2012	60	0.6	0.5
India	414	Vishakhapatnam	17.68	83.28	1937–2011	75	0.8	0.5
India	1369	Gangra	21.95	88.02	1974–2006	33	1.2	1.6
India	1270	Haldia	22.03	88.10	1971–2012	42	2.8	0.9
India	543	Diamond Harbour	22.20	88.17	1948–2012	65	4	0.7
India	369	Calcutta/Kolkata	22.55	88.30	1932–1999	68	7.4	1.3
Thailand	446	Ko Taphao Noi	7.83	98.43	1940–2015	76	1.3	0.9
Malaysia	1676	Pulau Langkawi	6.43	99.76	1986–2015	30	3.4	1.9
Malaysia	1595	Pulau Pinang	5.42	100.35	1986–2014	29	3.9	1.9

Locations, time spans, and trends of RLR PSMSL tide gauges and SONEL GPS stations. Error corresponds to 95% margin of error for the linear trend. The symbol ⋆ corresponds to nonsignificant trend (P-value > 0.1). *GPS*, Global Positioning System; *PSMSL*, Permanent Service for Mean Sea Level; *RSL*, relative sea level; *SONEL*, Système d'Observation du Niveau des Eaux Littorales.

Shared by India and Bangladesh, in the north of BoB, the Sundarbans region is the world's largest contiguous mangrove forest that covers ~10,000 km^2 of the GBM delta, with 60% in Bangladesh and 40% in India (Iftekhar and Saenger, 2008). This area, directly threatened by SL rise and alteration of freshwater flux, is recognized as a global priority for biodiversity conservation, especially regarding the Royal Bengal Tiger (Loucks et al., 2010). Brown and Nicholls (2015) reviewed available data, literature, and documentary sources and created a database of subsidence rates in the Bengal delta. They concluded an average subsidence rate of 2.8 mm $year^{-1}$ in Sundarbans region, the lowest rate observed in GBM delta. Loucks et al. (2010), using high-resolution elevation data and a scenario of SL increasing (by 28 cm), warned that in 50 years the Sundarbans tigers could join the Arctic's polar bears on the list of victims of climate change–induced habitat loss. Rahman et al. (2011), using Landsat images, showed that the Sundarbans coastline is currently in net erosion and was losing on an average about 5 km^2 $year^{-1}$ over 1973–2010 (~170 km^2, i.e., ~2%). Payo Garcia et al. (2016), through a numerical model with different SL rise scenarios (rise by 46 or 75 cm) and taking a net subsidence of ±2.5 mm $year^{-1}$, estimated that between 1% and 6% of Bangladesh Sundarbans area could be lost by 2100. The results obtained in this framework suggest that erosion, rather than inundation, may remain the dominant land loss driver by 2100. Pethick and Orford (2013) showed a rapid rise in RSL in the Sundarbans area. They used the only three available TGs, provided by Institute of Water Modelling of Bangladesh: Hiron Point (34 year), Mongla (20 year), and Khulna (72 year). In the RLR PSMSL data set, we could find the Hiron Point and Khepupara TG records, but they turned out to be too short (<24 years) for long-term analysis. Pethick and Orford (2013) found strong RSL trends of: ~8 mm $year^{-1}$ at Hiron Point (at the mouth of Pussur Estuary), ~6 mm $year^{-1}$ at Mongla, and ~3 mm $year^{-1}$ at Khulna (located 120 km inland). Moreover, they argued that the mean high water level was increasing at a much faster rate (14–17 mm $year^{-1}$) and a large part of the signal can be attributed to tide amplification, constricted by embankments.

Along other low-lying coastal regions of Bangladesh, high population density, inadequate infrastructure, and low adaptive capacity have made the urban residents highly vulnerable to climate change (Milliman et al., 1989; Choudhury et al., 1997; Warrick and Ahmad, 2012). Over 28% of the total population (~48 million) live in urban agglomerations (World Bank indicators (Mundial, 2014)). This percentage, which was below 5% in 1974, is expected to reach to 45% in 2030. At least 50% of the urban population

(~23 million) live in three major cities: Dhaka, Khulna, and Chittagong, where the land elevation, in whole or in part, is less than 10 m above SL. Hanson et al. (2011) estimated that more than 11 million people will be exposed to coastal flooding in 2070s at Dhaka, ~4 million at Khulna, and ~3 million at Chittagong.

Higgins et al. (2014), using InSAR satellite–based technique and GPS over 2007–11, mapped the subsidence within GBM delta in a region covering ~10,000 km^2 of irrigated cropland surrounding Dhaka city. The subsidence rate is about 10 mm year^{-1} around Dhaka and may reach 18 mm year^{-1} elsewhere in the area. Brown and Nicholls (2015) reported subsidence rates in the range −1 to 44 mm year^{-1}, with a mean of ~3 mm year^{-1}. These rates are associated with four principal processes: (1) tectonics, (2) sediment compaction, (3) sedimentation, and (4) human activities such as groundwater extraction, drainage, and embankment building.

Some studies tried to estimate the effect of future RSL rise on Bangladesh coast. Huq et al. (1995), among others, estimated that a 1-m rise can flood ~17% of land area and lead to displacement of more than 13 million people. Arfanuzzaman et al. (2016) estimated that with a 71 cm rise (with respect to 1980–99 levels) up to 25% of Bangladesh wetlands could be lost by 2100. Ruane et al. (2013) studied the impact of climate changes through different parameters on agricultural production in Bangladesh. They show that the agriculture production in southern Bangladesh is severely affected by SL rise. The projections of production lost due to coastal inundation, associated with 27 cm of SL rise, could reach 20% in southern Bangladesh (and 40% with 62 cm SL rise).

Finally, there is no clear consensus about the response of the GBM delta to natural and human forcings over decadal to century timescales. Moreover, all the studies on climate change impacts focused on coastal flooding by applying a simplified SL rise scenario, yet an uncertainty of 10 cm of RSL rise may result in major consequences for local people (Lee, 2013). Despite the crucial importance of this problem, very few studies have focused on assessing the actual RSL rates along the Bangladesh coast.

Singh (2002) estimated RSL trends from three TG records (22-year, 1977–98) provided by the Bangladesh Inland Water Transport (BIWTA): in the west, at Hiron Point ~4 mm year^{-1}, in the center, at Char Changa ~6 mm year^{-1} and, in the east, at Cox's Bazar ~8 mm year^{-1}. They argued that difference between these three RSL trends is probably due to local land subsidence in the eastern Bangladesh region (around Cox's Bazar). Lee (2013) used the Hiron Point TG record over 1990–2009 to reconstruct, by using ensemble empirical mode decomposition technique, the past RSL over 1950–2009 and found an RSL trend of ~8 mm year^{-1}. Sarwar (2013) used TG records, collected from the Bangladesh Water Development Board, BIWTA and the Metric PSMSL data set and provided a comprehensive analysis of SL changes in the region. They considered 13 TG records having at least 14 years of data, but a lot of discrepancies appeared in the trend analysis.

Along the Myanmar coast and within the Irrawady delta, where ~11 million people live, only one long-term RLR TG record is available at PSMSL: the Yangon TG operated during 1916–62. Unfortunately, about 47% observations are missing in this record. There are also three old stations but with the record length under 10 years (Akyab, Moulmein, and Amherst). It is an encouraging fact that since 2006 these locations have been reinstrumented, and current data are now available from the Metric PSMSL data set. The delta coast seems to be more or less in equilibrium, and sediment deposition currently balances subsidence and SL rise (Hedley et al., 2010). This can be explained by fewer numbers of large dams relative to its Asian neighbors. However, this situation is now rapidly changing with extensive damming projects in the basin. At Yangon, Hanson et al. (2011) project that, by 2070s, more than 5 million people could be exposed to coastal flooding.

A large part of the recent RSL trends estimated in the eastern BoB can be attributed to the ASL rise. Over 1993–2014, the rate of ASL trend is in the range 3–5 mm year^{-1} along the GBM coast and in the eastern part of the BoB and 1.8–5 mm year^{-1} along the eastern coast of India (Fig. 7.3). Over 1960–2014, the SL reconstruction gives an ASL trend of 2.5–3 mm year^{-1} along the GBM coast and the eastern part of the BoB, and in 1.8–2.5 mm year^{-1} along the east coast of India (Fig. 7.4), which is greater than the 20th century GMSL.

This finding was previously reported by Church et al. (2004) who found the fastest rate of ASL rise (4–5 mm year^{-1}) in the northeastern Indian Ocean over the period 1955–2003. Han et al. (2010), combining in situ and satellite observations with climate model simulations, identified a significant SL rise since 1960s in Indian Ocean (except in its southern tropical region). They demonstrated that changing surface winds, linked to the strengthening of the Indian Ocean Walker and Hadley circulations, drive this pattern. However, a recent decadal reversal in the upper-ocean temperature trends is observed in the North Indian Ocean (north of 5°S (Nieves et al., 2015)). An increase in the sea surface height decadal rate of ~6 mm year^{-1} was estimated between the period of 1993–2003 and that of 2004–13 from analysis of satellite altimetry data (Thompson et al., 2016). Thompson et al. (2016)

showed, through numerical model simulations, that this reversal has resulted from the combined effects of changing upper-ocean heat redistribution and the cross-equatorial heat transport, both being associated with decadal changes of surface winds.

5.2 Arabian Sea, Persian Gulf, and Maldives

The Arabian Sea is a region, in the northwest part of Indian Ocean, at strikingly intense geopolitical and economic crossroads, notably, via marine trade route for oil and gas resources export. We find the major harbors of Kochi and Mumbai on the southwest coast of India, and further in the northwest, the largest and most frequented ports serving the Arabian Sea, and, in the northeast, the major port of Karachi in Pakistan. Mumbai and Karachi are two large global megacities (with more than 10 million inhabitants, Fig. 7.1). The city of Karachi had a high population growth rate of 5.3% over 1960–2010 (Singh, 2014). On average, over the same period, Asian megacities faced an annual population growth rate of 3.7% against a rate of 2.6% in the rest of the world (Singh, 2014). Mumbai, with a current population of about 20 million, expects to achieve a 35% growth rate by 2025, and in Karachi, the current population of 14 million is expected to see an increase of 45% by 2025 (Kourtit and Nijkamp, 2013). These cities already faced major challenges of flooding and aquifer salinization, amplified by regional SL rise. The situation is being further aggravated in the Indus delta along Pakistan's coast, in Sindh province. This river system, among the largest deltas on Earth, is dominated by human activity since 19th century and is presently affected by (1) artificial flood levees, (2) barrages and their irrigation canals, (3) sediment impoundment behind upstream reservoirs, and (4) interbasin diversion (Syvitski et al., 2013). Consequently, there is a drastic reduction of sediment flux by more than 90% (Giosan et al., 2006; Syvitski and Kettner, 2011), which increases coastal retreat, seawater intrusion, and flooding. Moreover, Ferrier et al. (2015) showed that, in the Indus delta over the past 100 years, as much as ~0.5 mm $year^{-1}$ of the SL trend can be linked to erosion and deposition of sediment since the last glacial–interglacial cycles. Another important process occurs in this specific region: the influence of the groundwater depletion, deforming the Earth's solid surface and depressing the geoid and slowing SL rise near areas of significant groundwater loss (Veit and Conrad, 2016). Veit and Conrad (2016) define important groundwater depletion regions in Northwest India, Northeast Pakistan, and in the Arabian Peninsula, with a consequential slowdown in SL rise by ~0.5 ± 0.1 mm $year^{-1}$ since 1930.

Their work suggests that RSL in this region is currently as much as ~50 mm lower than it would be in the absence of global groundwater depletion.

Emery and Aubrey (1989) investigated the relative long-term SL from the Indian TGs during 1878–1982. On the west coast of India, they selected three TGs from the PSMSL data set at Mumbai (also known as Bombay), Mangalore, and Kochi (or Cochin) with a time length sufficient to detect significant changes. The longest and most coherent is the record of Mumbai (105 years, 1878–1982) presenting a significant linear SL trend of $-0.9\,\text{mm year}^{-1}$, followed by the Kochi record (43 years, 1878–1982) with a trend of $1.3\,\text{mm year}^{-1}$ and the Mangalore series (24 years, 1953–76) that has a $-2.1\,\text{mm year}^{-1}$ trend. These trends show a strong discrepancy, probably due to differences in the record lengths. Unnikrishnan and Shankar (2007) conducted complete reanalysis of these records. They estimated significant RSL trends from PSMSL TGs having at least 40 years length. In Arabian Sea, the TGs of Aden (58 years, 1880–1969), Karachi (44 years, 1916–92), Mumbai (113 years, 1878–1993), and Cochin (54 years, 1939–2003) were selected. The RSL rise estimated from these stations is between 1.1 and $1.7\,\text{mm year}^{-1}$. We updated the RSL trends at Cochin to be $0.7\,\text{mm year}^{-1}$ and at Mumbai to be $1.5\,\text{mm year}^{-1}$ (Table 7.8), and ~1.8–$2.5\,\text{mm year}^{-1}$ from the SL reconstruction (Fig. 7.4). Over 1993–2014, the rate of SL rise over the Arabian Sea from satellite altimeter is ~1.5–$3.5\,\text{mm year}^{-1}$ (Fig. 7.2). Although slightly lower, these estimates are consistent with GMSL rates.

Alothman et al. (2014) focused on the long-term SL rise in the northwestern Persian Gulf. The average of 15 TGs records, obtained from PSMSL, produces an RSL rate of $2.4\,\text{mm year}^{-1}$ for the period 1979–2007. Using six GPS stations, they estimated a subsidence rate of $-0.7\,\text{mm year}^{-1}$ in this region, in part due to excessive pumping in agricultural areas and wetting of unstable soils (Amin and Bankher, 1997).

The Maldives, located from 7°N to 0.5°S in the northeastern Arabian Sea, consists of 1190 small islands with 80% of the land area to be less than 1 m above SL (Khan et al., 2002). These atoll islands are morphologically sensitive to floods, tsunamis, and SL changes (Kench et al., 2006). Several studies detected a recent trend of SL rise at the Maldives (Khan et al., 2002; Woodworth, 2005; Church et al., 2006; Palanisamy et al., 2014). Palanisamy et al. (2014) compared two longest TGs: Malé and Gan (~20 years of length, available from PSMSL dataset), with satellite altimetry and past SL reconstruction. They inferred a significant rate of ASL rise at these two sites of ~$1.4 \pm 0.4\,\text{mm year}^{-1}$ over 1950–2009. This rate is slightly lower than the GMSL rate over the same period. However, it only represents the climatic

Table 7.8 Indian Ocean—Arabian Sea, Persian Gulf, and Maldives

Tide gauge from PSMSL

Country	ID	Name	LAT	LON	Date	Length (year)	RSL trend (mm year^{-1})	Error (mm year^{-1})
Tanzania	1600	Zanzibar	−6.15	39.18	1985–2013	29	⋆	⋆
Yemen	44	Aden	12.79	44.97	1916–1967	52	2.3	0.5
India	596	Kandla	23.02	70.22	1954–1996	43	2.6	0.8
India	43	Mumbai/Bombay	18.92	72.83	1878–1993	116	0.7	0.1
India	438	Cochin	9.97	76.27	1939–2007	69	1.5	0.4
Mauritius	1673	Port Louis II	−20.15	57.50	1987–2016	30	4.1	1.7
Mauritius	1672	Rodrigues Is.	−19.66	63.42	1987–2016	30	5.9	2.1

GPS—ULR6 from SONEL

Country	ID	LAT	LON	Date	Length (year)	Tide gauge	Distance	Vertical velocity (mm year^{-1})	Error (mm year^{-1})
Tanzania	ZNZB	−6.22	39.21	2010–2013	4	Zanzibar	7 km	⋆	⋆
Mauritius	VACS	−20.30	57.50	2008–2012	5	Port Louis II	15 km	−0.8	0.4

Locations, time spans and trends of RLR PSMSL tide gauges and SONEL GPS stations. Error corresponds to 95% margin of error for the linear trend. The symbol ⋆ corresponds to nonsignificant trend (P-value > 0.1). *GPS*, Global Positioning System; *PSMSL*, Permanent Service for Mean Sea Level; *RSL*, relative sea level; *SONEL*, Système d'Observation du Niveau des Eaux Littorales.

component of SL changes, and therefore does not take into account local subsidence that can amplify the RSL change, i.e., directly felt by the population. Now, it is crucial more than ever to estimate with accuracy the rate of vertical land motion at these sites because the ongoing and future SL rise subjects the population of the low-lying Maldives to enhanced vulnerability.

The nation of Mauritius, in the southwest, lives on a group of islands consisting of the main islands of Mauritius, Rodrigues, and Agalega and the archipelago of Saint Brandon. Two TG records with 30-year length are available from the RLR PSMSL data set in the capital of Mauritius Port Louis, where we found an RSL rate of ~4 mm year^{-1}, and the Rodrigues Island with an RSL trend of ~6 mm year^{-1} (Table 7.8). These high rates are confirmed by the ASL trends from the SL reconstruction (3–4 mm year^{-1}, Fig. 7.4) and from altimetry (5–7 mm year^{-1}, Fig. 7.3).

Globally, long-term, interannual, and decadal changes in the SL of the Arabian Sea have rarely been a subject of specific studies, probably due to the lack of historical quality data; the focus has primarily been on the regional physical oceanography of the northern Indian Ocean or that of the BoB.

An important feature was highlighted by Clarke and Liu (1994) who pointed that the interannual SL signal along the Indian west coast, from the equator to Mumbai, is generated by zonal interannual winds blowing along the equator. Shankar and Shetye (1999) demonstrated that the interdecadal SL variations recorded by the Mumbai TG closely follow the monsoon rainfall over the Indian subcontinent. They explained this by the changes in salinity in coastal waters, due to the seasonal fluctuations in river runoff, related to the strength of the monsoon and to the dynamics of ocean currents along the Indian coast.

Shankar et al. (2010) pointed to a much weaker interannual variability, in terms of low frequency, of the Indian west coast compared with the east coast. Aparna et al. (2012) demonstrated that the dominant climatic signals, IOD and ENSO, do not display any coherent response along the eastern Arabian Sea, in contrast to the BoB. Suresh et al. (2013) showed that the Indian west coast intraseasonal SL variations are mostly remotely forced by the winds from equatorial region, and Suresh et al. (2016) demonstrated that winds near Sri Lanka drive 60% of Indian west coast and eastern Arabian Sea seasonal SL. The Mumbai TG, the unique century-long TG record in the Indian Ocean, was used by Becker et al. (2014) to detect human influence on SL rise. They provided statistical evidence, from the power-law statistics framework, that 64% (i.e., ~0.7 mm year^{-1}) of the observed SL trend at Mumbai over the 20th century could be induced by externally driven changes in the Indian Ocean currents.

5.3 Indian Ocean Relative Sea Level Hotspots: Summary

- ***Bangladesh coast*** is an SL hotspot because of high density of coastal population that experiences devastating impact of cyclones on interannual timescale, and RSL rise is enhanced by land subsidence on the decadal scale.
- ***Irrawaddy delta*** is another SL hotspot with 11 million people living in the region. A combination of 3–5 mm $year^{-1}$ in the ASL rise with land subsidence of 6 mm $year^{-1}$ (Syvitski et al., 2009) leads to an RSL rise of more than 10 mm $year^{-1}$.
- ***Mauritius Island*** is a site potentially threatened by an RSL rise of 4–6 mm $year^{-1}$ over the past 30 years, and an indication of the ASL rising 2–3 times faster than the 20th century GMSL.

6. CONCLUSION

This chapter brings together SL observations, and analyzes similarities and differences in past RSL changes along the tropical coasts. We first reviewed the concept of RSL and the drivers of its regional variations. We defined the RSL hotspots and described the different types of observations used to estimate it. Second, we have identified a number of RSL hotspots per oceanic basin. We highlighted the vulnerability of the tropical deltaic coasts, more specifically those of Asia, and a current knowledge gap for priority-populated areas such as Brazil, Indonesia, Philippines, and Bangladesh. Obviously, this hotspot list is far from being exhaustive because most of these regions are still not sufficiently well instrumented with quality TGs and collocated GPS stations. While waiting for obtaining in the future precise and accurate long-term SL in situ measurements, new space missions are expected to provide unprecedentedly precise observations of SL changes along the tropical coasts (e.g., the satellite missions Saral/Altika, Sentinel-3/6, Jason-CS, SWOT).

Understanding and forecasting of the RSL critical thresholds along low-lying heavily populated tropical coastlines are among the most vital societal issues. High priority should be given to the development of integrated, multidisciplinary approaches to understanding the imprint of different geophysical coastal processes on the present-day RSL changes. Assessment of coastal vulnerability, to take appropriate measures to protect populations, can only be determined if the RSL threshold, and even more its uncertainty, is properly estimated.

ACKNOWLEDGMENTS

This work was funded by the Belmont Forum project BAND-AID (ANR-13-JCLI-0002, http://Belmont-BanDAiD.org or http://Belmont-SeaLevel.org). It was also supported by the French research agency (Agence Nationale de la Recherche; ANR) under the STORISK project (NR-15-CE03-0003). The authors are grateful to A. Cazenave for helpful insights on the tropical sea level and to G. Wöppelmann for useful comments on the last version of the manuscript. We thank the PSMSL, ESA-CCI, and SONEL teams for making tide gauge records, altimetric and GPS data, as well as corrections and accuracies, quickly and easily available for the community. We acknowledge B. Meyssignac, from LEGOS/CNES, for supplying the past sea level reconstruction data set.

REFERENCES

Abam, T.K.S., 2001. Regional hydrological research perspectives in the Niger Delta. Hydrol. Sci. J. 46 (1), 13–25.

Abidin, H.Z., Andreas, H., Gamal, M., Gumilar, I., Napitupulu, M., Fukuda, Y., et al., 2010. Land subsidence characteristics of the Jakarta Basin (Indonesia) and its relation with groundwater extraction and sea level rise. Groundwater Response to Changing Climate. IAH Sel. Pap. Hydrogeol. 16, 113–130.

Abidin, H.Z., Andreas, H., Gumilar, I., Sidiq, T.P., Gamal, M., 2015. Environmental impacts of land subsidence in urban areas of Indonesia. In: FIG Working Week Retrieved from https://www.fig.net/resources/proceedings/fig_proceedings/fig2015/papers/ts04i/TS04I_abidin_andreas_et_al_7568.pdf.

Ablain, M., Cazenave, A., Larnicol, G., Balmaseda, M., Cipollini, P., Faugère, Y., et al., 2015. Improved sea level record over the satellite altimetry era (1993–2010) from the Climate Change Initiative project. Ocean Sci. 11 (1), 67–82. https://doi.org/10.5194/os-11-67-2015.

Adelekan, I.O., 2009. Vulnerability of poor urban coastal communities to climate change in Lagos, Nigeria. In: Fifth Urban Research Symposium, pp. 28–30 Retrieved from http://siteresources.worldbank.org/INTURBANDEVELOPMENT/Resources/336387-1256566800920/6505269-1268260567624/Adelekan.pdf.

Albert, S., Leon, J.X., Grinham, A.R., Church, J.A., Gibbes, B.R., Woodroffe, C.D., 2016. Interactions between sea-level rise and wave exposure on reef island dynamics in the Solomon Islands. Environ. Res. Lett. 11 (5), 054011.

Alothman, A.O., Bos, M.S., Fernandes, R.M.S., Ayhan, M.E., 2014. Sea level rise in the north-western part of the Arabian Gulf. J. Geodyn. 81, 105–110.

Amin, A., Bankher, K., 1997. Causes of land subsidence in the kingdom of Saudi Arabia. Nat. Hazards. 16 (1), 57–63. https://doi.org/10.1023/A:1007942021332.

Aparna, S.G., McCreary, J.P., Shankar, D., Vinayachandran, P.N., 2012. Signatures of Indian Ocean Dipole and El Niño–Southern oscillation events in sea level variations in the Bay of Bengal. J. Geophys. Res. Oceans (1978–2012) 117 (C10). Retrieved from http://onlinelibrary.wiley.com/doi/10.1029/2012JC008055/full.

Arfanuzzaman, M., Mamnun, N., Islam, M.S., Dilshad, T., Syed, M.A., 2016. Evaluation of adaptation practices in the agriculture sector of Bangladesh: an ecosystem based assessment. Climate 4 (1), 11.

Aubrey, D.G., Emery, K.O., Uchupi, E., 1988. Changing coastal levels of South America and the Caribbean region from tide-gauge records. Tectonophysics 154 (3), 269–284.

Ballu, V., Bouin, M.-N., Siméoni, P., Crawford, W.C., Calmant, S., Boré, J.-M., et al., 2011. Comparing the role of absolute sea-level rise and vertical tectonic motions in coastal flooding, Torres Islands (Vanuatu). Proc. Natl. Acad. Sci. U.S.A. 108 (32), 13019–13022.

Becker, M., Meyssignac, B., Letetrel, C., Llovel, W., Cazenave, A., Delcroix, T., 2012. Sea level variations at tropical Pacific islands since 1950. Glob. Planet. Change. 80–81, 85–98. https://doi.org/10.1016/j.gloplacha.2011.09.004.
Becker, M., Karpytchev, M., Lennartz-Sassinek, S., 2014. Long-term sea level trends: natural or anthropogenic? Geophys. Res. Lett. 41 (15), 5571–5580. https://doi.org/10.1002/2014GL061027.
Bellard, C., Leclerc, C., Courchamp, F., 2014. Impact of sea level rise on the 10 insular biodiversity hotspots. Glob. Ecol. Biogeogr. 23 (2), 203–212.
Blum, M.D., Roberts, H.H., 2009. Drowning of the Mississippi Delta due to insufficient sediment supply and global sea-level rise. Nat. Geosci. 2 (7), 488–491.
Blum, M.D., Roberts, H.H., 2012. The Mississippi delta region: past, present, and future. Annu. Rev. Earth Planet. Sci. 40, 655–683.
Brammer, H., 2014. Bangladesh's dynamic coastal regions and sea-level rise. Clim. Risk Manag. 1, 51–62. https://doi.org/10.1016/j.crm.2013.10.001.
Brown, S., Kebede, A.S., Nicholls, R.J., 2011. Sea-level rise and impacts in Africa, 2000 to 2100. School of Civil Engineering and the Environment University of Southampton, UK. Retrieved from https://www.fig.net/resources/proceedings/fig_proceedings/fig2015/papers/ts04i/TS04I_abidin_andreas_et_al_7568.pdf.
Brown, S., Nicholls, R.J., 2015. Subsidence and human influences in mega deltas: the case of the Ganges–Brahmaputra–Meghna. Sci. Total Environ. 527, 362–374.
Buenfil-López, L.A., Rebollar-Plata, M., Muñoz-Sevilla, N.P., Juárez-León, B., 2012. Sea-level rise and subsidence/uplift processes in the Mexican South Pacific Coast. J. Coast. Res. 1154–1164. https://doi.org/10.2112/JCOASTRES-D-11-00118.1.
Catalao, J., Raju, D., Fernandes, R.M.S., 2013. Mapping vertical land movement in Singapore using InSAR GPS. In: ESA Special Publication, vol. 722, p. 54 Retrieved from https://ftp.space.dtu.dk/pub/Ioana/papers/s264_3cata.pdf.
Cazenave, A., Le Cozannet, G., 2013. Sea level rise and its coastal impacts. Earths Future. Retrieved from http://onlinelibrary.wiley.com/doi/10.1002/2013EF000188/abstract.
Chaussard, E., Amelung, F., Abidin, H., Hong, S.-H., 2013. Sinking cities in Indonesia: ALOS PALSAR detects rapid subsidence due to groundwater and gas extraction. Remote Sens. Environ. 128, 150–161.
Cheng, X., Xie, S.-P., Du, Y., Wang, J., Chen, X., Wang, J., 2016. Interannual-to-decadal variability and trends of sea level in the South China Sea. Clim. Dyn. 46 (9–10), 3113–3126. https://doi.org/10.1007/s00382-015-2756-1.
Choudhury, A.M., Haque, M.A., Quadir, D.A., 1997. Consequences of global warming and sea level rise in Bangladesh. Mar. Geodesy 20 (1), 13–31.
Church, J.A., White, N.J., Coleman, R., Lambeck, K., Mitrovica, J.X., 2004. Estimates of the regional distribution of sea level rise over the 1950-2000 period. J. Clim. 17 (13), 2609–2625.
Church, J.A., White, N.J., Hunter, J.R., 2006. Sea-level rise at tropical Pacific and Indian Ocean islands. Glob. Planet. Change 53 (3), 155–168.
Church, J.A., Clark, P.U., Cazenave, A., Gregory, J.M., Jevrejeva, S., Levermann, A., et al., 2013. Sea level change. Clim. Change 1137–1216.
Clarke, A.J., 2014. El Niño physics and El Niño predictability. Annu. Rev. Mar. Sci. 6, 79–99.
Clarke, A.J., Liu, X., 1994. Interannual Sea level in the northern and eastern Indian Ocean. J. Phys. Oceanogr. 24 (6), 1224–1235. https://doi.org/10.1175/1520-0485(1994)024<1224:ISLITN>2.0.CO;2.
Cohen, M.C., Lara, R.J., 2003. Temporal changes of mangrove vegetation boundaries in Amazonia: application of GIS and remote sensing techniques. Wetl. Ecol. Manag. 11 (4), 223–231.

Dai, A., Qian, T., Trenberth, K.E., Milliman, J.D., 2009. Changes in continental freshwater discharge from 1948 to 2004. J. Clim. 22 (10), 2773–2792. https://doi.org/10.1175/2008JCLI2592.1.

Dasgupta, S., Laplante, B., Meisner, C., Wheeler, D., Yan, J., 2009. The impact of sea level rise on developing countries: a comparative analysis. Clim. Change 93 (3–4), 379–388.

de Mesquita, A.R., dos Franco, A.S., Harari, J., de França, C.A.S., 2013. On sea level along the Brazillian coast. Rev. Bras. Geofis. 31 (5), 33–42. https://doi.org/10.22564/rbgf.vol31n5-2013.

Deng, W., Wei, G., Xie, L., Ke, T., Wang, Z., Zeng, T., Liu, Y., 2013. Variations in the Pacific Decadal Oscillation since 1853 in a coral record from the northern south China sea. J. Geophys. Res. Oceans 118 (5), 2358–2366.

Ding, X., Zheng, D., Chen, Y., Chao, J., Li, Z., 2001. Sea level change in Hong Kong from tide gauge measurements of 1954–1999. J. Geodesy 74 (10), 683–689.

Douglas, B.C., 2001. Sea level change in the era of the recording tide gauge. Int. Geophys. 75, 37–64.

Ducarme, B., Venedikov, A.P., de Mesquita, A.R., de Franca, C.A.S., Costa, D.S., Blitzkow, D., et al., 2007. New analysis of a 50 years tide gauge record at Cananéia (SP-Brazil) with the VAV tidal analysis program. In: Tregoning, D.P., Rizos, D.C. (Eds.), Dynamic Planet. Springer, Berlin Heidelberg, pp. 453–460. Retrieved from http://link.springer.com/chapter/10.1007/978-3-540-49350-1_66.

Duvat, V.K.E., Pillet, V., 2017. Shoreline changes in reef islands of the Central Pacific: Takapoto Atoll, Northern Tuamotu, French Polynesia. Geomorphology. 282, 96–118. https://doi.org/10.1016/j.geomorph.2017.01.002.

Duvat, V., Magnan, A., Pouget, F., 2013. Exposure of atoll population to coastal erosion and flooding: a South Tarawa assessment, Kiribati. Sustain. Sci. 8 (3), 423–440. https://doi.org/10.1007/s11625-013-0215-7.

Edelman, A., Gelding, A., Konovalov, E., McComiskie, R., Penny, A., Roberts, N., et al., 2014. State of the Tropics 2014 Report (Report). James Cook University, Cairns. Retrieved from http://stateofthetropics.org/.

Emery, K.O., Aubrey, D.G., 1986. Relative sea-level changes from tide-gauge records of eastern Asia mainland. Mar. Geol. 72 (1), 33–45.

Emery, K.O., Aubrey, D.G., 1989. Tide gauges of India. J. Coast. Res. 489–501.

Emery, K.O., Aubrey, D.G., 1991. Sea Levels, Land Levels, and Tide Gauges. Springer. New York etc. Retrieved from http://www.getcited.org/pub/102899569.

Enfield, D.B., 1989. El Niño, past and present. Rev. Geophys. 27 (1), 159–187.

Erban, L.E., Gorelick, S.M., Zebker, H.A., 2014. Groundwater extraction, land subsidence, and sea-level rise in the Mekong Delta, Vietnam. Environ. Res. Lett. 9 (8), 084010. https://doi.org/10.1088/1748-9326/9/8/084010.

Ericson, J.P., Vörösmarty, C.J., Dingman, S.L., Ward, L.G., Meybeck, M., 2006. Effective sea-level rise and deltas: causes of change and human dimension implications. Glob. Planet. Change 50 (1), 63–82.

Fenoglio-Marc, L., Schöne, T., Illigner, J., Becker, M., Manurung, P., Khafid, 2012. Sea level change and vertical motion from satellite altimetry, tide gauges and GPS in the Indonesian region. Mar. Geodesy 35 (Suppl. 1), 137–150.

Ferrier, K.L., Mitrovica, J.X., Giosan, L., Clift, P.D., 2015. Sea-level responses to erosion and deposition of sediment in the Indus River basin and the Arabian Sea. Earth Planet. Sci. Lett. 416, 12–20.

Fiedler, J.W., Conrad, C.P., 2010. Spatial variability of sea level rise due to water impoundment behind dams. Geophys. Res. Lett. 37 (12). Retrieved from http://onlinelibrary.wiley.com/doi/10.1029/2010GL043462/full.

França, M.C., Francisquini, M.I., Cohen, M.C., Pessenda, L.C., Rossetti, D.F., Guimaraes, J.T., Smith, C.B., 2012. The last mangroves of Marajó Island—eastern Amazon: impact of climate and/or relative sea-level changes. Rev. Palaeobot. Palynol. 187, 50–65.

French, G.T., Awosika, L.F., Ibe, C.E., 1995. Sea-level rise and Nigeria: potential impacts and consequences. J. Coast. Res. 224–242.

Fujihara, Y., Hoshikawa, K., Fujii, H., Kotera, A., Nagano, T., Yokoyama, S., 2015. Analysis and attribution of trends in water levels in the Vietnamese Mekong Delta. Hydrol. Process. Retrieved from http://onlinelibrary.wiley.com/doi/10.1002/hyp.10642/pdf.

Giosan, L., Constantinescu, S., Clift, P.D., Tabrez, A.R., Danish, M., Inam, A., 2006. Recent morphodynamics of the Indus delta shore and shelf. Cont. Shelf Res. 26 (14), 1668–1684. https://doi.org/10.1016/j.csr.2006.05.009.

Gratiot, N., Anthony, E.J., Gardel, A., Gaucherel, C., Proisy, C., Wells, J.T., 2008. Significant contribution of the 18.6 year tidal cycle to regional coastal changes. Nat. Geosci. 1 (3), 169–172.

Guo, J., Hu, Z., Wang, J., Chang, X., Li, G., 2015. Sea level changes of China seas and neighboring ocean based on satellite altimetry missions from 1993 to 2012. J. Coast. Res. 73 (sp1), 17–21.

Gupta, H., Kao, S.-J., Dai, M., 2012. The role of mega dams in reducing sediment fluxes: a case study of large Asian rivers. J. Hydrol. 464, 447–458.

Hallegatte, S., Ranger, N., Mestre, O., Dumas, P., Corfee-Morlot, J., Herweijer, C., Wood, R.M., 2011. Assessing climate change impacts, sea level rise and storm surge risk in port cities: a case study on Copenhagen. Clim. Change 104 (1), 113–137.

Hamlington, B.D., Leben, R.R., Nerem, R.S., Han, W., Kim, K.-Y., 2011. Reconstructing sea level using cyclostationary empirical orthogonal functions. J. Geophys. Res. Oceans. 116 (C12), C12015. https://doi.org/10.1029/2011JC007529.

Han, G., Huang, W., 2009. Low-frequency sea-level variability in the South China Sea and its relationship to ENSO. Theor. Appl. Climatol. 97 (1–2), 41–52.

Han, W., Meehl, G.A., Rajagopalan, B., Fasullo, J.T., Hu, A., Lin, J., et al., 2010. Patterns of Indian Ocean sea-level change in a warming climate. Nat. Geosci. 3 (8), 546–550.

Han, W., Vialard, J., McPhaden, M.J., Lee, T., Masumoto, Y., Feng, M., De Ruijter, W.P., 2014. Indian Ocean decadal variability: a review. Bull. Am. Meteorol. Soc. 95 (11), 1679–1703.

Hanson, S., Nicholls, R., Ranger, N., Hallegatte, S., Corfee-Morlot, J., Herweijer, C., Chateau, J., 2011. A global ranking of port cities with high exposure to climate extremes. Clim. Change. 104 (1), 89–111. https://doi.org/10.1007/s10584-010-9977-4.

He, L., Li, G., Li, K., Shu, Y., 2014. Estimation of regional sea level change in the Pearl River Delta from tide gauge and satellite altimetry data. Estuar. Coast. Shelf Sci. 141, 69–77.

Hedley, P.J., Bird, M.I., Robinson, R.A., 2010. Evolution of the Irrawaddy delta region since 1850. Geogr. J. 176 (2), 138–149.

Higgins, S.A., Overeem, I., Steckler, M.S., Syvitski, J.P., Seeber, L., Akhter, S.H., 2014. InSAR measurements of compaction and subsidence in the Ganges-Brahmaputra Delta, Bangladesh. J. Geophys. Res. Earth Surf. 119 (8), 1768–1781.

Hinkel, J., Brown, S., Exner, L., Nicholls, R.J., Vafeidis, A.T., Kebede, A.S., 2012. Sea-level rise impacts on Africa and the effects of mitigation and adaptation: an application of DIVA. Reg. Environ. Change 12 (1), 207–224.

Holgate, S.J., Matthews, A., Woodworth, P.L., Rickards, L.J., Tamisiea, M.E., Bradshaw, E., et al., 2013. New data systems and products at the permanent service for mean sea level. J. Coast. Res. 288, 493–504. https://doi.org/10.2112/JCOASTRES-D-12-00175.1.

Huq, S., Ali, S.I., Rahman, A.A., 1995. Sea-level rise and Bangladesh: a preliminary analysis. J. Coast. Res. 44–53.

Iftekhar, M.S., Saenger, P., 2008. Vegetation dynamics in the Bangladesh Sundarbans mangroves: a review of forest inventories. Wetl. Ecol. Manag. 16 (4), 291–312.

IPCC AR5, 2013. Climate change 2013: the physical science basis. In: Working Group I Contribution to the Fifth Assessment Report of the Intergovernmental Panel on Climate Change. Summary for Policymakers (IPCC, 2013). Retrieved from http://www.climatechange2013.org/images/report/WG1AR5_Frontmatter_FINAL.pdf.

Ivins, E.R., Dokka, R.K., Blom, R.G., 2007. Post-glacial sediment load and subsidence in coastal Louisiana. Geophys. Res. Lett. 34 (16). Retrieved from http://onlinelibrary.wiley.com/doi/10.1029/2007GL030003/full.

Jallow, B.P., Toure, S., Barrow, M.M., Mathieu, A.A., 1999. Coastal zone of the Gambia and the Abidjan region in Côte d'Ivoire: sea level rise vulnerability, response strategies, and adaptation options. Clim. Res. 12 (2–3), 129–136.

Jurkowski, G., Ni, J., Brown, L., 1984. Modern uparching of the Gulf coastal plain. J. Geophys. Res. Solid Earth 89 (B7), 6247–6255.

Kench, P.S., McLean, R.F., Brander, R.W., Nichol, S.L., Smithers, S.G., Ford, M.R., et al., 2006. Geological effects of tsunami on mid-ocean atoll islands: the Maldives before and after the Sumatran tsunami. Geology 34 (3), 177–180.

Kench, P.S., Thompson, D., Ford, M.R., Ogawa, H., McLean, R.F., 2015. Coral islands defy sea-level rise over the past century: records from a central Pacific atoll. Geology 43 (6), 515–518.

Kesel, R.H., 2003. Human modifications to the sediment regime of the Lower Mississippi River flood plain. Geomorphology 56 (3), 325–334.

Khan, T.M.A., Quadir, D.A., Murty, T.S., Kabir, A., Aktar, F., Sarker, M.A., 2002. Relative sea level changes in Maldives and vulnerability of land due to abnormal coastal inundation. Mar. Geodesy 25 (1–2), 133–143.

Kolker, A.S., Allison, M.A., Hameed, S., 2011. An evaluation of subsidence rates and sea-level variability in the northern Gulf of Mexico. Geophys. Res. Lett. 38 (21). Retrieved from http://www.agu.org/journals/gl/gl1121/2011GL049458/2011gl049458-t01.txt.

Kourtit, K., Nijkamp, P., 2013. In praise of megacities in a global world. Reg. Sci. Policy Pract. 5 (2), 167–182.

Lambeck, K., Woodroffe, C.D., Antonioli, F., Anzidei, M., Gehrels, W.R., Laborel, J., Wright, A.J., 2010. Paleoenvironmental Records, Geophysical Modelling, and Reconstruction of Sea Level Trends and Variability on Centennial and Longer Timescales. Wiley-Blackwell. Retrieved from http://www.academia.edu/download/29707925/LambecketalJL.pdf.

Le Cozannet, G., Garcin, M., Petitjean, L., Cazenave, A., Becker, M., Meyssignac, B., et al., 2013. Exploring the relation between sea level rise and shoreline erosion using sea level reconstructions: an example in French Polynesia. J. Coast. Res. 65. Retrieved from http://ics2013.org/papers/Paper3699_rev.pdf.

Le Cozannet, G., Raucoules, D., Wöppelmann, G., Garcin, M., Da Sylva, S., Meyssignac, B., et al., 2015. Vertical ground motion and historical sea-level records in Dakar (Senegal). Environ. Res. Lett. 10 (8), 084016.

Lee, H.S., 2013. Estimation of extreme sea levels along the Bangladesh coast due to storm surge and sea level rise using EEMD and EVA. J. Geophys. Res. Oceans. 118 (9), 4273–4285. https://doi.org/10.1002/jgrc.20310.

Lemos, A.T., Ghisolfi, R.D., 2011. Long-term mean sea level measurements along the Brazilian coast: a preliminary assessment. Pan Am. J. Aquat. Sci. 5 (2), 331–340.

Letetrel, C., Karpytchev, M., Bouin, M.-N., Marcos, M., Santamaría-Gómez, A., Wöppelmann, G., 2015. Estimation of vertical land movement rates along the coasts of the Gulf of Mexico over the past decades. Continent. Shelf Res. 111, 42–51.

Llovel, W., Cazenave, A., Rogel, P., Lombard, A., Nguyen, M.B., 2009. Two-dimensional reconstruction of past sea level (1950–2003) from tide gauge data and an Ocean General Circulation Model. Clim. Past. 5 (2), 217–227. https://doi.org/10.5194/cp-5-217-2009.

Losada, I.J., Reguero, B.G., Méndez, F.J., Castanedo, S., Abascal, A.J., Mínguez, R., 2013. Long-term changes in sea-level components in Latin America and the Caribbean. Glob. Planet. Change 104, 34–50.

Loucks, C., Barber-Meyer, S., Hossain, M.A.A., Barlow, A., Chowdhury, R.M., 2010. Sea level rise and tigers: predicted impacts to Bangladesh's Sundarbans mangroves. Clim. Change 98 (1), 291–298.

Lovelock, C.E., Cahoon, D.R., Friess, D.A., Guntenspergen, G.R., Krauss, K.W., Reef, R., et al., 2015. The vulnerability of Indo-Pacific mangrove forests to sea-level rise. Nature. Retrieved from http://www.nature.com/nature/journal/vaop/ncurrent/full/nature15538.html.

Mansur, A.V., Brondízio, E.S., Roy, S., Hetrick, S., Vogt, N.D., Newton, A., 2016. An assessment of urban vulnerability in the Amazon Delta and Estuary: a multi-criterion index of flood exposure, socio-economic conditions and infrastructure. Sustain. Sci. 1–19.

McCann, W.R., 2006. Estimating the Threat of Tsunamogenic Earthquakes and Earthquake Induced-Landslide Tsunami in the Caribbean. World Scientific Publishing, Singapore. Retrieved from https://books.google.fr/books?hl=fr&lr=&id=OVz0DNfCdYMC&oi=fnd&pg=PA43&dq=W.McCann+2006+tsunami&ots=sSXZ0uC3jl&sig=8OJQob1YijoYkwKqgputZHVzVVc.

McGranahan, G., Balk, D., Anderson, B., 2007. The rising tide: assessing the risks of climate change and human settlements in low elevation coastal zones. Environ. Urban. 19 (1), 17–37.

McLean, R., Kench, P., 2015. Destruction or persistence of coral atoll islands in the face of 20th and 21st century sea-level rise? Wiley Interdiscip. Rev. Clim. Change 6 (5), 445–463.

Mcleod, E., Hinkel, J., Vafeidis, A.T., Nicholls, R.J., Harvey, N., Salm, R., 2010. Sea-level rise vulnerability in the countries of the Coral Triangle. Sustain. Sci. 5 (2), 207–222.

Mei-e, R., 1993. Relative sea-level changes in China over the last 80 years. J. Coast. Res. 229–241.

Melet, A., Almar, R., Meyssignac, B., 2016. What dominates sea level at the coast: a case study for the Gulf of Guinea. Ocean Dyn. 66 (5), 623–636.

Merrifield, M.A., 2011. A shift in western tropical Pacific sea level trends during the 1990s. J. Clim. 24 (15), 4126–4138.

Merrifield, M.A., Thompson, P.R., Lander, M., 2012. Multidecadal sea level anomalies and trends in the western tropical Pacific. Geophys. Res. Lett. 39 (13). Retrieved from http://onlinelibrary.wiley.com/doi/10.1029/2012GL052032/full.

Mesquita, A.R., 2003. Sea-level variations along the Brazilian Coast: a short review. J. Coast. Res. 21–31.

Meyssignac, B., Becker, M., Llovel, W., Cazenave, A., 2012a. An assessment of two-dimensional past sea level reconstructions over 1950–2009 based on tide-gauge data and different input sea level grids. Surv. Geophys. 1–28.

Meyssignac, B., Salas y Melia, D., Becker, M., Llovel, W., Cazenave, A., 2012b. Tropical Pacific spatial trend patterns in observed sea level: internal variability and/or anthropogenic signature? Clim. Past 8 (2), 787–802.

Milliman, J., Haq, B.U., 1996. Sea-Level Rise and Coastal Subsidence: Causes, Consequences, and Strategies, vol. 2. Springer Science & Business Media. Retrieved from https://books.google.fr/books?hl=fr&lr=&id=oXrPSnlFswQC&oi=fnd&pg=PR13&dq=milliman+haq+1996&ots=rzxOvOOmnK&sig=5F2wpoUq177_92Zfh5CGpbLzWe0.

Milliman, J.D., Broadus, J.M., Gable, F., 1989. Environmental and economic implications of rising sea level and subsiding deltas: the Nile and Bengal examples. Ambio 340–345.

Milly, P.C.D., Cazenave, A., Famiglietti, J.S., Gornitz, V., Laval, K., Lettenmaier, D.P., et al., 2010. Terrestrial water-storage contributions to sea-level rise and variability. Underst. Sea Level Rise Var. 226–255.

Milne, G.A., Gehrels, W.R., Hughes, C.W., Tamisiea, M.E., 2009. Identifying the causes of sea-level change. Nat. Geosci. 2 (7), 471–478. https://doi.org/10.1038/ngeo544.
Mimura, N., Nurse, L., McLean, R., Agard, J., Briguglio, L., Lefale, P., et al., 2007. Small islands. Clim. Change 687–716.
Mitchum, G.T., Wyrtki, K., 1988. Overview of Pacific sea level variability. Mar. Geodesy 12 (4), 235–245.
Mitrovica, J.X., Tamisiea, M.E., Davis, J.L., Milne, G.A., 2001. Recent mass balance of polar ice sheets inferred from patterns of global sea-level change. Nature 409 (6823), 1026–1029.
Mittermeier, R.A., Turner, W.R., Larsen, F.W., Brooks, T.M., Gascon, C., 2011. Global biodiversity conservation: the critical role of hotspots. In: Biodiversity Hotspots. Springer, pp. 3–22. Retrieved from http://link.springer.com/chapter/10.1007/978-3-642-20992-5_1.
Mochizuki, T., Kimoto, M., Watanabe, M., Chikamoto, Y., Ishii, M., 2016. Interbasin effects of the Indian Ocean on Pacific decadal climate change. Geophys. Res. Lett. 43 (13). 2016GL069940 https://doi.org/10.1002/2016GL069940.
Moon, J.-H., Song, Y.T., Lee, H., 2015. PDO and ENSO modulations intensified decadal sea level variability in the tropical Pacific. J. Geophys. Res. Oceans 120 (12), 8229–8237.
Moriconi-Ebrard, F., Harre, D., Heinrigs, P., 2016. Urbanisation Dynamics in West Africa 1950–2010. Organisation for Economic Co-operation and Development, Paris. Retrieved from http://www.oecd-ilibrary.org/content/book/9789264252233-en.
Morton, R.A., Bernier, J.C., Barras, J.A., 2006. Evidence of regional subsidence and associated interior wetland loss induced by hydrocarbon production, Gulf Coast region, USA. Environ. Geol. 50 (2), 261–274.
Muehe, D., 2006. Erosion in the Brazilian coastal zone: an overview. J. Coast. Res. 43–48.
Muehe, D., 2010. Brazilian coastal vulnerability to climate change. Pan Am. J. Aquat. Sci. 5 (2), 173–183.
Muehe, D., Neves, C.F., 1995. The implications of sea-level rise on the Brazilian coast: a preliminary assessment. J. Coast. Res. 54–78.
Mundial, B., 2014. World development indicators 2014. Relac. Int. Retrieved from http://sedici.unlp.edu.ar/bitstream/handle/10915/38454/Documento_completo.pdf?sequence=1.
Nali, J.O., Rigo, D., 2011. Urban floods: assessing the effects of sea level rise and mitigation measures. In: Presented at the 12nd International Conference on Urban Drainage, Porto Alegre/Brazil Retrieved from https://web.sbe.hw.ac.uk/staffprofiles/bdgsa/temp/12th%20ICUD/PDF/PAP005452.pdf.
Nandy, S., Bandopadhyay, S., 2011. Trend of sea level change in the Hugli estuary, India. Indian J. Geo Mar. Sci. 40 (6), 802–812.
Neves, C.F., Muehe, D., 1995. Potential impacts of sea-level rise on the metropolitan region of Recife, Brazil. J. Coast. Res. 116–131.
Nicholls, R.J., Cazenave, A., 2010. Sea-level rise and its impact on coastal zones. Science 328 (5985), 1517–1520.
Nicholls, R.J., Mimura, N., 1998. Regional issues raised by sea-level rise and their policy implications. Clim. Res. 11 (1), 5–18.
Nicholls, R.J., Hoozemans, F.M., Marchand, M., 1999. Increasing flood risk and wetland losses due to global sea-level rise: regional and global analyses. Glob. Environ. Change 9, S69–S87.
Nicholls, R.J., Marinova, N., Lowe, J.A., Brown, S., Vellinga, P., De Gusmao, D., et al., 2011. Sea-level rise and its possible impacts given a 'beyond 4 C world' in the twenty-first century. Philos. Trans. R. Soc. Lond. A Math. Phys. Eng. Sci. 369 (1934), 161–181.
Nieves, V., Willis, J.K., Patzert, W.C., 2015. Recent hiatus caused by decadal shift in Indo-Pacific heating. Science 349 (6247), 532–535.

Nurse, L.A., Mclean, R.F., Agard, J., Briguglio, L.P., Duvat-magnan, V., Pelesikoti, N., et al., 2014. Small islands. In: Barros, V.R., Field, C.B., Dokken, D.J., Mastrandrea, M.D., Mach, K.J., Bilir, T.E., Chatterjee, M., Ebi, K.L., Estrada, Y.O., Genova, R.C., Girma, B., Kissel, E.S., Levy, A.N., MacCracken, S., Mastrandrea, P.R., White, L.L. (Eds.), Climate Change 2014: Impacts, Adaptation, and Vulnerability. Part B: Regional Aspects. Contribution of Working Group II to the Fifth Assessment Report of the Intergovernmental Panel on Climate Change. Cambridge University Press, p. 1613. Retrieved from https://hal.archives-ouvertes.fr/hal-01090732/.

Overeem, I., Syvitski, J.P.M., 2009. Dynamics and Vulnerability of Delta Systems. GKSS Research Centre, LOICZ Internat. Project Office, Inst. for Coastal Research. Retrieved from http://46.37.191.43/~futureearth/wp-content/uploads/2016/02/LOICZ-RS35.pdf.

Palanisamy, H., Becker, M., Meyssignac, B., Henry, O., Cazenave, A., 2012. Regional sea level change and variability in the Caribbean sea since 1950. J. Geod. Sci. 2 (2), 125–133.

Palanisamy, H., Cazenave, A., Meyssignac, B., Soudarin, L., Wöppelmann, G., Becker, M., 2014. Regional sea level variability, total relative sea level rise and its impacts on islands and coastal zones of Indian Ocean over the last sixty years. Glob. Planet. Change. 116, 54–67. https://doi.org/10.1016/j.gloplacha.2014.02.001.

Palanisamy, H., Cazenave, A., Delcroix, T., Meyssignac, B., 2015. Spatial trend patterns in the Pacific Ocean sea level during the altimetry era: the contribution of thermocline depth change and internal climate variability. Ocean Dyn. 65 (3), 341–356.

Payo Garcia, A., Mukhopadhyay, A., Hazra, S., Ghosh, T., Ghosh, S., Brown, S., et al., 2016. Projected changes in area of the Sundarban mangrove forest in Bangladesh due to SLR by 2100. Clim. Change. 139 (2), 279–291. https://doi.org/P10.1007/s10584-016-1769-z.

Peltier, W.R., 2004. Global glacial isostasy and the surface of the ice-age Earth: the ICE-5G (VM2) model and GRACE. Annu. Rev. Earth Planet. Sci. 32, 111–149.

Peng, D., Palanisamy, H., Cazenave, A., Meyssignac, B., 2013. Interannual sea level variations in the South China Sea over 1950–2009. Mar. Geodesy 36 (2), 164–182.

Perez, R.T., Amadore, L.A., Feir, R.B., 1999. Climate change impacts and responses in the Philippines coastal sector. Clim. Res. 12 (2–3), 97–107.

Pethick, J., Orford, J.D., 2013. Rapid rise in effective sea-level in southwest Bangladesh: its causes and contemporary rates. Glob. Planet. Change. 111, 237–245. https://doi.org/10.1016/j.gloplacha.2013.09.019.

Phien-wej, N., Giao, P.H., Nutalaya, P., 2006. Land subsidence in Bangkok, Thailand. Eng. Geol. 82 (4), 187–201. https://doi.org/10.1016/j.enggeo.2005.10.004.

Piecuch, C.G., Ponte, R.M., 2015. Inverted barometer contributions to recent sea level changes along the northeast coast of North America. Geophys. Res. Lett. 42 (14), 5918–5925.

Ponte, R.M., 1994. Understanding the relation between wind-and pressure-driven sea level variability. J. Geophys. Res. Oceans 99 (C4), 8033–8039.

Pugh, D., Woodworth, P., 2014. Sea-Level Science: Understanding Tides, Surges, Tsunamis and Mean Sea-Level Changes. Cambridge University Press. Retrieved from https://books.google.fr/books?hl=fr&lr=&id=QiBGAwAAQBAJ&oi=fnd&pg=PR7&dq=%22bangladesh%22+sea+level+tide+gauge&ots=2v_zrsxpfy&sig=1Y6RGByE_iiZv4G0epFQc-jkjbE.

Rahman, A.F., Dragoni, D., El-Masri, B., 2011. Response of the Sundarbans coastline to sea level rise and decreased sediment flow: a remote sensing assessment. Remote Sens. Environ. 115 (12), 3121–3128. https://doi.org/10.1016/j.rse.2011.06.019.

Rao, K.N., Subraelu, P., Rao, T.V., Malini, B.H., Ratheesh, R., Bhattacharya, S., et al., 2008. Sea-level rise and coastal vulnerability: an assessment of Andhra Pradesh coast, India through remote sensing and GIS. J. Coast. Conserv. 12 (4), 195–207.

Raucoules, D., Le Cozannet, G., Wöppelmann, G., De Michele, M., Gravelle, M., Daag, A., Marcos, M., 2013. High nonlinear urban ground motion in Manila (Philippines) from 1993 to 2010 observed by DInSAR: implications for sea-level measurement. Remote Sens. Environ. 139, 386–397.

Ray, R.D., Douglas, B.C., 2011. Experiments in reconstructing twentieth-century sea levels. Prog. Oceanogr. 91 (4), 496–515. https://doi.org/10.1016/j.pocean.2011.07.021.

Reguero, B.G., Losada, I.J., Díaz-Simal, P., Méndez, F.J., Beck, M.W., 2015. Effects of climate change on exposure to coastal flooding in Latin America and the Caribbean. PLoS One. 10 (7), e0133409. https://doi.org/10.1371/journal.pone.0133409.

Riva, R.E., Bamber, J.L., Lavallée, D.A., Wouters, B., 2010. Sea-level fingerprint of continental water and ice mass change from GRACE. Geophys. Res. Lett. 37 (19). Retrieved from http://onlinelibrary.wiley.com/doi/10.1029/2010GL044770/full.

Roden, G.I., 1963. Sea level variations at Panama. J. Geophys. Res. 68 (20), 5701–5710.

Rodolfo, K.S., Siringan, F.P., 2006. Global sea-level rise is recognised, but flooding from anthropogenic land subsidence is ignored around northern Manila Bay, Philippines. Disasters 30 (1), 118–139.

Rong, Z., Liu, Y., Zong, H., Cheng, Y., 2007. Interannual sea level variability in the South China Sea and its response to ENSO. Glob. Planet. Change 55 (4), 257–272.

Ruane, A.C., Major, D.C., Winston, H.Y., Alam, M., Hussain, S.G., Khan, A.S., et al., 2013. Multi-factor impact analysis of agricultural production in Bangladesh with climate change. Glob. Environ. Change 23 (1), 338–350.

Saglio-Yatzimirsky, M.-C., 2013. Megacity Slums: Social Exclusion, Space and Urban Policies in Brazil and India, vol. 1. World Scientific. Retrieved from https://books.google.fr/books?hl=fr&lr=&id=PDi6CgAAQBAJ&oi=fnd&pg=PR7&dq=megacity+slums&ots=fSZaZ0mCAp&sig=gC6_HziOvuzohQJ6QaGeKqQ4iXI.

Sallenger, Jr., A. H., Doran, K. S., Howd, P. A., 2012. Hotspot of accelerated sea-level rise on the Atlantic coast of North America. Nat. Clim. Change. https://doi.org/10.1038/nclimate1597.

Santamaría-Gómez, A., Gravelle, M., Dangendorf, S., Marcos, M., Spada, G., Wöppelmann, G., 2017. Uncertainty of the 20th century sea-level rise due to vertical land motion errors. Earth Planet. Sci. Lett. 473, 24–32.

Saramul, S., Ezer, T., 2014. Spatial variations of sea level along the coast of Thailand: impacts of extreme land subsidence, earthquakes and the seasonal monsoon. Glob. Planet. Change 122, 70–81.

Sarwar, M.G.M., 2013. Sea-level rise along the coast of Bangladesh. In: Disaster Risk Reduction Approaches in Bangladesh. Springer, pp. 217–231. Retrieved from http://link.springer.com/chapter/10.1007/978-4-431-54252-0_10.

Sarwar, M.G.M., Woodroffe, C.D., 2013. Rates of shoreline change along the coast of Bangladesh. J. Coast. Conserv. 17 (3), 515–526. https://doi.org/10.1007/s11852-013-0251-6.

Shankar, D., Shetye, S.R., 1999. Are interdecadal sea level changes along the Indian coast influenced by variability of monsoon rainfall? J. Geophys. Res. Oceans (1978–2012) 104 (C11), 26031–26042.

Shankar, D., Aparna, S.G., McCreary, J.P., Suresh, I., Neetu, S., Durand, F., et al., 2010. Minima of interannual sea-level variability in the Indian Ocean. Prog. Oceanogr. 84 (3–4), 225–241. https://doi.org/10.1016/j.pocean.2009.10.002.

Shearman, P., Bryan, J., Walsh, J.P., 2013. Trends in deltaic change over three decades in the Asia-Pacific region. J. Coast. Res. 290, 1169–1183. https://doi.org/10.2112/JCOASTRES-D-12-00120.1.

Short, A.D., da Klein, A.H.F., 2016. Brazilian beach systems: review and overview. In: Brazilian Beach Systems. Springer, pp. 573–608. Retrieved from http://link.springer.com/chapter/10.1007/978-3-319-30394-9_20.

Singh, O.P., 2002. Predictability of sea level in the Meghna estuary of Bangladesh. Glob. Planet. Change. 32 (2–3), 245–251. https://doi.org/10.1016/S0921-8181(01)00152-7.

Singh, R.B., 2014. Urban Development Challenges, Risks and Resilience in Asian Mega Cities. Springer.

Sinha, P.C., Rao, Y.R., Dube, S.K., Murty, T.S., 1997. Effect of sea level rise on tidal circulation in the Hooghly Estuary, Bay of Bengal. Mar. Geodesy. 20 (4), 341–366. https://doi.org/10.1080/01490419709388114.

Soumya, M., Vethamony, P., Tkalich, P., 2015. Inter-annual sea level variability in the southern South China Sea. Glob. Planet. Change 133, 17–26.

Stammer, D., 2008. Response of the global ocean to Greenland and Antarctic ice melting. J. Geophys. Res. Oceans 113 (C6). Retrieved from http://onlinelibrary.wiley.com/doi/10.1029/2006JC004079/full.

Stammer, D., Cazenave, A., Ponte, R.M., Tamisiea, M.E., 2013. Causes for contemporary regional sea level changes. Annu. Rev. Mar. Sci. 5, 21–46.

Strassburg, M.W., Hamlington, B.D., Manrung, R.R., Lumban-Gaol, J., Nababan, B., Kim, K.-Y., 2015. Sea level trends in Southeast Asian seas. Clim. Past 11 (5). Retrieved from http://digitalcommons.odu.edu/ccpo_pubs/138/.

Suresh, I., Vialard, J., Lengaigne, M., Han, W., McCreary, J., Durand, F., Muraleedharan, P.M., 2013. Origins of wind-driven intraseasonal sea level variations in the North Indian Ocean coastal waveguide. Geophys. Res. Lett. 40 (21), 2013GL058312. https://doi.org/10.1002/2013GL058312.

Suresh, I., Vialard, J., Izumo, T., Lengaigne, M., Han, W., McCreary, J., Muraleedharan, P.M., 2016. Dominant role of winds near Sri Lanka in driving seasonal sea level variations along the west coast of India. Geophys. Res. Lett. 43 (13), 7028–7035.

Syvitski, J.P., 2008. Deltas at risk. Sustain. Sci. 3 (1), 23–32.

Syvitski, J.P., Kettner, A., 2011. Sediment flux and the Anthropocene. Philos. Trans. R. Soc. Lond. A Math. Phys. Eng. Sci. 369 (1938), 957–975.

Syvitski, J.P., Kettner, A.J., Overeem, I., Hutton, E.W., Hannon, M.T., Brakenridge, G.R., et al., 2009. Sinking deltas due to human activities. Nat. Geosci. 2 (10), 681–686.

Syvitski, J.P., Kettner, A.J., Overeem, I., Giosan, L., Brakenridge, G.R., Hannon, M., Bilham, R., 2013. Anthropocene metamorphosis of the Indus Delta and lower floodplain. Anthropocene 3, 24–35.

Tamisiea, M.E., 2011. Ongoing glacial isostatic contributions to observations of sea level change. Geophys. J. Int. 186 (3), 1036–1044.

Tamisiea, M.E., Mitrovica, J.X., 2011. The moving boundaries of sea level change: understanding the origins of geographic variability. Oceanography. Retrieved from http://agris.fao.org/agris-search/search.do?recordID=DJ2012090245.

Thompson, P.R., Piecuch, C.G., Merrifield, M.A., McCreary, J.P., Firing, E., 2016. Forcing of recent decadal variability in the equatorial and North Indian Ocean. J. Geophys. Res. Oceans 121 (9), 6762–6778.

Tkalich, P., Vethamony, P., Luu, Q.-H., Babu, M.T., 2013. Sea Level Trend and Variability in the Singapore Strait. Retrieved from http://drs.nio.org/drs/handle/2264/4270.

Törnqvist, T.E., Wallace, D.J., Storms, J.E., Wallinga, J., Van Dam, R.L., Blaauw, M., et al., 2008. Mississippi Delta subsidence primarily caused by compaction of Holocene strata. Nat. Geosci. 1 (3), 173–176.

Torres, R.R., Tsimplis, M.N., 2013. Sea-level trends and interannual variability in the Caribbean Sea. J. Geophys. Res. Oceans 118 (6), 2934–2947.

Trisirisatayawong, I., Naeije, M., Simons, W., Fenoglio-Marc, L., 2011. Sea level change in the Gulf of Thailand from GPS-corrected tide gauge data and multi-satellite altimetry. Glob. Planet. Change 76 (3), 137–151.

Tseng, Y.-H., Breaker, L.C., Chang, E.T.-Y., 2010. Sea level variations in the regional seas around Taiwan. J. Oceanogr. 66 (1), 27–39.

UN-HABITAT, 2014. State of African Cities 2014, Re-Imagining Sustainable Urban Transitions, vol. 004/14E. UN-Habitat. Retrieved from http://unhabitat.org/books/state-of-african-cities-2014-re-imagining-sustainable-urban-transitions/.

Unnikrishnan, A.S., Shankar, D., 2007. Are sea-level-rise trends along the coasts of the north Indian Ocean consistent with global estimates? Glob. Planet. Change 57 (3), 301–307.

Veit, E., Conrad, C.P., 2016. The impact of groundwater depletion on spatial variations in sea level change during the past century. Geophys. Res. Lett. 43 (7), 3351–3359.

Wada, Y., Beek, L.P., Weiland, F.C.S., Chao, B.F., Wu, Y.-H., Bierkens, M.F., 2012. Past and future contribution of global groundwater depletion to sea-level rise. Geophys. Res. Lett. 39 (9). Retrieved from http://onlinelibrary.wiley.com/doi/10.1029/2012GL051230/full.

Wada, Y., Lo, M.-H., Yeh, P.J.-F., Reager, J.T., Famiglietti, J.S., Wu, R.-J., Tseng, Y.-H., 2016. Fate of water pumped from underground and contributions to sea-level rise. Nat. Clim. Change 6 (8), 777–780.

Warrick, R.A., Ahmad, Q.K., 2012. The Implications of Climate and Sea-Level Change for Bangladesh. Springer Science & Business Media. Retrieved from https://books.google.fr/books?hl=fr&lr=&id=Q3DvCAAAQBAJ&oi=fnd&pg=PP8&dq=%22bangladesh%22+sea+level+tide+gauge&ots=lnrOsBYEe7&sig=Nz-eDXAkLzim6RTvGPW3wBKZZB8.

Webb, A.P., Kench, P.S., 2010. The dynamic response of reef islands to sea-level rise: evidence from multi-decadal analysis of island change in the Central Pacific. Glob. Planet. Change 72 (3), 234–246.

Wilson, S.G., Fischetti, T.R., 2010. Coastline Population Trends in the United States: 1960 to 2008. US Department of Commerce, Economics and Statistics Administration, US Census Bureau.

Wilson, C.A., Goodbred, S.L., January 5, 2015. Construction and Maintenance of the Ganges-Brahmaputra-Meghna Delta: Linking Process, Morphology, and Stratigraphy [review-article]. From http://www.annualreviews.org/doi/10.1146/annurev-marine-010213-135032.

Wolanski, E., 2006. The Environment in Asia Pacific Harbours. Springer. Retrieved from http://link.springer.com/content/pdf/10.1007/1-4020-3655-8.pdf.

Wolstencroft, M., Shen, Z., Törnqvist, T.E., Milne, G.A., Kulp, M., 2014. Understanding subsidence in the Mississippi Delta region due to sediment, ice, and ocean loading: insights from geophysical modeling. J. Geophys. Res. Solid Earth 119 (4). 2013JB010928 https://doi.org/10.1002/2013JB010928.

Woodworth, P.L., 2005. Have there been large recent sea level changes in the Maldive Islands? Glob. Planet. Change 49 (1), 1–18.

Woodworth, P.L., Aman, A., Aarup, T., 2007. Sea level monitoring in Africa. Afr. J. Mar. Sci. 29 (3), 321–330.

Wöppelmann, G., Marcos, M., 2016. Vertical land motion as a key to understanding sea level change and variability. Rev. Geophys. 54 (1), 2015RG000502. https://doi.org/10.1002/2015RG000502.

Wöppelmann, G., Miguez, B.M., Bouin, M.-N., Altamimi, Z., 2007. Geocentric sea-level trend estimates from GPS analyses at relevant tide gauges world-wide. Glob. Planet. Change 57 (3), 396–406.

Wöppelmann, G., Míguez, B.M., Créach, R., 2008. Tide gauge records at Dakar, Senegal (Africa): towards a 100-years consistent sea-level time series? In: European Geosciences Union, General Assembly 2008 (Vienna, Austria, 13–18th April 2008). Retrieved from http://eseas.sonel.org/IMG/pdf/EGU2008_GW.pdf.

Wu, T.W., Song, L.S., Li, W.L., Wang, Z.W., Zhang, H.Z., Xin, X.X., et al., 2014. An overview of BCC climate system model development and application for climate change studies. J. Meteorol. Res. 28 (1), 34–56. https://doi.org/10.1007/s13351-014-3041-7.

Wunsch, C., Stammer, D., 1997. Atmospheric loading and the oceanic "inverted barometer" effect. Rev. Geophys. 35 (1), 79–107.
Wyrtki, K., 1973. Teleconnections in the equatorial Pacific Ocean. Science 180 (4081), 66–68.
Wyrtki, K., 1975. El Niño—the dynamic response of the equatorial Pacific Oceanto atmospheric forcing. J. Phys. Oceanogr. 5 (4), 572–584.
Xu, B., Feng, G., Li, Z., Wang, Q., Wang, C., Xie, R., 2016. Coastal subsidence monitoring associated with land reclamation using the point target based SBAS-InSAR method: a case study of Shenzhen, China. Remote Sens. 8 (8), 652. https://doi.org/10.3390/rs8080652.
Yanagi, T., Akaki, T., 1994. Sea Level variation in the Eastern Asia. J. Oceanogr. 50 (6), 643–651.
Zhang, X., Church, J.A., 2012. Sea level trends, interannual and decadal variability in the Pacific Ocean. Geophys. Res. Lett. 39 (21). Retrieved from http://onlinelibrary.wiley.com/doi/10.1029/2012GL053240/full.

CHAPTER 8

Exploring Tropical Variability and Extremes Impacts on Population Vulnerability in Piura, Peru: The Case of the 1997–98 El Niño

Ivan J. Ramírez[1,2]

[1]Department of Geography and Environmental Sciences, University of Colorado Denver, Denver, CO, United States; [2]Consortium for Capacity Building/INSTAAR, University of Colorado Boulder, Boulder, CO, United States

Contents

1. INTRODUCTION

Climate variability and extreme events and their impacts on human, ecological, and physical systems have serious implications for public health, food security, and economic development. According to the World Meteorological Organization, impacts on society from 1970 to 2012 from hydrological and meteorological extreme events (hereinafter referred to as hydrometeorological extremes), such as droughts, temperature and rainfall extremes, storms, floods, mass movement wet (e.g., landslides), flash floods, and related disasters, include 1.94 million deaths and estimated economic losses of US$ 2.4 trillion (2014: 3). In the continent of South America, alone, for example,

Tropical Extremes: Natural Variability and Trends
ISBN 978-0-12-809248-4
https://doi.org/10.1016/B978-0-12-809248-4.00008-X

hydrometeorological extremes contributed to 54,995 lives lost and US$ 71.8 billion in economic losses. Floods were the primary cause of impact on human mortality (80.0%) and economy (63.0%) in South America (WMO, 2014: 18). According to the Intergovernmental Panel on Climate Change (IPCC) (2012), disaster losses have increased since the 1950s, and developing countries have felt the greatest burden of impact (as a proportion of gross domestic product). Furthermore, the trends of some hydrometeorological extremes have changed in character (e.g., longer droughts, extreme precipitation) but to varying degrees of confidence because of data limitations and confounding human-related factors. There is high confidence, however, that the rise in climate-related disaster impacts and losses are attributable to increasing societal exposure and vulnerability (IPCC, 2012).

Within this context of hydrometeorological risk to societies, there is a great need to better understand population vulnerability to hydrometeorological extremes (Climate and Development Knowledge Network [CDKN], 2014), including how societal vulnerability intersects with and influences the impacts of climate, water, and weather-related hazards. Societal vulnerability refers to the socioeconomic, political, cultural, institutional, and environmental factors that mediate exposure and risk of peoples and places (Lavell et al., 2012), at various scales from individuals to households and communities. It also reflects the societal context, which underlies health risks in a population (i.e., social determinants of health (Marmot, 2005)). Of particular import is understanding vulnerability in the tropical environments of the Global South[1], where communities face greater exposure and risk to climate variability and hydrometeorological extremes (IPCC, 2012).

In the Global South, an important source of climate variability and hydrometeorological risk is the El Niño–Southern Oscillation (ENSO) pattern. During ENSO, ocean–atmosphere interactions that are basin-wide or regional, influence local to global weather patterns, increasing the probability of climate, water, and weather anomalies around the world (McPhaden et al., 2006; Zebiak et al., 2015; Glantz, 2015). Consequently, these linkages of quasiperiodic climatic changes, which occur every 3–8 years, have

[1] Global South refers to the concept of the "North–South divide," which describes the inequality gap in terms of economic and social development between countries that reside in the Northern Hemisphere (NH) (rich) and Southern Hemisphere (SH) (poor) (https://www.rgs.org/NR/rdonlyres/6AFE1B7F-9141-472A-95C1-52AA291AA679/0/60sGlobalNorthSouthDivide.pdf). The concept is imperfect because countries such as Australia and New Zealand are geographically located in the SH but are categorized as Global North because of their economies. As well, it does not account for the emerging economies of the BRIC (Brazil, Russia, India, and China) countries. Nevertheless, it is a useful concept to highlight uneven development that generally persists between the two hemispheres.

untoward effects on population health and well-being (Nicholls, 1991; Kovats et al., 2003; World Health Organization [WHO], 2016).

ENSO teleconnections and associated hydrometeorological extremes affect society and population health through various direct and indirect exposure pathways and impacts on what the World Health Organization refers to as "the basic determinants of health" (2009). Basic determinants are related to and part of the social determinants of health which refers to a broader critical concept in public health, which implicates social, economic, and political structures as the cause of health inequalities (http://www.who.int/social_determinants/en/). They refer to basic needs which are largely unmet in large segments of populations (Gasper, 2005: 1), living in countries where the burden of climate change–related impacts on public health is disproportionate (WHO, 2009). Basics include air, water, shelter, and food, as well as freedom from disease. The initial determinants highlight the fact that during El Niños and La Niñas, i.e., the warm and cold phases of ENSO, hydrometeorological extremes have the potential to create high impacts on the natural, social, and built environments in which people live, work, and subsist. Consequently, the basic necessities of populations, including the natural resources, which they depend on and infrastructures (e.g., water, sanitation, public health, and energy), which prevent disease exposure degrade or become disrupted (WHO, 2016). The latter determinant (i.e., freedom from disease) refers to the proliferation of infectious diseases when infrastructures collapse and ecologies, both coastal and inland, are altered because of climate variability and hydrometeorological extremes spawned by ENSO (Ramirez, 2015). Thus, during periods of ENSO, particularly the El Niño phase, population exposure to a variety of climate-related hazards and health risks may increase (WHO, 2009, 2016), particularly in vulnerable regions.

In this context, the purpose of this chapter is to explore the impacts of El Niño-related hydrometeorological extremes on society and population health in northwestern South America. Using a cross-disciplinary approach, this research presents a case study that focuses on impacts and vulnerability associated with the 1997–98 El Niño in Piura, Peru. This place-based study illustrates the complexity of tropical vulnerability and risk in the Global South and highlights the importance of understanding intersections among climate variability, hydrometeorological extremes, and society. In doing so, such place-based knowledge may provide lessons for disaster risk reduction (DRR) and capacity building in public health and development planning, while suggesting ways to enhance societal resilience to present and future hydrometeorological extremes.

Following this introduction is the conceptual approach and framework used to explore the impacts of El Niño-related hydrometeorological extremes on society and population health. The next sections discuss regional and national vulnerability to El Niño-related hydrometeorological extremes and the application of the framework to the case study of Piura in northern Peru, which illustrates how hydrometeorological hazards intersect with societal vulnerabilities and population health, including infectious disease risk, in a semiarid ecosystem adjacent to the Pacific Ocean. In conclusion, the chapter closes with lessons for public health and DRR in the context of hydrometeorological extremes.

2. CONCEPTUAL APPROACH

This study explores the effects of tropical variability and extremes on society and population health using a cross-disciplinary lens, which Glantz (2003) refers to as "climate affairs." Climate affairs is a system-based multidisciplinary approach to understanding the many facets of climate, including change, variability, and extremes, and how societies interact with climate and ocean impacts on human, ecological, and physical systems (Glantz and Adeel, 2001; Glantz, 2003; Ramirez et al., 2013). In simplest terms, the approach is a guide to understanding climate from a holistic and broader perspective that includes not only physical science but also social dimensions and even considerations of climate ethics (Glantz and Jamieson, 2000; Glantz, 2003; Jamieson, 2014), which often underlie climate vulnerability.

Fig. 8.1 shows the basic elements or pillars of a climate affairs approach, which provides a flexible framework to examine El Niño-related hydrometeorological extremes and its impacts on and interactions with society (see Box 8.1 for climate affairs background). As a research lens, it offers insights about climate and ocean phenomena, which may otherwise be overlooked in a more traditional scientific analysis. For example, in a 1997 colloquium at the National Center for Atmospheric Research entitled "A Systems Approach to ENSO,"[2] Glantz introduces the notion of "climate affairs" (implicitly) by focusing on not only ENSO science but also ecosystem, societal, and policy dimensions of air–sea interactions in the tropical Pacific Ocean. In another example, an "affairs" approach disentangles how unsustainable human activities (e.g., agriculture) and perceptions/views of climate

[2] This colloquium was organized in response to the developing 1997–98 El Niño. http://www.ilankelman.org/glantz/Glantz1997SystemsENSO.pdf.

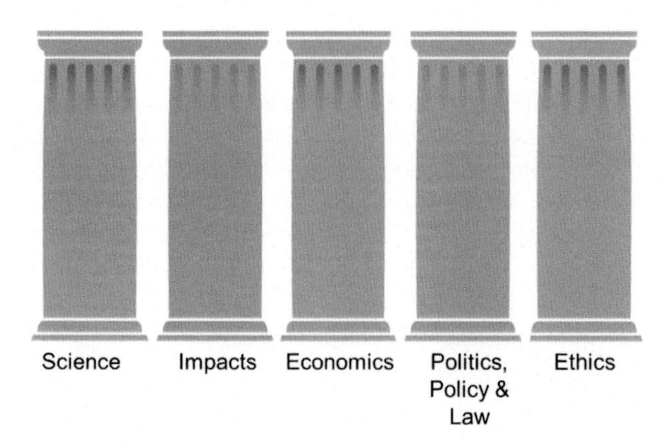

Figure 8.1 Pillars of climate affairs approach based on Glantz (2003).

intersect with natural variability to generate drought vulnerability in fragile landscapes such as the Great Plains in the US and the northeast region of Brazil (Glantz, 1994). More recently, the approach was employed to rethink epidemic cholera's connection to El Niño in Latin America, by considering multiple dimensions of ENSO, including how society defines events and how socioeconomics underlies root causes of vulnerability, rather than simply correlating air temperature changes with disease incidence (Ramirez et al., 2013).

The application of climate affairs to understand impacts and vulnerability to El Niño-related hydrometeorological extremes in Piura, Peru is visualized in Fig. 8.2 (climate–ecosystems–society [CES] model). The model shown is adapted from a conceptualization from Ramírez (2012) influenced by vulnerability analysis framework of Turner et al. (2003). It is grounded in studies of climate–health (Kovats et al., 2003; Ebi et al., 2006; Confalonieri et al., 2007; WHO, 2009; Maantay and Becker, 2012; Balbus et al., 2016), human–environment vulnerability (Cutter et al., 2003; Turner et al., 2003; Wisner et al., 2004, 2012), and climate–society interactions (Kates, 1985; Caviedes, 1984; Ribot, 1995, 2009; Glantz, 2001, 2003, 2015; Zebiak et al., 2015). Thus, it is flexible to address various climate-related issues, including public health, infrastructure, agriculture, food, water, energy, and development impacts.

The CES model illustrates the interrelationships among climate, ecosystems, and society. Within the triad, each component is linked and dependent on the other, and therefore the model does not imply a simple "cause and effect." Rather, the model suggests a coupling of human–environment

Box 8.1 Pillars of Climate Affairs

Climate affairs is multidimensional and has five basic pillars to understanding climate–society interactions. The first pillar highlights the importance of climate science, and hence climate affairs is grounded in understandings of the physical dynamics of climate and ocean systems, as well as variability, fluctuations, changes, and hydrometeorological extremes at multiple spatial and temporal scales. In addition, it recognizes the value of traditional and regional/local knowledge or ways of understanding climate and environmental phenomenon (Ramirez et al., 2013). The second pillar is climate impacts, which refers to impacts of climate in shaping and changing ecosystems, which, in turn, impact various dimensions of society; equally important, and not in isolation are human impacts on ecosystems and climate and oceans. Thus, to understand why societies are vulnerable to hydrometeorological extremes, one must consider not only the variability of climate in a region but also how human activities alter and degrade ecosystems and increase exposure to environmental health hazards, including infectious disease agents. The third pillar is economics, economic costs of disasters, and responses to hydrometeorological extremes, which may include the losses of life, property, and livelihoods, as well as estimating the societal value of climate information and services (https://www.wmo.int/pages/prog/dra/eguides/en/3-national-relationships/3-4-socio-economic-benefit-analysis). The fourth pillar is politics, policy, and law, which draws attention to the "processes pursued by different actors to achieve a policy objective desired by one group or nation, often at the expense of the objectives of other groups or nation" (Glantz, 2003: 123). It also refers to the "regulations, laws, and nonbinding resolutions" that address climate-related issues, including multisectoral coordinating organizations that focus on responses to El Niño (Peru's Multisectoral Committee for the Study of El Niño [ENFEN]) (http://www.met.igp.gob.pe/variabclim/enfen/), and multilateral mandates, such as the UNISDR's Sendai Framework for DRR (http://www.unisdr.org/we/coordinate/sendai-framework) or the UN Framework Convention on Climate Change (http://unfccc.int/meetings/bonn_nov_2017/session/10376.php). The final pillar is ethics and equity, which one could argue is actually the motivational foundation of the pillars, driving the need for a multidisciplinary and equitable approach to climate risk management. Ethics and equity refers to the differential impacts of climate impacts and interactions, including distributional issues of responsibility and compensation, existing inequalities between the Global North and South, inequalities at the local community level, human rights, and environmental and social injustice (Glantz, 2003: 165–174). In summary, climate affairs places hydrometeorological extremes within a wider context of factors to explain societal vulnerability and resilience to climate processes.

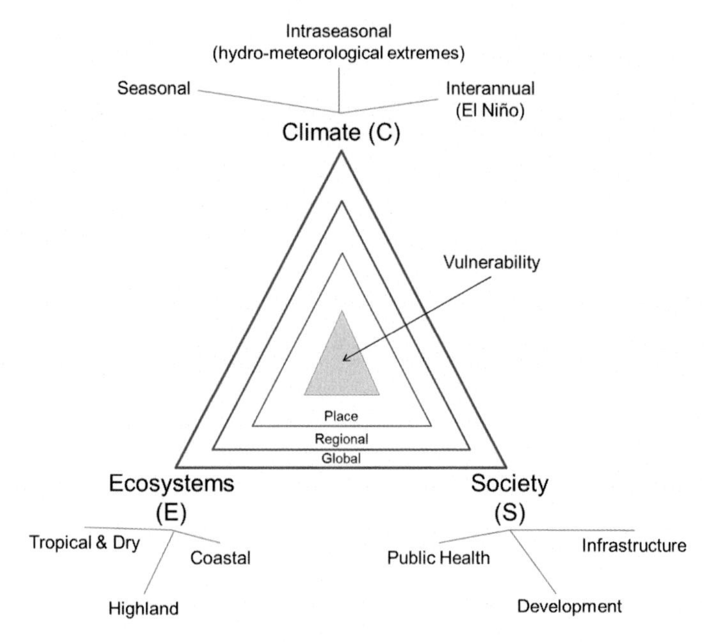

Figure 8.2 Heuristic model of climate–ecosystem–society (CES) interactions in Piura, Peru.

systems that explains impacts and vulnerability that arise from interactions that are place-specific and occur at various spatial scales (e.g., from global and regional to local) (Turner et al., 2003). Although not visualized, the model also includes a temporal dimension, which recognizes that societal vulnerability to disasters is historically contingent and relevant (Wisner et al., 2004: 54). Thus, the past provides insights and potential lessons for future climate-related hazards. At a secondary level are subcomponents of climate (temporal scales, processes, and events), ecosystems (biomes based on Köppen's climate classification), and society (economic and institutional sectors of society or cultural subgroups), which can be adapted to different societal contexts, including intersecting inequalities (e.g., socioeconomic and gender), to understand climate vulnerability (Oxfam, 2010; CARE, 2016).

In terms of climate impacts, the CES model considers multiple pathways in which atmosphere and ocean processes affect society and population health. In one direction (i.e., CES), climate and ocean processes are important factors that affect societal well-being. This occurs within varying social and environmental conditions that modify pathways that expose society to a variety of environmental and climatic hazards (Ramírez, 2012; Balbus et al., 2016). These mechanisms may range from global to regional climate and ocean processes to local hydrometeorological extremes, which influence exposure pathways such as impacts from extreme heat, changes in agroecosystems and hydrology, reduced air, food, and water quality, changes in disease ecologies, and population displacement (Confalonieri et al., 2007; Balbus et al., 2016). In the case of El Niño, as described earlier, rainfall and temperature teleconnections in tropical regions impact basic determinants of population health and jointly influence infectious disease transmission. El Niño may also impact society through environmental changes in oceans (e.g., sea surface temperatures [SST]), which affect marine and coastal ecologies, which in turn impact food chains, food webs, and ecosystem services that many organisms, animals, and humans alike are dependent on.

Conversely, the model in the other direction (i.e., society–ecosystems–climate) accounts for societal impacts on ecosystems and climate, which in turn affect societal well-being. In other words, humans mediate the relationships between climate processes and societal impacts, including consequences to public health (Wilson, 2001). For example, human activities (e.g., urban development, farming, and other land use/cover changes) alter terrestrial and aquatic environments, which disrupt ecosystems and the organisms, animals, and communities that reside within it. Furthermore, human activities have the capacity to alter the atmosphere and change global and regional climates, which in turn influence patterns of hydrometeorological extremes (IPCC, 2012), and subsequent impacts on society. One example of an impact on society is an outbreak of a mosquito-borne disease such as malaria, influenced by a deforested landscape, which changes the ecology among human populations, mosquito vectors, and plasmodium parasites, increasing human–disease interactions and exposure (Vittor et al., 2009). A second example is an outbreak of a diarrheal disease such as cholera, influenced by urbanization and polluted coastal waters, increasing human exposure to a number of waterborne pathogens, including *vibrio cholerae* bacteria (Penrose et al., 2010; Chowdhury et al., 2011).

In the two examples, hydrometeorological extremes may play an important role by influencing malaria and cholera ecology (e.g., rainfall and temperature anomalies) and local hydrology (e.g., floods and drought), and thus influence exposure pathways and aggravate social conditions and public health. However, as the examples highlight, humans also play a concurrent role through activities that degrade the environment and generate vulnerability. In addition, such human activities are influenced by social, cultural, economic, and political factors, including development agendas and public policy, which shape the conditions in which people live, i.e., social determinants of health (Marmot, 2005). According to Wisner et al. (2004, 2012), structural factors are the root causes of societal vulnerability to disasters, including climate-related events, and reflect the distribution of power in a society. They determine people's access to opportunities and resources, including food, income, and public services, which generate unequal exposures to unsafe living conditions, climate-related hazards, and consequently, an array of health and societal impacts.

In the following sections, regional and national vulnerabilities to El Niño are discussed, followed by the application of the CES model to local vulnerabilities (climate, social, health, infrastructure, and gender) in Piura, Peru during the 1997–98 El Niño. The 1997–98 episode while not absolute (meaning it cannot be applied to all El Niño impacts) serves as a case study of a significant extreme event, which was well documented, with a multitude of lessons (see Glantz, 2001: 4–43; http://www.bvsde.paho.org/bvsacd/cd68/ElNiño.pdf).

3. REGIONAL VULNERABILITY TO EL NIÑO

Vulnerability to El Niño is a reality for many coastal economies, particularly along the Andean corridor in northwestern South America. To begin, the geographic proximity of these countries to the eastern tropical Pacific Ocean places them at the center of El Niño's environmental influence, which brings seasonally anomalous tropical storms, coastal sea-level variations, torrential rains and floods, and droughts to the region. El Niño disrupts human, ecological, and physical systems when it recurs every few years. Furthermore, underlying these physical changes and potential environmental hazards associated with El Niño and La Niña episodes is the human development context in which climate impacts occur (Ribot, 1995, 2009; Glantz, 2001; Trigoso Rubio, 2007). According to Moreno (2006), populations in Latin America are markedly vulnerable to climate-related hazards because of tropical ecology (environmental suitability for pathogens and vectors) and importantly, social and infrastructure deprivation, including food and water insecurities, which increase human exposure and susceptibility to climate-sensitive diseases. As an example, Table 8.1 shows the human development and health setting for the region in the years preceding the 1997–98 El Niño. As the indicators show, basic needs, such as access to improved water and sanitation services, were not met for some percentage

Table 8.1 Human development and health indicators in northwestern South America

Indicator	Colombia	Ecuador	Peru
GNI per capita (US$) 1997[a]	2650	2210	2250
No access to improved drinking water sources (%) 1997[a]	10	22	22
No access to improved sanitation facilities (%) 1997[a]	27	34	40
Maternal mortality ratio (per 100,000 live births) 1995[c]	81	130	220
Diarrheal disease mortality rate (per 100,000 children <5 year) 1995[b]	27	65	61
Pneumonia and influenza mortality rate (per 100,000 children <5 year) 1995[b]	32	73	153

Data Sources: [a]PAHO/WHO, 2017. Communicable Diseases and Health Analysis/Health Information and Analysis. PLISA Database. Health Situation in the Americas: Basic Indicators. Washington, DC, United States of America. http://www.paho.org/data/index.php/en/indicators.html; [b]PAHO, 2002. Health in the America, 2002 Edition Volume 1. Scientific and Technical Publication No. 587. PAHO, Washington, DC; [c]WHO, 2015. Global Health Observatory Data Repository. WHO, Geneva. http://apps.who.int/gho/data/view.main.1390?lang=en.

of the population, which may explain preexisting mortality rates of diarrheal disease, pneumonia, and influenza among children (less than 5 years), as well as maternal deaths in each respective country. Such indicators underscore the vulnerability of the region to health problems particularly those associated with social inequalities.

Within this context of societal vulnerability, Andean countries face compounding health risks in the face of El Niño-related hydrometeorological extremes. As Table 8.2 shows, Colombia saw a significant rise in dengue incidence in 1998, whereas Peru and Ecuador reported substantial malaria and cholera cases, compared with the previous year. All three diseases, while categorically climate-sensitive, may also be termed "poverty-sensitive" health problems because of links to social determinants of health (Teklehaimanot and Meija, 2008; Talavera and Perez, 2009). In addition, the region experienced significant economic impacts associated with hydrometeorological extremes and related disasters. The greatest economic impacts were reported in Ecuador and Peru, where losses were estimated at ~6380 (US$ millions).

4. EL NIÑO VULNERABILITY IN PERU

Peru, in particular, has a memorable history with El Niño and impacts from hydrometeorological extremes (Lagos and Buizer, 1992; Rodriguez et al., 2005; Lagos et al., 2008), including warm events in 1925, 1972–73, 1982–83

Table 8.2 Health and economic impacts associated with the 1997–98 El Niño

Indicator	Colombia	Ecuador	Peru
Malaria cases % change[a]	3.3	53.1	37.1
Cholera cases % change[b]	−70.7	5676.9	1236.2
Dengue cases % change[a]	160.1	6.7	−27.2
El Niño losses (US$ millions) 1997–98[c]	564	2882	3498

Data Sources: [a]PAHO/WHO, 2017. Communicable Diseases and Health Analysis/Health Information and Analysis. PLISA Database. Health Situation in the Americas: Basic Indicators. Washington, DC, United States of America. http://www.paho.org/data/index.php/en/indicators.html; [b]PAHO, 2008. Number of Cholera Case in the Americas, 1990-2008; [c]CEPAL (Economic Commission for Latin America and the Caribbean), 1999. Efectos macroeconomicos del fenomeno El Niño de 1997-98 (Macroeconomic effects of the 1997-98 El Niño). https://www.cepal.org/publicaciones/xml/0/40870/EL_NIO1997-98_ECON_ANDINAS.pdf (in Spanish).

and 1997–98, 2015–16, and most recently, the "Costero" event of 2017 (Ramírez and Briones, 2017; Takahashi, 2017)[3]. In Peru, El Niño teleconnections are felt across all natural regions (low-lying coast to highlands and jungle) and geographies (north to south), but to varying degrees, dependent on the characteristics of the event and timing. The northern coast, which is well known for El Niño sensitivity (Ramirez, 2015), may experience above-average air temperatures and torrential rains along with storm surges and inland coastal flooding, whereas the south and Central Andes (Lagos et al., 2008; Peru's Servicio Nacional de Meteorologia e Hidrologia [National Meteorological and Hydrological Services] [SENAMHI], 2009) and Amazon may tend to expect drought (Marengo et al., 2008). In the central coast where the capitol (Lima) is located, above-average air temperatures are reported (SENAMHI, 2004); although heavy rains and flooding were reported there incidentally during the extreme events of 1997–98, 2015–16, and the El Niño "Costero" in 2017.

It is also important to consider that teleconnections depend on the season of the year (Diaz et al., 2001) and are typically generalized in terms of El Niño and La Niña relationships from December–February and June–August. Teleconnections are also influenced by the magnitude and spatial distribution of SST anomalies (Goddard et al., 2001), which may span the basin or concentrate in the central or eastern Pacific (Takahashi, 2017), and affect how impacts manifest regionally and worldwide. Teleconnections are strongest during the Southern Hemisphere (SH) summer (i.e., December–March) when SST anomalies are largest in the tropical Pacific (National

[3] At the time of writing, an unusual event referred to as an El Niño "Costero" (coastal) surprised Peru and Ecuador because of its rapid evolution and severe societal, economic, and health impacts (Fraser, 2017).

Oceanic and Atmospheric Administration [NOAA], 2005). However, teleconnections are not symmetric between the two phases of ENSO (Meehl, 2002); anomalies associated with La Niñas are considered weaker relative to those experienced during El Niños (Hoerling, 2002; Zebiak, 2002).

In Perú, the effects of El Niño on society are felt across various sectors, including public health, infrastructure, agriculture, fisheries, water, energy, transportation, education, industry, and tourism (Consejo Nacional del Ambiente [CONAM], 2002). Environmental changes associated with El Niños disrupt coastal upwelling processes, which, along with invading warm waters, adversely affect marine ecosystems (Barber and Chavez, 1983; Tarazona and Valle, 1999; Escribano et al., 2004; Chavez et al., 2008). Fish populations, for example, respond by migrating north and south, going to greater ocean depths, or swimming closer to shore. For some species, such as anchovy, which is an important commodity in Peru (Caviedes, 1984; Carr and Broad, 2000), the impacts can be fatal particularly when the effects of extreme environmental conditions (e.g., anomalous warming waters) are coupled with human activities. For example, in 1973, the collapse of the anchovy industry was associated with an El Niño of moderate extremes and overharvesting (Caviedes and Fik, 1992; Glantz, 2001: 232). For farmers, rainfall extremes along with temperature variability affect crop yields and livestock (Lagos, 1998; Woodman, 1998; CONAM, 2002). For example, potato yields in the Central Andes are vulnerable to droughts and El Niño's (Orlove et al., 2000). Rainfall extremes may also spawn insect (e.g., rain abundance) and plant diseases (e.g., rain deficit) that damage crops (Cisneros and Mujica, 1999). Climate impacts on fisheries and agriculture can endanger livelihoods and food security.

In the energy sector, El Niño-related flooding damages infrastructure that causes power outages, which in turn affects human systems, such as water and sanitation services, refrigeration, local businesses, and hospitals (Valverde, 1998; CONAM, 2002). During such events, water contamination can occur and foods can spoil. In general, infrastructure damage (i.e., the washing out of bridges, roads, and communication lines) contributes to social suffering, not only because of the costs in terms of economic losses (i.e., costs to rebuild and disrupting market access) but also because of the subsequent effects on human health. In addition to drowning and physical injuries, in Peru, the breakdown of infrastructure (e.g., energy along with water and sanitation systems) due to hydrometeorological hazards increases risk and exposure to an array of infectious diseases that include water- and mosquito-borne, as well as skin-, eye-, and respiratory-related infections (Caviedes, 1984; Gueri, 1984; PAHO, 1998a; Sandoval, 1999; WHO, 2016).

Furthermore, hydrometeorological extremes and hazards may induce migration when people lose their homes to floods, as an example of hazard, and become forcibly displaced (McMichael, 2013; Yonetani, 2016), temporarily or long term. Consequently, migration may lead to the spread of infectious diseases during travel or once people settle in temporary housing or resettle in a new area. For example, persons without prior exposure to a mosquito-borne disease, e.g., malaria, may migrate to an endemic area. Alternatively, malaria-infected populace may migrate to an area without the disease and introduce the disease to a susceptible community without prior exposure (Martens and Hall, 2000).

In addition, El Niño impacts stymie public health efforts during and following hydrometeorological-related disasters in vulnerable places. Besides the effects on population health, disasters affect the health sector's capacity to respond (WHO, 2016). Extreme flooding events may isolate communities when roads and bridges are damaged and destroyed (Caviedes, 1984; PAHO, 1998b; Velasco-Zapata and Broad, 2001: 194). During the 1997–98 El Niño, torrential rains led to flooding, landslides, mudflows, which impacted Peru's infrastructure (Velasco-Zapata and Broad, 2001: 193), including 7277 km of roads and 60 bridges destroyed (Penaherrera del Aguila, 1999: 160). During such disasters, impacts on transportation systems and roads prevent public health workers from reaching those injured and treating them. Furthermore, these impacts disrupt the flow of goods between cities, which can aggravate food access, and cause undernutrition, as was the case in 1983. Following heavy rainfall associated with the 1982–83 El Niño, the highway from the Port of Paita to the city of Piura in northern Peru was flooded and submerged, isolating the communities (Velasco-Zapata and Broad, 2001: 194) and preventing them from seeking assistance and receiving supplies and necessities. Furthermore, ill-prepared (i.e., incapable of sustaining high water volume) bridges in Piura, Peru, collapsed when the Piura River overflowed its banks, further isolating residents in Piura from the neighboring municipality, Castilla. Overall, El Niños and hydrometeorological-related disasters have the potential to disrupt human well-being in Perú, increase social vulnerability, and in turn affect population health.

5. LOCAL VULNERABILITY TO EL NIÑO: THE CASE OF PIURA

5.1 Climate Vulnerability

Piura, located on the northern coast of Peru, is one of the most sensitive subregions to El Niño (Caviedes, 1984; Woodman, 1998; Ordinola, 2002;

Rodriguez et al., 2005). Geographically, it is a place situated between the strong influence of coastal upwelling and the Peruvian Current in the west and the Andes in the east. A map of ecological zones (Fig. 8.3) in Piura shows that the region has a mix of arid to semiarid desert characteristics on the low-lying coast that transitions to a variable tropical highland environment (mix of desert scrub, woodland, dry, and moist forests) as you reach the Andes in the northeast, bordering Ecuador (CONAM and AACHCP, 2005).

Historically, Piura is where the term "El Niño" originated. It was first named by Peruvian fishermen (on the northern coast) who noticed a periodic warming of ocean waters coinciding with Christmas (Carillo, 1892; Caviedes, 1984). Since the late 15th century, the region has been synonymous with quasiperiodic hydrometeorological extremes, particularly torrential rains (Eguiguren, 1894; Sears, 1895). In the last 50 years, several El Niño events have been associated with catastrophic impacts in Piura (Cordova Aguilar, 1996; Ministry of Health [MINSA] Piura, 2001; Ramírez and Briones, 2017; Takahashi, 2017). From 1970 to 2003, floods contributed to 80% of climate-related disasters in the Piura River basin (CONAM and AACHCP, 2005). One of the most significant episodes was the 1997–98 event, which many termed, the "El Niño of the century."

The 1997–98 El Niño was historic in terms of the magnitude of SST anomalies, spatial extent, evolution, and importantly, its impacts on societies and ecosystems. This extraordinary event began in the middle to late Northern Hemisphere spring in 1997 and lasted 13–19 months, which depends on the index one uses (See Tables 8.3 and 8.4 below for comparison of SST anomalies and El Niño events). According to the ONI index, El Niño began in May of 1997 and ended in May of 1998 (NOAA, 2017). The Peruvian index (ICEN) on the other hand indicates that El Niño began in March of 1997 but ended in September of 1998 (ENFEN, 2012). Although there is some disagreement about the timing of the evolution of El Niño, both agree that it was an extraordinary event in terms of SST anomalies.

Over the course of its evolution, the 1997–98 El Niño expanded across the central and eastern Pacific Ocean and increased in magnitude and geographic span of anomalies eastward and westward across the basin (Fig. 8.4). The event peaked in November and December of 1997, reaching SST anomalies as high as 3.8°C (Niño 1 + 2 region) and 2.3°C (Niño 3.4 region). However, while it began to gradually decay thereafter, high SST anomalies persisted until July of 1998, particularly in the equatorial and coastal upwelling regions near the coasts of northern Peru and southern Ecuador.

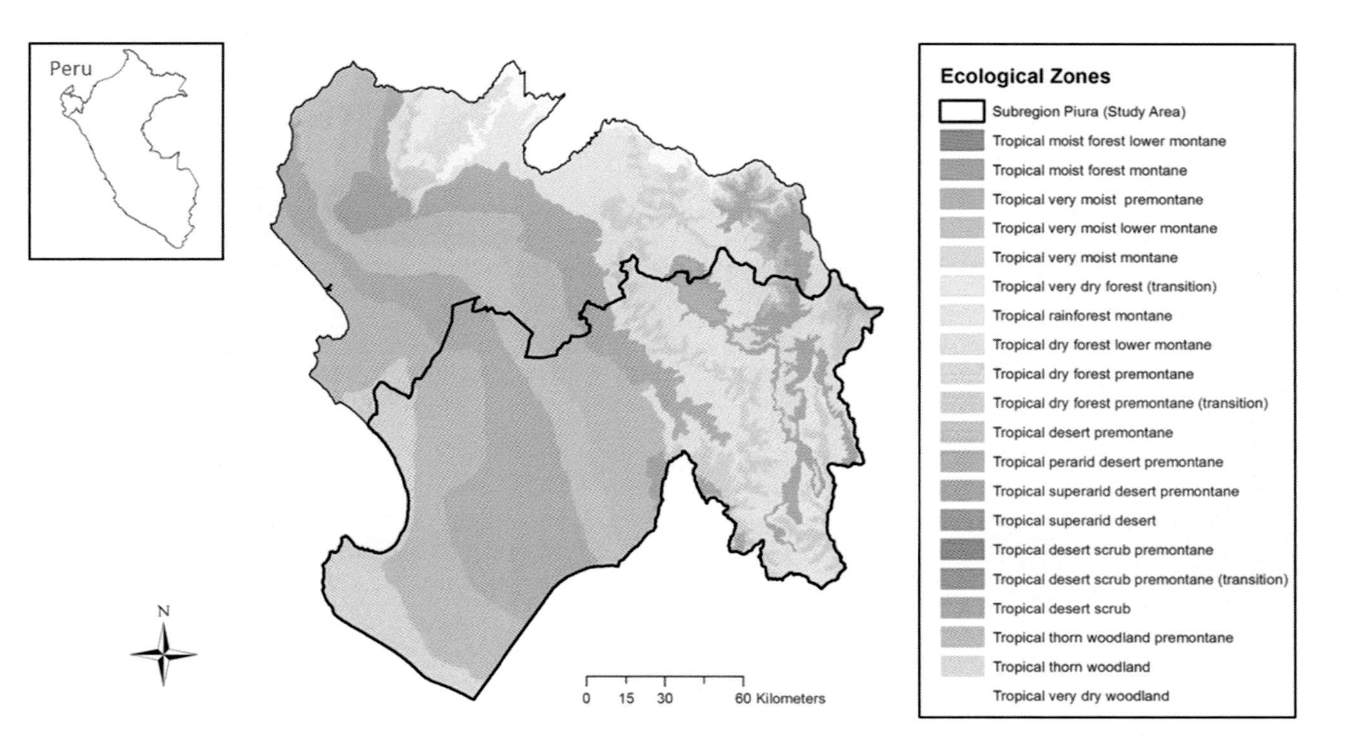

Figure 8.3 Map of ecological zones in the Department of Piura, Peru (which includes the subregion used in this case study). Zones are based on precipitation, temperature, evapotranspiration, and vegetation variables, according to the Holdridge classification (1967). *(Data Source: Universidad de Piura (University of Piura [UDEP]), 2008. Datos Climaticos (Climate Datasets). (Piura, Peru).)*

Table 8.3 Comparison of El Niño (red) and La Niña (blue) events according to the ONI index for 1997–98

Year	DJF	JFM	FMA	MAM	AMJ	MJJ	JJA	JAS	ASO	SON	OND	NDJ
1997	−0.5	−0.4	−0.2	0.1	0.6	1	1.4	1.7	2	2.2	2.3	2.3
1998	2.1	1.8	1.4	1	0.5	−0.1	−0.7	−1	−1.2	−1.2	−1.3	−1.4

Data Source: NOAA, 2017. Cold and Warm Episodes by Season. http://origin.cpc.ncep.noaa.gov/products/analysis_monitoring/ensostuff/ONI_v5.php.

Table 8.4 Comparison of El Niño (red) and La Niña (blue) events according to the ICEN (El Niño Costero) index for 1997–98

Year	DJF	JFM	FMA	MAM	AMJ	MJJ	JJA	JAS	ASO	SON	OND	NDJ
1997	−0.8	−0.1	0.6	1.4	2.2	3.0	3.5	3.8	3.7	3.8	3.8	3.8
1998	3.4	3.0	2.8	2.8	2.6	2.2	1.6	1.1	0.6	0.1	−0.1	−0.3

Data Source: ENFEN (Peru's Multisectoral Committee for the Study of El Niño), 2012. Definicion operacional de los eventos El Niño y La Niña y sus magnitudes en la costa del Peru (Operational definition of El Niño and La Niña events and their magnitudes along the coast of Peru). http://www.imarpe.pe/imarpe/archivos/informes/imarpe_comenf_not_tecni_enfen_09abr12.pdf.

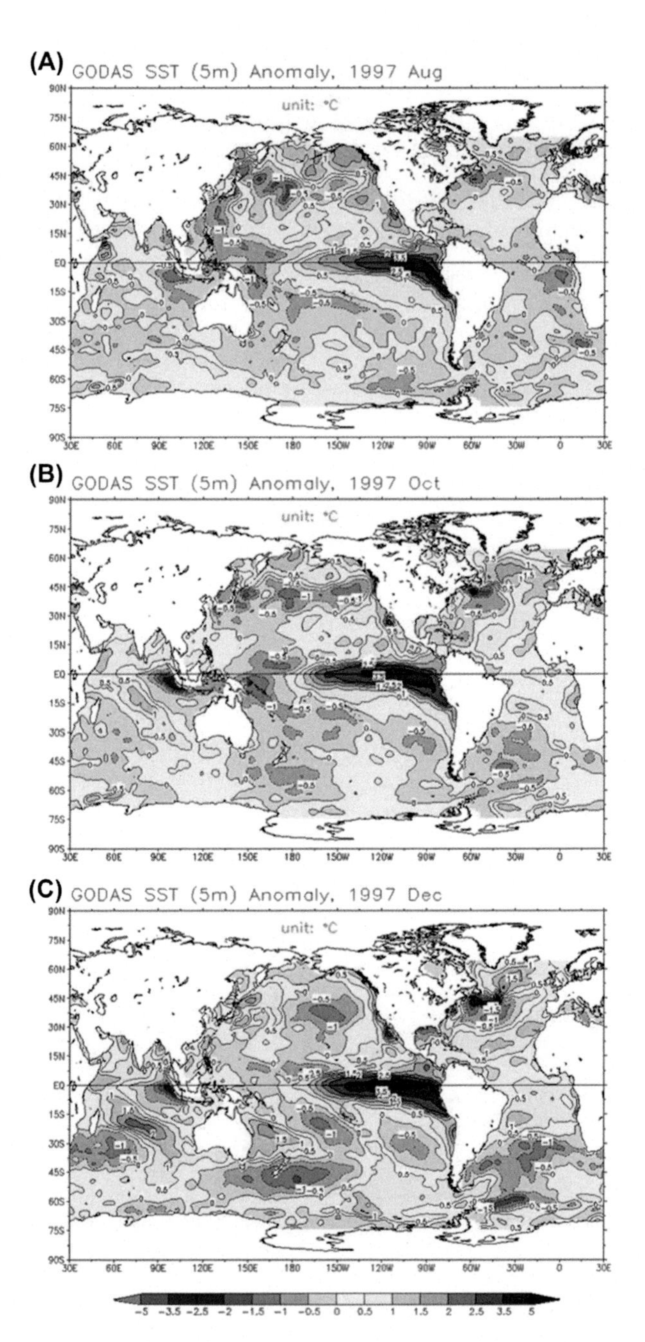

Figure 8.4 Maps of sea surface temperature anomalies across the tropical Pacific Ocean for (A) August 1997 (B) October 1997 (C) December 1997. *(Data Source: http://www.cpc.ncep.noaa.gov/products/GODAS/monthly.shtml.)*

In Piura, the average SST anomaly in the port of Paita, for example, was 2.9°C from July of 1997 to May of 1998 (see Table 8.5). During this El Niño period, air temperature anomalies along the coast of Peru, particularly in the north, were also several degrees above normal. In the city of Piura (capitol of the region), for example, air temperature anomalies began rising above 1°C in March of 1997, peaked in September at 4°C and did not subside below 1°C until August of 1998 (Fig. 8.5). In total, the population of Piura endured 17 months of above-average temperatures. Following the

Table 8.5 Sea surface temperature anomalies in the port of Paita (meteorological station) in Piura for 1997–98

Year	J	F	M	A	M	J	J	A	S	O	N	D
1997	−0.7	−0.2	0.7	0.6	1.5	1.9	2.4	3.4	3.6	2.7	3.4	3.3
1998	3.2	3.1	3.0	2.2	2.0	0.3	0.2	0.0	−0.1	−0.4	−0.5	−0.5

Data Source: Universidad de Piura (University of Piura [UDEP]), 2008. Datos Climaticos (Climate Datasets). (Piura, Peru).

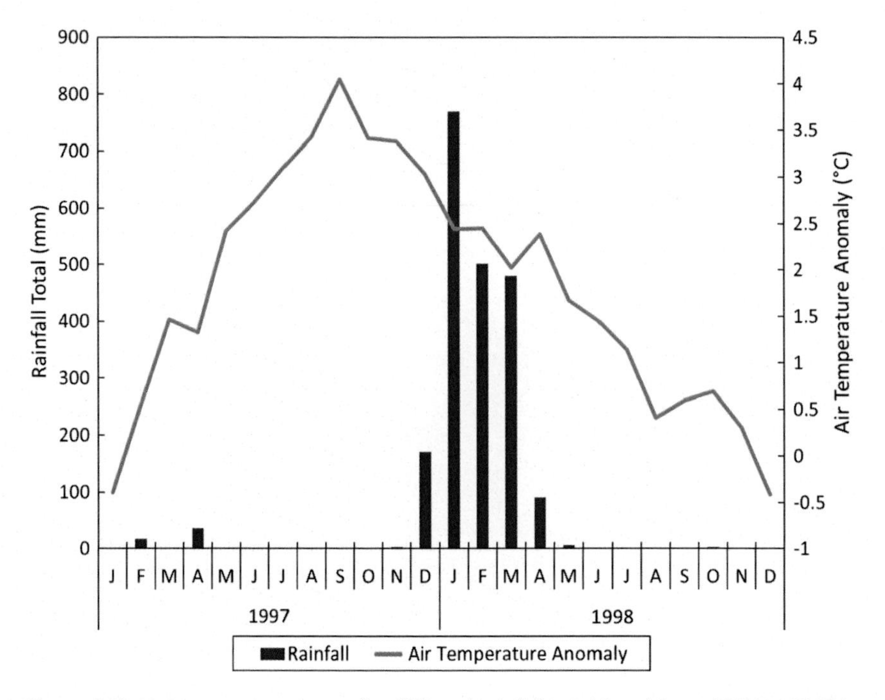

Figure 8.5 Air temperature anomalies (°C) and rainfall total (mm) from 1997 to 1998 in the city of Piura (Miraflores meteorological station). *(Data Source: Universidad de Piura (University of Piura [UDEP]), 2008. Datos Climaticos (Climate Datasets). (Piura, Peru).)*

anomalously warm SH winter and spring were extreme rains, which began in December of 1997 and ended in April of 1998. In total, the city of Piura received 2012 mm of rain within a 5-month period (Fig. 8.5), the equivalent of 9.9 times the annual average (base period—1971–2000) of rainfall expected.

5.2 Socioeconomic Vulnerability

In 1998, the subregion of Piura (n = 33 districts), which is the focus of this case study, had a population of 847,257. As Fig. 8.6 shows, most of the population lived in districts in the low-lying coast, which is highly urbanized (~80% and above). Increasing urbanization in the area has created pressures on the existing infrastructures within the region. An overall view of social vulnerability in Piura using a Basic Needs Unmet index (Necesidades Básicas Insatisfechas [NBI], Fig. 8.7) reflects social disparities of both coastal and highland populations. The NBI index, which is one way in which the Peruvian government maps poverty and social inequality, is composed of a number of variables that reflect access to water, sanitation, electricity, education, type of housing, dependency ratio, etc. (Peru's National Institute for Statistics and Information [INEI], 1994a). It maps the number of basic needs that households do not have access to. The data presented in the figure are mapping districts where at least one basic need is unmet throughout the subregion.

Fig. 8.8 highlights several basic needs percentages in Piura. An estimated 82% of districts were without piped municipal water (directly in their households, based on the 1993 census) for more than 50% of the population. Furthermore, 79% of districts for more than 50% of the population were lacking access to a toilet in their households (Fig. 8.8B). Without potable water and adequate sanitation, the practice of hygiene, which is a basic measure to prevent exposure to water-related diseases, is quite challenging. Thus, the socioecological context of Piura is one of water stress (living in an arid environment) coupled with socioeconomic inequalities, which vary geographically as one travels from the coast to the highlands. Water stress is therefore defined not only by the ecological characteristics of the subregion but also by population access to safe drinking water in households, which is unevenly distributed across Piura. Relatedly, water and sanitation disparities are indicative of other social deprivations, which comprise of and affect the social determinants of health. For example, 88% of districts were without electricity for more than 50% of their population, and 40% of districts had illiteracy rates of more than 30% (Fig. 8.9). The latter indicators underscore

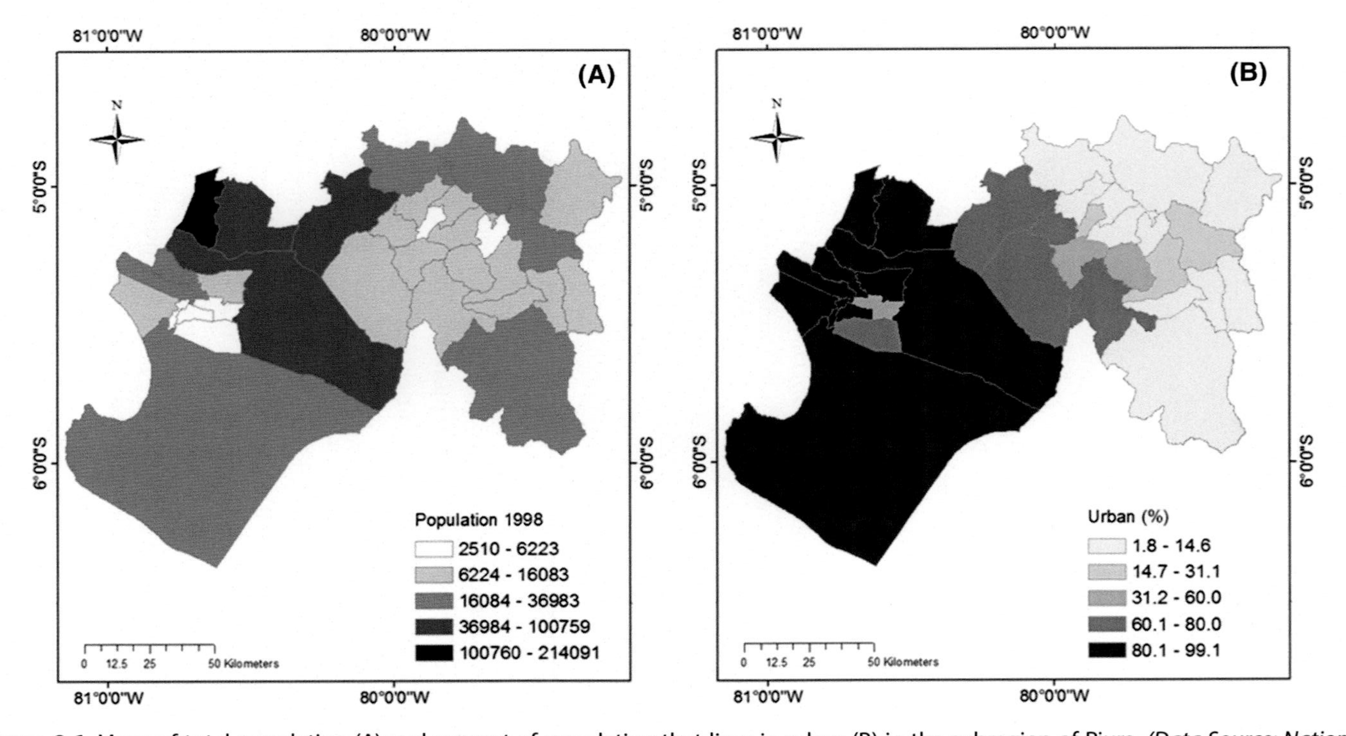

Figure 8.6 Maps of total population (A) and percent of population that lives in urban (B) in the subregion of Piura. *(Data Source: National Institute for Statistics and Information, Peru [INEI], 1994b. Piura: Compendio Estadistico 1993-1994 (Piura: Summary of Statistics 1993-1994). Lima, Peru. (in Spanish); National Institute for Statistics, Information, Peru [INEI], 2000. Encuesta Demográfica y de Salud Familiar 2000 (Study of Family Demography and Health 2000). http://desa.inei.gob.pe/endes/ (in Spanish).)*

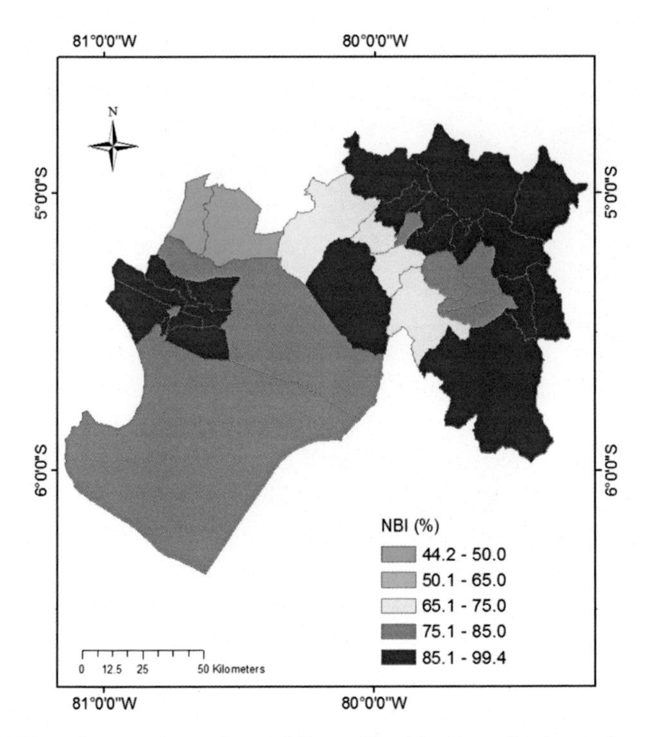

Figure 8.7 Map of percentage of population with at least one basic need unmet in the subregion of Piura. *(Data Source: National Institute for Statistics and Information, Peru [INEI], 1994b. Piura: Compendio Estadistico 1993-1994 (Piura: Summary of Statistics 1993-1994). Lima, Peru. (in Spanish).)*

energy and education poverty, which impedes access to public services and amenities at the household level (http://www.who.int/indoorair/publications/fuelforlife/en/), access to adequate health care, and correlate with poor health outcomes in developing countries (Grosse and Auffrey, 1989) and in developed countries (Marcus, 2006).

5.3 Health Vulnerability

Evidence that socioeconomic vulnerability contributed to health vulnerability in Piura is apparent when observing rates of preventable infectious diseases in Piura prior to the 1997–98 El Niño event. Several endemic diseases that are environment-related and poverty-sensitive health problems included acute diarrheal diseases, respiratory-related infections, mosquito-borne diseases, such as malaria and dengue, as well as infant mortality and maternal mortality. The following are examples for 1997: dehydration associated with acute

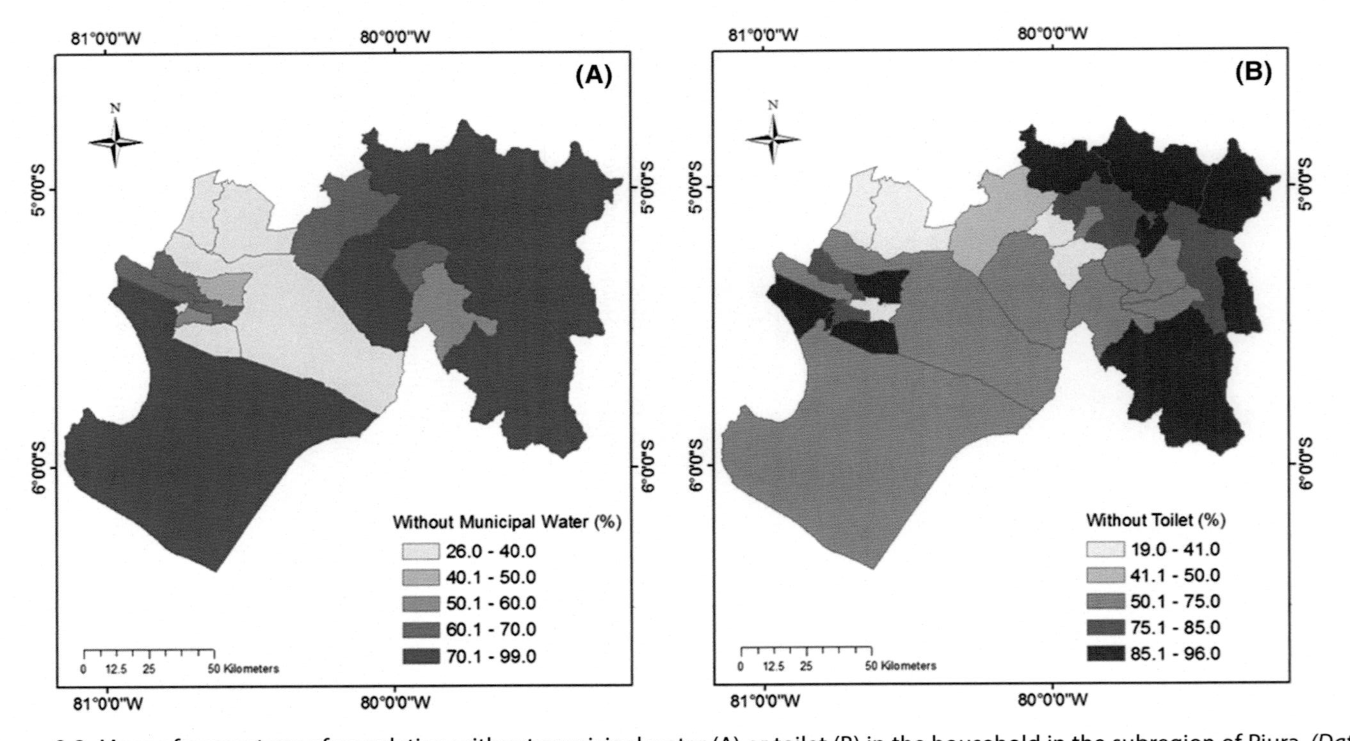

Figure 8.8 Maps of percentage of population without municipal water (A) or toilet (B) in the household in the subregion of Piura. *(Data Source: National Institute for Statistics and Information, Peru [INEI], 1994b. Piura: Compendio Estadistico 1993-1994 (Piura: Summary of Statistics 1993-1994). Lima, Peru. (in Spanish).)*

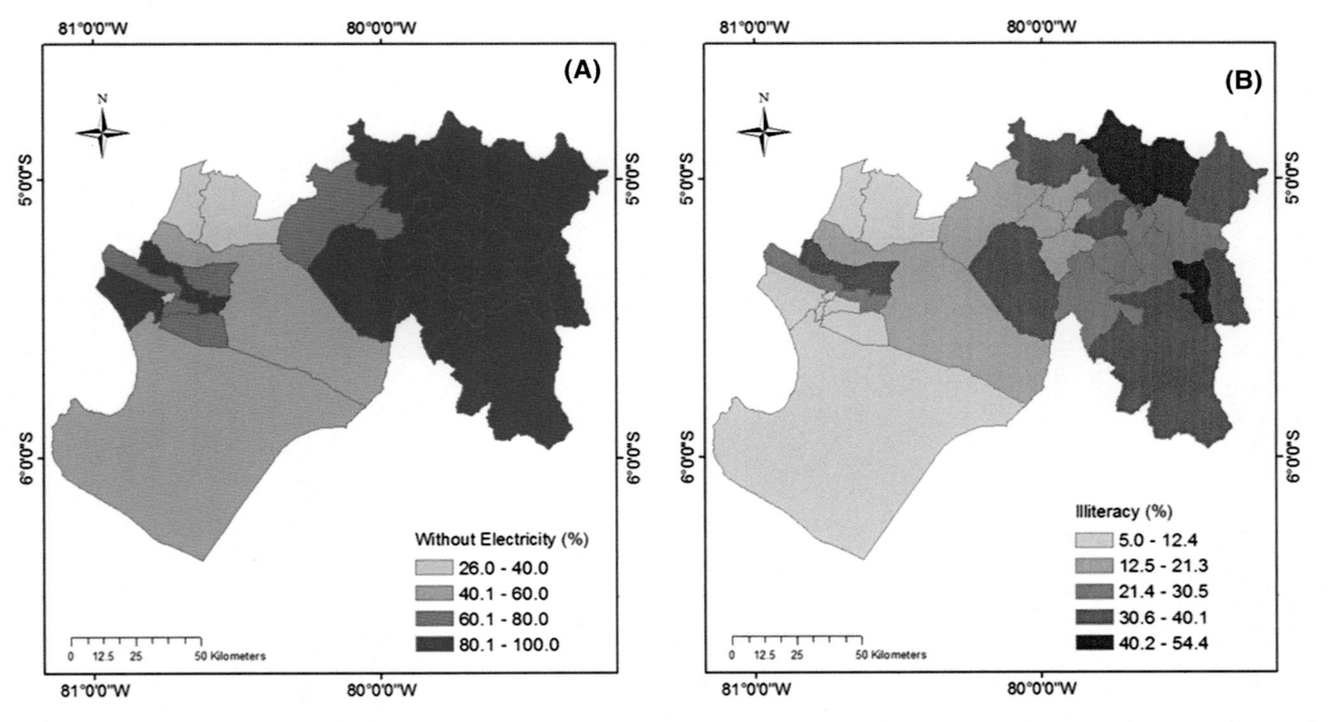

Figure 8.9 Maps of percentage of population without electricity (A) or that is illiterate (B) in the household in the subregion of Piura. *(Data Source: National Institute for Statistics and Information, Peru [INEI], 1994b. Piura: Compendio Estadistico 1993-1994 (Piura: Summary of Statistics 1993-1994). Lima, Peru. (in Spanish).)*

diarrheal disease (percent of total acute diarrheal disease)—31.3%, malaria—3.4 (per 100,000), cholera—0.6 (per 100,000), infant mortality—16.5 (per 1000), and maternal mortality—88.5 (per 1000) (MINSA, 2001). During and in the aftermath of El Niños, particularly strong events, these conditions among the population tend to become aggravated. Fig. 8.10 shows select health outcomes in Piura before and after the two strongest El Niños of the 20th century. As the figure shows, respiratory, diarrheal, and malaria cases increased markedly from 1 year (onset year of El Niño) to the next year (peak and period where torrential rains occur, usually in the first semester of the year). Figs. 8.11 and 8.12 illustrate how these emergent epidemics coincide in time, during the first 12 weeks of the year 1998. Singer (2009) refers to this ensemble of infectious disease epidemics in time and across space associated with environmental change as an "ecosyndemic." The concept is linked to the

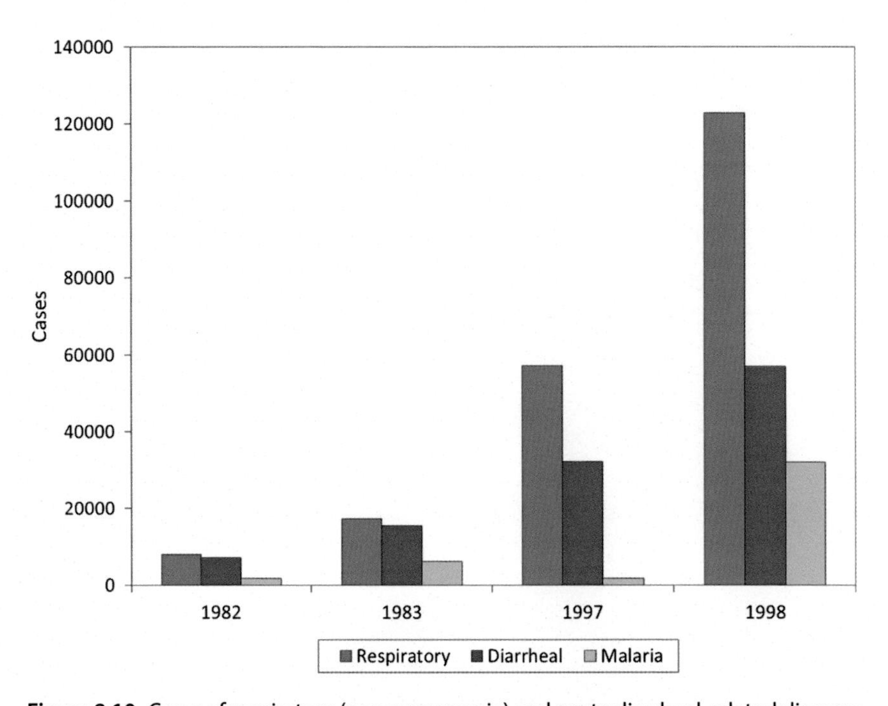

Figure 8.10 Cases of respiratory (non-pneumonia) and acute diarrheal-related diseases (non-cholera) and malaria in the year before and year of El Niño in 1982–83 and 1997–98. *(Data Source: Ministerio de Salud en Piura [Ministry of Health], (MINSA), 2001. Enfermedades emergentes – reemergentes y modificaciones en el perfil epidemiologico del noroeste Peruano post Fenomeno El Niño. (Emerging and reemerging diseases and impacts on epidemiology in northern Peru, post-El Niño). Powerpoint Presentation. Regional Direction of Piura, Region Grau (in Spanish).)*

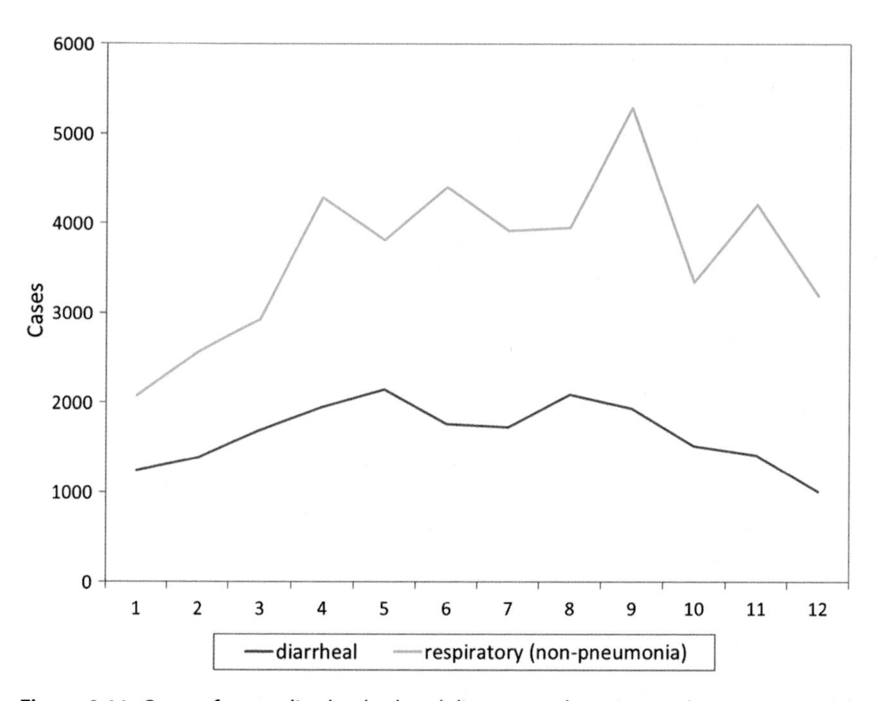

Figure 8.11 Cases of acute diarrheal-related diseases and respiratory (non-pneumonia) during the first 12 weeks of 1998. *(Data Source: Sandoval, P.S., 1999. Evaluación de Daños y Acciones del Fenómeno El Niño (Evaluation of damages caused by the El Niño phenomenon and actions taken). Oficina de Planificación, Dirección de Salud Regional (Planning Office, Regional Health Post), Piura, Peru (in Spanish).)*

emergent topic of disease and health syndemics (Singer, 2009), which refers to the clustering of more than one disease or health problem in a community with the potential for interactions that reinforce exposure and transmission. Thus, in the time of El Niños, the health sector in Piura and Peru in general must contend with an excess burden of disease and health afflictions, including an increase in infant and maternal mortality rates associated with postdisaster effects (i.e., impacts on basic determinants of health, physical, and psychosocial trauma) (MINSA, 2001).

5.4 Infrastructure Vulnerability

In addition to and likely in part a cause of the health effects are impacts on the homes and infrastructure of communities. Fig. 8.13 is a map of the population affected by El Niño-related disasters in 1998, which suggest loss of homes and population displacement. It also implies economic disruption. Table 8.6 shows estimates of impacts on infrastructure measured as

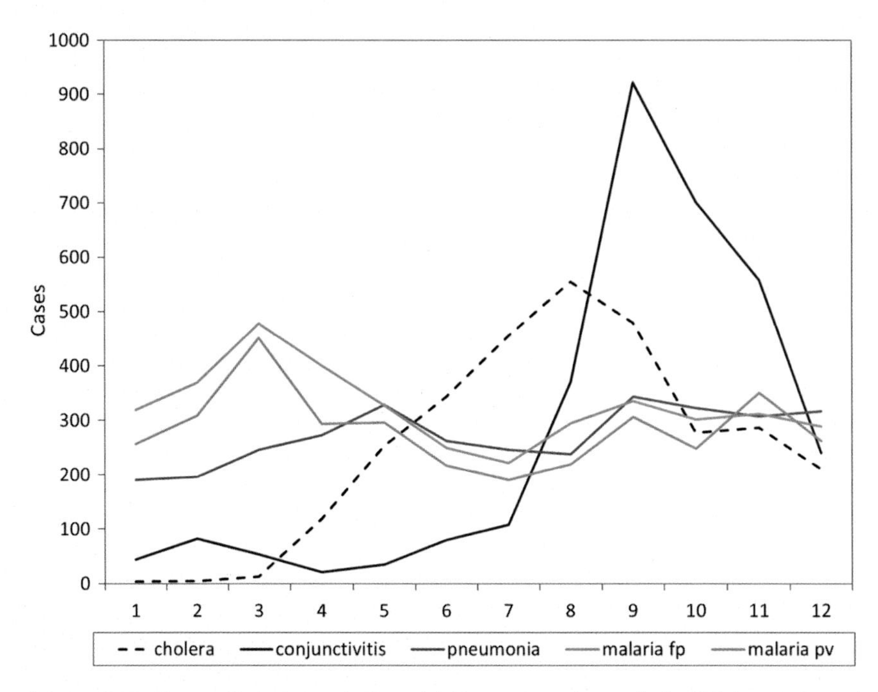

Figure 8.12 Cases of cholera, conjunctivitis, pneumonia, malaria (falciparum), and malaria (vivax) during the first 12 weeks of 1998. *(Data Source: Sandoval, P.S., 1999. Evaluación de Daños y Acciones del Fenómeno El Niño (Evaluation of damages caused by the El Niño phenomenon and actions taken). Oficina de Planificación, Dirección de Salud Regional (Planning Office, Regional Health Post), Piura, Peru (in Spanish).)*

economic losses across multiple sectors associated with the 1997–98 El Niño-related disasters. For example, estimated losses (US$ millions) were experienced in agriculture (50), sanitation (18), and health (0.5). Although the health sector may appear to be the least affected, the damages incurred by hydrometeorological extremes on health infrastructure impede societal responses, as described earlier, to health and social emergencies in Piura. Furthermore, incomes fell following the 1997–98 El Niño, increasing food insecurity, particularly in rural areas (Reyes, 2002). These impacts on well-being along with destruction of homes and other impacts (e.g., agriculture) contributed to wider impacts on people's lives and well-being, and their abilities to cope with the effects of disasters and ecological changes.

5.5 Gender Vulnerability

An aspect of El Niño-related hydrometeorological extremes, which is often neglected in impact assessments and planning, is the gender

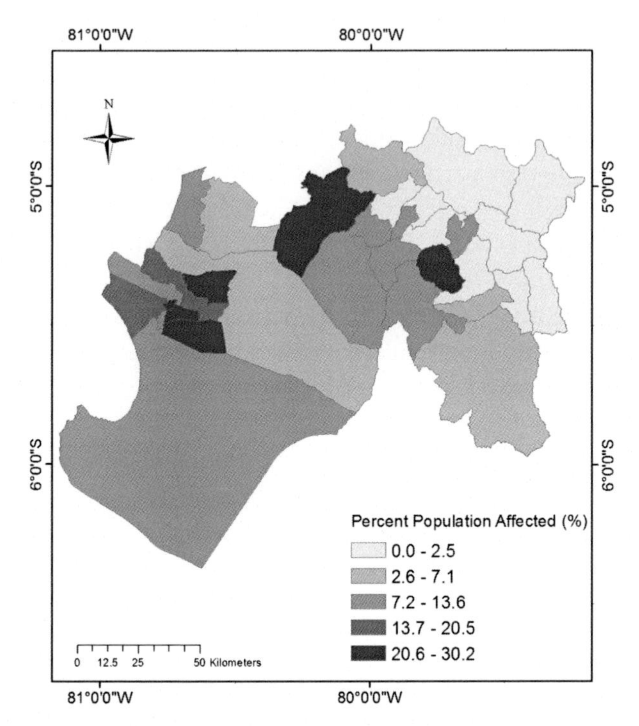

Figure 8.13 Percentage of the population affected by El Niño-related disasters (damage and destruction of homes) in 1998. *(Data Source: National Institute for Statistics and Information, Peru [INEI], 1998. Evaluacion de daños y valorizacion de perdidas ocasionadas por los efectos del fenomeno de El Niño (Evaluation of damages and estimation of economic losses associated with the effects of El Niño). 30 May, 1998. Piura, Peru. (in Spanish).)*

Table 8.6 Estimated reported losses during the 1997–98 El Niño

Sectors	US$ losses (millions)
Agriculture	50,344,919
Fisheries	2,225,064
Industry and tourism	5,536,632
Energy and mining	5,588,841
Transportation	143,597,107
Sanitation	18,630,820
Education	11,972,829
Health	467,368
Other sectors	10,143,695
Total	248,507,276

Data Sources: National Institute for Statistics and Information, Peru [INEI], 1998. Evaluacion de daños y valorizacion de perdidas ocasionadas por los efectos del fenomeno de El Niño (Evaluation of damages and estimation of economic losses associated with the effects of El Niño). 30 May, 1998. Piura, Peru. (in Spanish).

dimension of vulnerability (Reyes, 2002). Like preexisting social and health vulnerabilities, gender inequality at multiple levels in society poses a significant challenge to the public health sector and disaster risk management during and following exposure to climate-related hazards. Inequality for women, particularly those that are socioeconomically poor, and their children, means lack of access or differential access to basic needs, such as food, water, shelter, as well as information and services, resources, and control over assets that would enable someone or a community to thrive in a society and respond to hydrometeorological threats. Gender vulnerability is particularly marked in communities, such as those found in the tropics of Latin America, Asia, and Africa, where livelihoods and survival depend on natural resources (CARE, 2016). Women in these communities are closely linked to the environment and are sensitive to environmental changes in various sectors in society (e.g., water, agriculture, and energy). The impacts of El Niño influence regional and local hydrology and therefore, impact water resources and related sectors. In rural Piura, gender inequality intersects with other vulnerabilities producing disproportionate health effects on women and their children, including exposure to the array of infectious diseases described earlier (i.e., cholera, malaria, dengue, acute respiratory, and diarrheal diseases), and increase in malnutrition (Reyes, 2002). Pregnant women are particularly vulnerable because of lack of food security and increased susceptibility to infections (e.g., malaria) before, during, and after gestation (http://www.who.int/gender-equity-rights/knowledge/gender-health-malaria.pdf?ua=1). Overall, hydrometeorological-related extremes and their hazards potentially exacerbate existing population vulnerabilities, including social and health inequities.

6. CONCLUDING THOUGHTS

The impacts of tropical variability and hydrometeorological extremes pose a significant challenge for South American countries, particularly in the semiarid northwest tropics, where coastal and highland populations face greater exposure to El Niño-related climate hazards within a broader context of socioecological risks. In this chapter, the case study of Piura, Peru illustrated the complexity of hydrometeorological risk by examining the interactions of climate, ecosystem, and society components in an El Niño-sensitive place (hot spot) during the extreme event of 1997–98. The analysis showed, as others have (Trigoso Rubio, 2007; Ramirez, 2015; Sorensen et al., 2017), that vulnerability to El Niño-related hydrometeorological extremes is not only a matter of physical climate risk but also one of preexisting socioeconomic, health, infrastructure, and gender conditions, which set the stage for unsafe environments and cumulative human suffering (e.g., disaster losses and public health impacts). Thus, a "problem" climate (e.g., water stress and quasiperiodic climate extremes) does not beget hazards and disasters in Piura on its own. The untoward effects on societal well-being are the product of the coupled interactions of a "problem" climate with a "problem" society (e.g., uneven development, human-mediated ecosystem change, and social inequalities)[4], which increase exposure and sensitivity of populations to potential harms from hydrometeorological extremes.

Beyond the economic costs of damages to the built environment, public services, industries, and in general livelihoods, for example, is an excess burden of disease morbidity (i.e., ecosyndemics), in addition to loss of life, along with gender inequalities, all of which exacerbate societal vulnerability. Thus, in conceiving public policy and strategies for adaptation and resilience to increasing hydrometeorological risk (climate change) and the quasiperiodic threats of El Niño (variability), decision makers must be reminded to take into account the potential intersections of various vulnerabilities (i.e., climate, social, and disease) that coexist in their respective constituent communities and affect the public's health and well-being. Therefore, action warrants integrated thinking that promotes "hydrometeorological extreme readiness" in the context of El Niño (Glantz et al., 2018). This suggests that policy-makers not only need to develop and improve early warning systems, which utilize climate and hydrometeorological information to respond to foreseeable threats but they should also incorporate social and public health information, which informs responders of the multistressed context that populations experience in El Niño hot spots, such as Piura. In doing so, policy-makers can better address hydrometeorological risk associated with chronic vulnerabilities, such as social and infrastructure depravation, gender inequity, and health disparities, build up existing resilience capacities, and move toward DRR that is part and parcel of El Niño "readiness" and sustainable development.

[4] The notions of a "problem" climate and "problem" society were inspired by the Glantz essay (2003: 242–245).

REFERENCES

Balbus, J., Crimmins, A., Gamble, J.L., Easterling, D.R., Kunkel, K.E., Saha, S., Sarofim, M.C., 2016. Ch. 1: Introduction: climate change and human health. In: The Impacts of Climate Change on Human Health in the United States: A Scientific Assessment. U.S. Global Change Research Program, Washington, DC, pp. 25–42. https://doi.org/10.7930/J0VX0DFW.

Barber, R.T., Chavez, F.P., 1983. Biological consequences of El Niño. Science 22, 1203–1210.

CARE, 2016. Gender Dynamics in a Changing Climate: How Gender and Adaptive Capacity Affect Resilience. CARE. http://careclimatechange.org/wp-content/uploads/2015/11/Gender-and-Adaptation-Learning-Brief.pdf.

Carillo, C., 1892. Disertación sobre las Corrientes Oceanicas y Estudios de la Corriente Peruana de Humboldt (Examination of ocean currents and the study of the Humboldt Current). Bol. Soc. Geogr. Lima 11, 84 (in Spanish).

Carr, M., Broad, K., 2000. Satellites, society, and the Peruvian fisheries during the 1997-1998 El Niño. In: Halpern, D. (Ed.), Satellites, Océanography and Society. Elsevier Science B.V., Atlanta, pp. 171–191.

Caviedes, C.N., 1984. El Niño 1982-83. Geogr. Rev. 74, 267–290. http://marineecology.wcp.muohio.edu/climate_projects_04/el_Niño/ecology_cesar.pdf.

Caviedes, C.N., Fik, T.J., 1992. The Peru-Chile eastern fisheries and climatic oscillation. In: Glantz, M.H. (Ed.), Climate Variability, Climate Change and Fisheries. Cambridge University Press, pp. 355–376.

CEPAL (Economic Commission for Latin America and the Caribbean), 1999. Efectos macroeconomicos del fenomeno El Niño de 1997-98 (Macroeconomic Effects of the 1997-98 El Niño). (in Spanish) https://www.cepal.org/publicaciones/xml/0/40870/EL_NIO1997-98_ECON_ANDINAS.pdf.

Chavez, F.P., Bertrand, A., Guevarro-Carrasco, R., Soler, P., Csirke, J., 2008. The northern Humboldt Current system: ocean dynamics, ecosystem processes, and fisheries. Prog. Oceanogr. 79, 1–15. http://www.imarpe.gob.pe/paita/documentos/Editorial_Conf_Humboldt.pdf.

Chowdhury, F., Rahman, M.A., Begum, Y.A., Khan, A.I., Faruque, A.S.G., Saha, N.C., Baby, N.I., et al., 2011. Impact of rapid urbanization on the rates of infection by Vibrio cholerae 01 and enterooxigenic escherichia coli in Dhaka, Bangladesh. PLoS Negl. Trop. Dis. 5, e999. https://www.ncbi.nlm.nih.gov/pmc/articles/PMC3071362/pdf/pntd.0000999.pdf.

Cisneros, F., Mujica, N., 1999. Impacto del cambia climatico en la agricultura: efectos del fenómeno El Niño en los cultivos de la costa central (Impacts from climate change on agricultura: the effect of El Niño on crops in the central coast). In: Marticorena, B. (Ed.), Peru: Vulnerabilidad Frente Al Cambio Climático: Aproximaciones a la experiencia con el fenómeno El Niño (Peru: Vulnerability to Climate Change: The experience of the El Niño Phenomenon as an Analogue). Consejo Nacional de Ambiente, pp. 115–136 (in Spanish).

Climate and Development Knowledge Network (CDKN), 2014. The IPCC's Fifth Assessment Report: What's in It for Latin America? Executive Summary. CDKN. https://cdkn.org/wp-content/uploads/2014/11/IPCC-AR5-Whats-in-it-for-Latin-America.pdf.

CONAM, Autoridad Autónoma de la Cuenca Hidrográfica [Local Authority of the Watershed Hydrology] Chira Piura (AACHCP), 2005. Evaluacion local integrada y estrategia de adaptación al cambio climático en la Cuenca del Rio Piura (Integrated Local Avaluación and Strategy for Adaptation to Climate Change in the Rio Piura Watershed). Peru http://bibliotecavirtual.minam.gob.pe/biam/handle/minam/696.

Confalonieri, U., Menne, B., Akhtar, R., Ebi, K., Hauengue, M., Kovats, R.S., Revich, B., Woodward, A., 2007. Human health. In: Parry, M.L., Canziani, O.F., Palutikof, J.P., van der Linden, P.J., Hanson, C.E. (Eds.), Climate Change 2007: Impacts, Adaptation and Vulnerability. Cambridge University Press, Cambridge, UK, pp. 391–431. Contribution of Working Group II to the Fourth Assessment Report of the Intergovernmental Panel on Climate Change https://www.ipcc.ch/pdf/assessment-report/ar4/wg2/ar4-wg2-chapter8.pdf.

Consejo Nacional del Ambiente (CONAM), December 13, 2002. Estrategia nacional de cambio climatico, Perú (National Climate Change Strategy). Version 8. (in Spanish) http://sinia.minam.gob.pe/index.php?idElementoInformacion=1.

Cordova Aguilar, H., 1996. Piura. Geospacios 9. Universidad de la Serena, Peru (in Spanish).

Cutter, S.L., Boruff, B.J., Shirley, W.L., 2003. Social vulnerability to environmental hazards. Soc. Sci. Q. 84, 242–261.

Diaz, H.F., Hoerling, M.P., Eishield, J.K., 2001. ENSO variability, teleconnections and climate change. Int. J. Climatol. 21, 1845–1862.

Ebi, K., Kovats, R.S., Menne, B., 2006. An approach for assessing human health vulnerability and public health interventions to adapt to climate change. Environ. Health Perspect. 114, 1930–1934. https://doi.org/10.1289/ehp.8430.

Eguiguren, D.V., 1894. Las lluvias en Piura (Rains in Piura). Bol. Soc. Geogr. Lima 4, 241–258 (in Spanish).

ENFEN (Peru's Multisectoral Committee for the Study of El Niño), 2012. Definicion operacional de los eventos El Niño y La Niña y sus magnitudes en la costa del Peru (Operational Definition of El Niño and La Niña Events and Their Magnitudes along the Coast of Peru). http://www.imarpe.pe/imarpe/archivos/informes/imarpe_comenf_not_tecni_enfen_09abr12.pdf.

Escribano, R., et al., 2004. Biological and chemical consequences of the 1997-1998 El Niño in the Chilean coastal upwelling system: a synthesis. Deep Sea Res. 11, 2389–2411.

Fraser, B., April 25, 2017. Surprise El Niño causes devastation but offers lessons for ecologists. Nat. News. http://www.nature.com/news/surprise-el-ni%C3%B1o-causes-devastation-but-offers-lessons-for-ecologists-1.21891.

Gasper, D., 2005. The Ethics of Development. Edinburgh University Press, Edinburg.

Glantz, M.H. (Ed.), 1994. Drought Follows the Plow. Cambridge University Press, New York.

Glantz, M.H., 2001. In: Currents of Change: Impacts of El Niño and La Niña on Climate and Society. Cambridge University Press, New York.

Glantz, M.H., 2003. In: Climate Affairs. Climate Affairs: A Primer. Island Press, Washington, DC.

Glantz, M.H., 2015. Shades of chaos: lessons learned about lessons learned about forecasting El Niño and its impacts. Int. J. Disaster Risk Sci. 6, 94–103. https://doi.org/10.1007/s13753-015-0045-6. https://link.springer.com/content/pdf/10.1007/s13753-015-0045-6.pdf.

Glantz, M.H., Adeel, Z., 2001. El Niño of the century: once burnt, twice shy? Glob. Environ. Change 11, 171–174.

Glantz, M.H., Jamieson, D., 2000. Societal response to Hurricane Mitch and intra- versus intergenerational equity issues: whose norms should apply? Risk Anal. 6, 869–882. https://pdfs.semanticscholar.org/d649/feaeacc952bb712fc3e70d4c13e7e08b2904.pdf.

Glantz, M.H., Naranjo, L., Baudoin, M., Ramírez, I.J., 2018. What does it mean to be El Niño Ready? Atmosphere 9, 94. https://doi.org/10.3390/atmos9030094.

Goddard, L., Mason, S.J., Zebiak, S.E., Ropelewski, C.F., Basher, R., Cane, M., 2001. Current approaches to seasonal-to-annual climate predictions. Int. J. Climatol. 21, 1111–1152.

Grosse, R.N., Auffrey, C., 1989. Literacy and health status in developing countries. Annu. Rev. Publ. Health. 10, 281–297. http://www.annualreviews.org/doi/pdf/10.1146/annurev.pu.10.050189.001433.

Gueri, A., 1984. Lessons learned: health effects of El Niño in Peru. Disasters Prep. Mitig. 19. PAHO http://helid.digicollection.org/en/d/Jdi019e/2.html.

Hoerling, M., 2002. The symmetry issue. In: Glantz, M.H. (Ed.), La Niña and its Impacts: Facts and Speculation. United Nations University Press, pp. 57–62.
Holdridge, L.R., 1967. Life Zone Ecology. Tropical Science Center, San Jose, Costa Rica. http://reddcr.go.cr/sites/default/files/centro-de-documentacion/holdridge_1966_-_life_zone_ecology.pdf.
Intergovernmental Panel on Climate Change (IPCC), 2012. Summary for policymakers. In: Field, C.B., Barros, V., Stocker, T.F., Qin, D., Dokken, D.J., Ebi, K.L., Mastrandrea, M.D., Mach, K.J., Plattner, G.K., Allen, S.K., Tignor, M., Midgley, P.M. (Eds.), Managing the Risks of Extreme Events and Disasters to Advance Climate Change Adaptation. A Special Report of Working Groups I and II of the IPCC. Cambridge University Press, Cambridge, pp. 3–21. https://www.ipcc.ch/pdf/special-reports/srex/SREX_Full_Report.pdf.
Jamieson, D., 2014. Reason in a Dark Time. Oxford University Press, New York.
Kates, R., 1985. The interaction of climate and society. In: Kates, R.W., Ausubel, J.H., Berberian, M. (Eds.), Climate Impact Assessment: Studies of the Interaction of Climate and Society. John Wiley & Sons, San Francisco, pp. 3–36.
Kovats, R.S., Bouma, M.J., Hajat, S., Worrall, E., Haines, A., 2003. El Niño and health. Lancet 362, 1481–1489.
Lagos, P., 1998. Use of the information and prediction of El Niño in the Perúvian agrarian sector. In: Glantz, M.H. (Ed.), Assessment of the Use of Remote Sensing and Other Information Related to ENSO: The Use of ENSO Information in Perú. Environmental and Societal Impacts Group, National Center for Atmospheric Research, Boulder, CO, pp. 29–35. NASA/NCAR/Perú Project Final Report.
Lagos, P., Buizer, J., 1992. El Niño and Peru: a nation's response to interannual climate variability. In: Majumdar, et al. (Ed.), Natural and Technological Disasters: Causes, Effects and Preventive Measures. The Pennsylvania Academy of Science, Erie, pp. 223–238.
Lagos, P., Silva, Y., Nickl, E., Mosquera, K., 2008. El Niño-related precipitation variability in Peru. Adv. Geosci. 14, 231–237. http://www.adv-geosci.net/14/231/2008/.
Lavell, A., Oppenheimer, M., Diop, C., Hess, J., Lempert, R., Li, J., Muir-Wood, R., Myeong, S., 2012. Climate change: new dimensions in disaster risk, exposure, vulnerability, and resilience. In: Field, C.B., Barros, V., Stocker, T.F., Qin, D., Dokken, D.J., Ebi, K.L., Kelman, I., Gaillard, J.C., Mercer, J. (Eds.) (2015). Int. J. Disaster Risk Sci., 6 https://doi.org/10.1007/s13753-015-0038-5.
Maantay, J., Becker, S., 2012. The health impacts of global climate change: a geographic perspective. Appl. Geogr. 33, 1–3.
Marcus, E.N., 2006. The silent epidemic- the health effects of illiteracy. N. Engl. J. Med. 355, 339–341. http://www.nejm.org/doi/full/10.1056/NEJMp058328#t=article.
Marengo, J.A., Nobre, C.A., Tomasella, J., 2008. The drought of Amazonia in 2005. J. Clim. 21, 495–516. https://doi.org/10.1175/2007JCLI1600.1.
Marmot, M., 2005. Social determinants of health inequalities. Lancet 365, 1099–1104.
Martens, P., Hall, H., 2000. Malaria on the move: human population movement and malaria transmission. Emerg. Infect. Dis. 6, 103–109. https://doi.org/10.3201/eid0602.000202.
McMichael, A.J., 2013. Globalization, climate change, and human health. N. Engl. J. Med. 368, 1335–1343. http://www.nejm.org/doi/pdf/10.1056/NEJMra1109341.
McPhaden, M.J., Zebiak, S.E., Glantz, M.H., 2006. ENSO as an integrating concept in earth science. Science 314, 1740–1745.
Meehl, G., 2002. Attribution of societal and environmental impacts to specific La Niña and El Niño events. In: Glantz, M.H. (Ed.), La Niña and Its Impacts: Facts and Speculation. United Nations University Press, pp. 63–67.
Ministerio de Salud en Piura [Ministry of Health], (MINSA), 2001. Enfermedades emergentes – reemergentes y modificaciones en el perfil epidemiologico del noroeste Peruano post Fenomeno El Niño (Emerging and Reemerging Diseases and Impacts on Epidemiology in Northern Peru, Post-El Niño). Powerpoint Presentation. Regional Direction of Piura, Region Grau (in Spanish).

Moreno, A.R., 2006. Climate change and human health in Latin America: drivers, effects, and policies. Reg. Environ. Change 6, 157–164.
National Institute for Statistics and Information, Peru (INEI), 1994a. Necesidades Basicas Insatisfechas (Basic Needs Unmet). Lima, Peru (in Spanish) http://proyectos.inei.gob.pe/web/biblioineipub/bancopub/Est/Lib0068/n00.htm.
National Institute for Statistics and Information, Peru (INEI), 1994b. Piura: Compendio Estadistico 1993-1994 (Piura: Summary of Statistics 1993-1994). Lima, Peru. 1994b, (in Spanish).
National Institute for Statistics and Information, Peru (INEI), 1998. Evaluacion de daños y valorizacion de perdidas ocasionadas por los efectos del fenomeno de El Niño (Evaluation of damages and estimation of economic losses associated with the effects of El Niño). 30 May, 1998. Piura, Peru (in Spanish).
National Institute for Statistics and Information, Peru [INEI], 2000. Encuesta Demográfica y de Salud Familiar 2000 (Study of Family Demography and Health 2000). (in Spanish) http://desa.inei.gob.pe/endes/.
National Oceanic and Atmospheric Administration (NOAA), 2005. Frequently Asked Questions about El Niño and La Niña. Climate Prediction Center. http://www.cpc.noaa.gov/products/analysis_monitoring/ensostuff/ensofaq.shtml.
Nicholls, N., 1991. Teleconnections and health. In: Glantz, M.H., Katz, R.W., Nicholl, N. (Eds.), Teleconnections Linking Worldwide Climate Anomalies. Cambridge University Press, New York, pp. 493–510.
NOAA, 2017. Cold and Warm Episodes by Season. http://origin.cpc.ncep.noaa.gov/products/analysis_monitoring/ensostuff/ONI_v5.php.
Ordinola, N., 2002. The consequences of cold events for Peru. In: Glantz, M.H. (Ed.), La Niña and Its Impacts: Facts and Speculation. United Nations University Press, Tokyo, pp. 146–150.
Orlove, B.S., Chiang, J.C.H., Cane, M.A., 2000. Forecasting Andean rainfall and crop yield from the influence of El Niño on Pleiades visibility. Nature 403, 68–71.
Oxfam, 2010. Gender, Disaster Risk Reduction, and Climate Change Adaptation: A Learning Companion. Oxfam. https://www.gdnonline.org/resources/OxfamGender&ARR.pdf.
PAHO, 1998. El Niño and Its Impact on Health. 122nd Meeting, Washington, DC June 1998, Provisional Agenda Item 4.4.
PAHO, 2002. Health in the America, 2002 Edition Volume 1. Scientific and Technical Publication No. 587. PAHO, Washington, DC.
PAHO, 2008. Number of Cholera Case in the Americas, 1990-2008.
PAHO/WHO, 2017. Communicable Diseases and Health Analysis/Health Information and Analysis. PLISA Database. Health Situation in the Americas: Basic Indicators. Washington, DC, United States of America http://www.paho.org/data/index.php/en/indicators.html.
Pan American Health Organization (PAHO), 1998b. Perú: Fenómeno "El Niño". Informe estratégico #2, OPS-Perú, Week 10. (in Spanish).
Penaherrera del Aguila, C., 1999. El Niño 1997-1998 y sus impactos en el territorio peruano (The 1997-1998 El Niño and its impacts on Peruvian territory). In: Marticorena, B. (Ed.), Peru: Vulnerabilidad Frente Al Cambio Climático: Aproximaciones a la experiencia con el fenómeno El Niño (Peru: Vulnerability to Climate Change: The Experience of the El Niño Phenomenon as an Analogue). Consejo Nacional de Ambiente, pp. 151–162 (in Spanish).
Penrose, K., Castro, M.C.d., Werema, J., Ryan, E.T., 2010. Informal urban settlements and cholera risk in Dar es Salaam, Tanzania. PLoS Negl. Trop. Dis. 4, e631. https://doi.org/10.1371/journal.pntd.0000631.
Ramírez, I.J., 2012. Cholera in a Time of El Niño and Vulnerability in Piura, Peru: A Climate Affairs Approach. (Doctoral dissertation). Proquest Dissertations and Theses 2012. Michigan State University. Publication Number, AAT 3490977.
Ramirez, I.J., 2015. Cholera resurgence in Piura, Peru: examining climate associations during the 1997-98 El Niño. GeoJournal 80, 129–143.

Ramírez, I.J., Briones, F., 2017. Understanding the El Niño "Costero" of 2017: the definition problem and challenges of climate forecasting and disaster responses. International Journal of Disaster Risk Science 8, 489–492. https://doi.org/10.1007/s13753-017-0151-8.

Ramirez, I.J., Grady, S., Glantz, M.H., 2013. Reexamining El Niño and cholera in Peru: a climate affairs approach. Weather Clim. Soc. 5, 148–161.

Reyes, R.R., 2002. Gendering responses to El Niño in rural Peru. Gend. Dev. 10, 60–69.

Ribot, J., 1995. The causal structure of vulnerability: its application to climate impact analysis. GeoJournal 35, 119–122.

Ribot, J., 2009. Vulnerability does not just fall from the sky: toward multi-scale pro-poor climate policy. In: Mearns, R., Norton, A. (Eds.), Dimensions of Climate Change: Equity and Vulnerability in a Warming World. The World Bank, Washington, DC. http://www.icarus.info/wp-content/uploads/2009/10/Ribot-Vulnerability-Final-Draft-for-Distribution.pdf.

Rodriguez, R., Macabres, A., Luckman, B., Evans, M., Masiokas, M., Ektvedt, T.M., 2005. "El Niño" events recorded in dry-forest species of the lowlands of northwest Peru. Dendrochronologia 22, 181–186.

Sandoval, P.S., 1999. Evaluación de Daños y Acciones del Fenómeno El Niño (Evaluation of damages caused by the El Niño phenomenon and actions taken). Oficina de Planificación, Dirección de Salud Regional (Planning Office, Regional Health Post), Piura, Peru (in Spanish).

Sears, A.F., 1895. The coastal desert of Perú. Bull. Am. Geogr. Soc. 28, 256–271.

SENAMHI, 2009. Escanarios de cambio climatico en la cuenca del Rio Urubamba para el ano 2100: resumen técnico (Climate Change Scenarios for the Urubamba River Basin in 2100: A Technical Summary). Lima, Perú (in Spanish).

Servicio Nacional de Meteorologia e Hidrologia [National Meteorological and Hydrological Services] (SENAMHI), Lambayeque, 2004. El evento El Niño Oscilacion Sur 1997-1998: su impacto en el departamento de Lambayeque (The 1997/98 ENSO Event: Impacts in the Department of Lambayeque). Chiclayo, Perú http://www.senamhi.gob.pe/?p=0702 (in Spanish).

Singer, M., 2009. Introduction to Syndemics: A Critical Systems Approach to Public and Community Health. John Wiley & Sons, San Francisco.

Sorensen, C.J., Borbor-Cordova, M.J., Calvello-Hynes, E., Diaz, A., Lemery, J., Stewart-Ibarra, A.M., 2017. Climate variability, vulnerability and natural disasters: a case study of Zika virus in Manabi, Ecuador following the 2016 earthquake. GeoHealth. 1. https://doi.org/10.1002/2017GH000104.

Takahashi, K., 2017. Fenomono El Niño: global vs 'Costero' (The El Niño phenomenon: Global vs Coastal). In: Generacion de informacion y monitoreo del Fenomeno El Niño – Boletin Tecnico. Instituto Geofisico del Peru. Ministerio del Ambiente, Peru. 4(4) (in Spanish).

Talavera, A., Perez, E., 2009. Is cholera disease associated with poverty? J. Infect. Dev. Ctries. 3, 408–411. https://doi.org/10.3855/jidc.410.

Tarazona, J., Valle, S., 1999. Impactos potenciales del cambio climatico global sobre el ecosistema marino Peruano (Potential impacts from global climate change on the Peruvian marine ecosystem). In: Marticorena, B. (Ed.), Peru: Vulnerabilidad Frente Al Cambio Climático: Aproximaciones a la experiencia con el fenómeno El Niño (Peru: Vulnerability to Climate Change: The Experience of the El Niño Phenomenon as an Analogue). Consejo Nacional de Ambiente, pp. 95–114 (in Spanish).

Teklehaimanot, A., Meija, P., 2008. Malaria and poverty. Ann. N.Y. Acad. Sci. 1136, 32–37. https://doi.org/10.1196/annals.1425.037. http://onlinelibrary.wiley.com/doi/10.1196/annals.1425.037/epdf.

Trigoso Rubio, E., 2007. Climate change impacts and adaptation in Peru: the case of Puno and Piura. Human development report 2007/2008. In: Human Development Report Office Occasional Paper. http://hdr.undp.org/en/content/climate-change-impacts-and-adaptation-peru.

Turner II, B.L., Kasperson, R.E., Matson, P.A., McCarthy, J.J., Corell, R.W., Christensen, L., Eckley, N., et al., 2003. Science and technology for sustainable development special feature: a framework for vulnerability analysis in sustainability science. Proc. Natl. Acad. Sci. U.S.A. 100, 8074–8079.

Universidad de Piura (University of Piura [UDEP]), 2008. Datos Climaticos (Climate Datasets) (Piura, Peru).

Valverde, A., 1998. Influence of El Niño in the systems of hydroelectric generation. In: Glantz, M.H. (Ed.), Assessment of the Use of Remote Sensing and Other Information Related to ENSO: The Use of ENSO Information in Perú. Environmental and Societal Impacts Group, National Center for Atmospheric Research, pp. 45–48. NASA/NCAR/Perú Project Final Report. Boulder, CO.

Velasco-Zapata, A., Broad, K., 2001. Peru country case study: impacts and responses to the 1997-98 El Niño event. In: Glantz, M.H. (Ed.), Once Burned, Twice Shy? Lessons Learned from the 1997-98 El Niño. United Nations University Press, Tokyo, pp. 186–199.

Vittor, A.Y., Pan, W., Gilman, R.H., Tielsch, J., Glass, G., Shields, T., Sánchez-Lozano, W., Pinedo, V.V., Salas-Cobos, E., Flores, S., Patz, J.A., 2009. Linking deforestation to malaria in the Amazon: characterization of the breeding habitat of the principal malaria vector Anopheles darling. Am. J. Trop. Med. Hyg. 81, 5–12.

WHO, 2015. Global Health Observatory Data Repository. WHO, Geneva. http://apps.who.int/gho/data/view.main.1390?lang=en.

WHO, 2016. El Niño and Health: A Global Overview – January 2016. http://www.who.int/hac/crises/el-Niño/who_el_Niño_and_health_global_report_21jan2016.pdf?ua=1.

Wilson, M.L., 2001. Ecology and infectious disease. In: Aron, J.L., Patz, J. (Eds.), Ecosystem Change and Public Health: A Global Perspective. Johns Hopkins University, Baltimore, pp. 283–324.

Wisner, B., Blaikie, P., Cannon, T., Davis, I., 2004. At Risk: Natural Hazards, People's Vulnerability and Disasters, second ed. Routledge, New York.

Wisner, B., Gaillard, J.C., Kelman, I., 2012. Framing disaster. In: The Routledge Handbook of Hazards and Disaster Risk Reduction. Routledge, New York. https://www.routledgehandbooks.com/doi/10.4324/9780203844236.ch3.

Woodman, R., 1998. El Fenómeno El Niño y el Clima en el Perú (The El Niño phenemonon and climate in Perú). In: El Perú en los Albores del Siglo XXI (Perú at the Turn of the 21st Century), pp. 201–242Ediciones del Congreso del Perú, Lima-Perú (in Spanish).

World Health Organization (WHO), 2009. Protecting Health from Climate Change: Science, Policy and People. WHO, Geneva.

World Meteorological Organization (WMO), 2014. Atlas of Mortality and Economic Losses from Weather, Climate, and Water Extremes (1970-2012). WMO-No. 1123 WMO, Geneva.https://public.wmo.int/en/resources/library/atlas-mortality-and-economic-losses-weather-and-climate-extremes-1970-2012.

Yonetani, M., 2016. Disaster-related displacement in a changing climate. World Meteorol. Organ. Bull. 65. https://public.wmo.int/en/resources/bulletin/disaster-related-displacement-changing-climate.

Zebiak, S., 2002. The identification of differences in forecasting El Niño and La Niña. In: Glantz, M.H. (Ed.), La Niña and Its Impacts: Facts and Speculation. United Nations University Press, Tokyo, pp. 74–77.

Zebiak, S.E., Orlove, B., Munoz, A.G., Vaughan, C., Hansen, J., Troy, T., Thomson, M.C., et al., 2015. Investigating El Niño-southern oscillation and society relationships. WIREs Clim. Change 6, 17–34. https://doi.org/10.1002/wcc.294.

CHAPTER 9

Tropics as Tempest

Bjorn Stevens[1], Gabor Drotos[2], Tobias Becker[1], Thorsten Mauritsen[1]
[1]Max Planck Institute for Meteorology, Hamburg, Germany; [2]Hungarian Academy of Sciences and Eötvös Loránd University, Budapest, Hungary

Contents

1. TRANQUIL OR TEMPESTUOUS?

War was once famously described as long periods of immense boredom punctuated by moments of shear terror. If literature is a guide, this is also an apt description of tropical weather. As a case in point, consider the description of the Hurricane in Zora Neale Hurston's famous novel, *Their Eyes Were Watching God*.

The wind came back with triple fury, and put out the light for the last time. They sat in company with the others in other shanties, their eyes straining against crude walls and their souls asking if He meant to measure their puny might against His. They seemed to be staring at the dark, but their eyes were watching God.

Few would venture to describe tropical climate in similar terms. In contrast, the view of the tropics as a climatically tranquil region—tempering the intemperate extratropics—is prevalent. The terror is gone. Only the boredom remains. This view, which is rooted in the analysis of the instrumental temperature record and proxies for temperature variations over the last 10 million years, might well be mistaken. When one pieces together the results of a number of recent studies, a different view of the tropics, as a region capable of globally reorganizing its circulation, and substantially altering the planetary energy budget and the distribution of freshwater, no longer seems implausible. The apparent tranquility of the tropics may well belie its tempest.

Tropical Extremes: Natural Variability and Trends
ISBN 978-0-12-809248-4
https://doi.org/10.1016/B978-0-12-809248-4.00009-1

2. PAST CHANGES IN THE DEEP TROPICS

There is little doubt that surface warming over the roughly 150 year period of the modern instrumental record is pronounced in the Northern Hemisphere extratropics, especially in the arctic and over land. Less clear is the degree of tropical warming. Because the temperature across the tropical troposphere is connected by moist convection to those regions in the tropics where the near-surface moist static energy maximizes (roughly speaking the regions where surface waters are warmest), a warming of the regions of warm surface waters is expected to be especially evident in the upper troposphere (Emanuel, 1986). Such a signal expresses the expectation that the vertical structure of warming scales with the sensitivity of a reversible saturated adiabat to the temperature at cloud base. Latent heating differentiates the moist from the saturated adiabats, and this scales with the saturation vapor pressure at cloud base, which is a strong function of temperature. Hence any change to the temperature near cloud base in the convecting regions will be amplified with height, so that the strongest signals should be evident in the upper tropical troposphere. Given this expectation, reports of modest or even statistically insignificant warming in the upper troposphere (near 200 hPa Christy et al., 2007) have been taken to be indicative of a lack of warming in the convecting regions. However, as understanding of the instrumental record has improved, a purported lack of warming (particularly in the upper tropical troposphere) has increasingly been identified with how the data are treated. After accounting for the lack of homogeneity in the radiosonde record, or the expected influence of stratospheric cooling on the vertically extensive weighting functions used to infer temperatures in the upper troposphere, a clearer signal of tropical warming emerges, albeit less pronounced than in models, but with a vertical structure that is consistent with the models and the theoretical understanding that underpins their construction (Fu et al., 2011; Santer et al., 2017). So these studies would maintain that, contrary to lingering perceptions, the warm tropical oceans are warming over the instrumental record and this warming is amplified with height.

More contraindicative for the perception of a tranquil tropics, is the ongoing revision of understanding of the proxy record of tropical sea surface temperatures (SSTs). The original (and still prevalent) view, of a more tranquil tropics, arises from measurements of the magnesium-to-calcium ratio (Mg/Ca) from foraminiferal shell in deep-sea sediments (Wara et al., 2005). The data seemed to suggest that in the mid-Pliocene Warm Period (around 3 Ma, million years before present), when global temperatures are estimated to have been 3K warmer than at present, the warm-water regions of the tropical oceans were not appreciably warmer than they are today. More recent analyses, which correct for changes in Mg/Ca in the background seawater, show that the warm pool region had been as much as 2K warmer during the mid-Pliocene Warm Period (Medina-Elizalde and Lea, 2010; O'Brien et al., 2014). Moreover, the gradual cooling of the Earth's surface that led to the onset of glaciation in the Quaternary is evident within the tropics well in advance of the first appearance of ice sheets, suggesting that changes in the tropics were not simply tempering much larger changes at high latitudes (Herbert et al., 2010). Although not unequivocal (Ravelo et al., 2014; Zhang et al., 2014), these studies are consistent with the evolution in thinking regarding tropical warming in the recent past and contribute to the impression that the warmest ocean waters, while not varying as much as those at high latitudes, do show considerable variations in response to forcing.

Less disputed is the idea that forcing may shape the structure of the tropical circulation. SST trends over the 20th century (Deser et al., 2010) are far from homogeneous with less warming or even a cooling over the equatorial eastern Pacific (cold tongue region) and the greatest warming over the California and Benguela Current systems. A similar pattern of warming is evident in the Pliocene. Irrespective of whether the warmest waters were warmer during the Pliocene warm period, the cold, upwelling, eastern boundary currents were much warmer, with some proxies indicating that waters in the California and Benguela Current systems were ≈10K warmer 3 million years ago (Herbert et al., 2010). Such large temperature changes would reshape the pattern of tropical SST and, because surface winds and hence near-surface moist static energy responds to SST gradients, the pattern of tropical convection (Lindzen and Nigam, 1987). In contrast to the pattern of Pliocene warming, a more equable distribution of SST more weakly constrains the distribution of precipitation, so that the tropical circulation system as a whole becomes more labile and hence susceptible to fluctuations. For instance, to the extent the warming of the eastern boundary current systems reduces zonal gradients, it is conceivable that large-scale convective systems associated with Madden–Julian oscillation (MJO)–like disturbances would propagate across the Pacific. This would allow such a circulation to more dramatically and directly impact South America and Africa, and more effectively (Palmer and Owen, 1986, through planetary waves) influence European and Middle Eastern climate. Weakening of the subtropical highs in association with warming over these current systems would affect the meridional energy transport of the stationary eddies (Nigam 1997), and this would in turn influence the position of the ITCZ and monsoon circulations. In summary, even if there is equivocation as to how much temperatures in the deep tropics change with globally averaged surface temperatures, there is ample evidence of large-scale changes in the structure of the tropical circulation with changes in radiative forcing.

3. TEMPESTUOUSNESS IN MODEL-BASED REPRESENTATIONS OF TROPICAL PRECIPITATION

The idea of an intemperate tropics also aligns with experience from modeling. Even the most comprehensive models have difficulty in reproducing observed patterns of tropical precipitation (Knutti and Sedlácek, 2012), or the distribution of cloud radiative effects, let alone large-scale patterns of variability. Examples include the double ITCZ problem, wherein models show a too pronounced band of precipitation stretching across the Pacific near the equator (Oueslati and Bellon, 2015). Similar issues arise in the Atlantic, as in coupled models the zonal gradient of SSTs can even have the wrong sign, and convection tends to be coastally trapped rather than maximizing in the middle of the ocean basin (Siongco et al., 2014). Models also struggle to get the distribution of cloud radiative effects correct in the tropics, tending to make the bright oceans (with large negative cloud radiative effects) too dark, and the dark oceans (with less negative cloud radiative effects) too bright (Nam et al., 2012; Hourdin et al., 2015). Finally the challenge models have in representing large-scale patterns of variability ranging from the MJO on interseasonal scales to monsoon circulations, to ENSO on interannual timescales, and to patterns of decadal variability in the Pacific are well documented. These differences are also apparent when models are run with prescribed SSTs (Crueger et al., 2013; Siongco et al., 2014), demonstrating that small differences in how atmospheric processes are represented can result in very large circulation differences. Such sensitivities are not what one would expect for a tranquil tropics.

The degree of sensitivity of the tropical circulation in comprehensive models is evident in calculations using specified distributions of SSTs, even in the absence of zonal asymmetries. Fig. 9.1 shows the distribution of tropical precipitation resulting from a prescribed SST on an aqua-planet. Output from five state-of-the art atmospheric general circulation models used for climate research or numerical weather prediction is shown. Two of the models show a rainband well removed from the equator with a

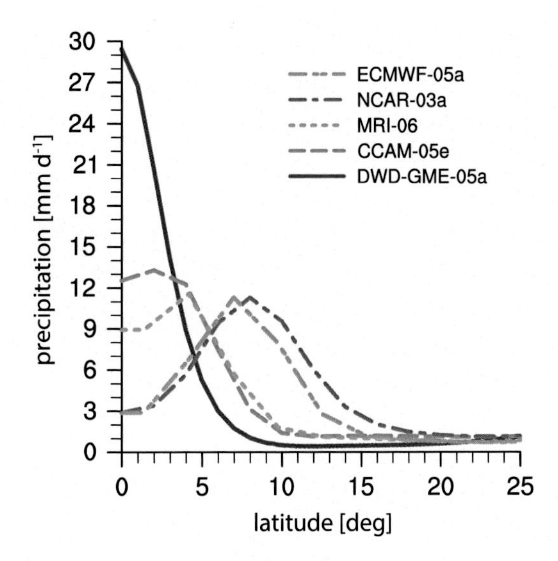

Figure 9.1 Distribution of precipitation with latitude from aqua-planet experiments with patterns of SST specified following the QOBS prescription. *(Model output is taken from the Aqua Planet Experiment (APE; Williamson et al., 2012) as described by Möbis, B., Stevens, B., 2012. Factors controlling the position of the intertropical convergence zone on an aquaplanet. J. Adv. Model. Earth Syst. 4 (4), 1–16.)*

pronounced minimum in precipitation in the inner tropics, equatorward of 5°. Two other models show a broadband of precipitation at the equator, with the maximum in precipitation slightly displaced toward the edge of the band. The remaining model has a very strong precipitation peak, a factor of three higher than the peak amount in the other models, centered at the equator. Möbis and Stevens (2012) and Oueslati and Bellon (2013) showed that many of these differences can be reproduced in a single model, simply by making small changes to how cumulus convection is parameterized—foremost the lateral mixing between the convective updraft and its environment.

These sensitivities are also manifested in how precipitation responds to warming. Fig. 9.2 plots the change in precipitation, as calculated by different models participating in the fifth phase of the coupled model intercomparison project, arising from a uniform 4K warming. Intermodel differences in the pattern of precipitation change are even larger than those of the base state. Some of the models (e.g., the MPI-ESM-LR or FGOALS-G2) show an apparent narrowing, or equatorward shift, of the ITCZ, whereas others (MIROC5 and to a lesser degree IPSL-CM5A-LR) show precipitation enhancement on the subtropical margins of the rainbands. Even if models

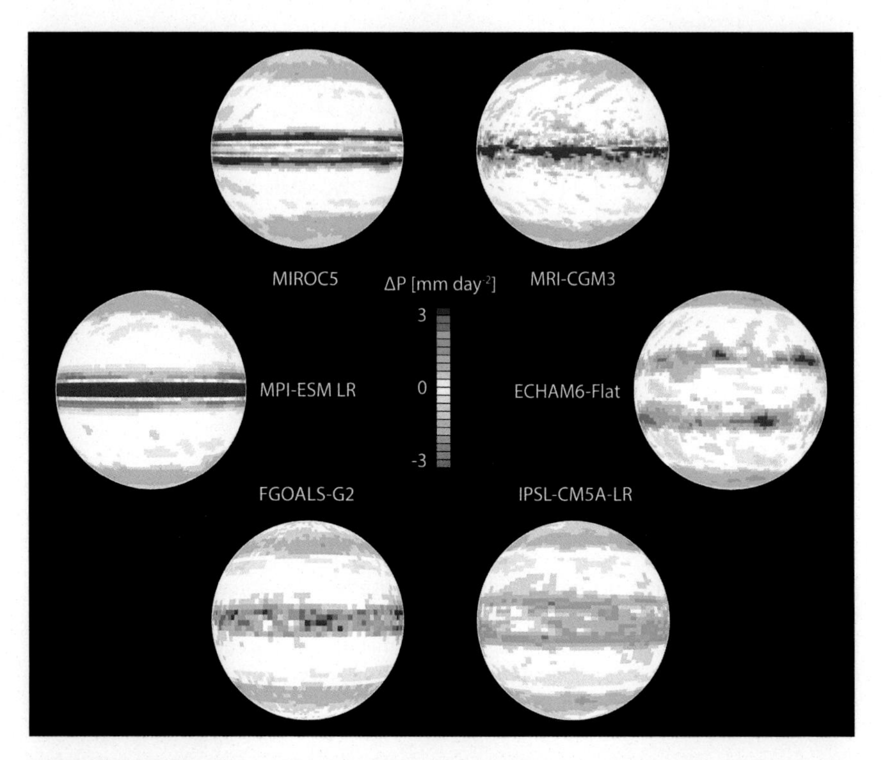

Figure 9.2 The change in precipitation as calculated using different models providing output from aqua-planet simulations as part of CMIP5. All models use the Qobs sea surface temperature (SST) (Blackburn et al., 2013), and Qobs plus 4K from which the precipitation change is calculated. *(This figure is adapted from Stevens, B., Bony, S., 2013. What are climate models missing? Science 340 (6136), 1053–1054 and also includes an additional simulation (ECHAM-Flat) showing the sensitivity to the base SST distribution.)*

agree in a particular feature, such as the equatorward intensification of the rainband, they may disagree on what accompanies it. For instance, in the MPI-ESM a strong strengthening of the rainband on the equator is accompanied by drying on the flanks of the ITCZ, whereas for the FGOALS-G2 model, the equatorial shift is less pronounced and the drying on the subtropical margins is more muted. Comparing the MPI-ESM-LR to ECHAM6-Flat is also instructive, as these calculations were performed by the same model, but for ECHAM6-Flat the underlying distribution of SST, which was then uniformly warmed, followed the Flat prescription,[1] which means that the distribution was less peaked at the equator. Hence the

[1] The standard aqua-planet simulations allow SST to vary with latitude φ as $\sin^2(\varphi) + \sin^4(\varphi)$, in the Flat prescription, the SST varies near the equator only as $\sin^4(\varphi)$.

underlying SST that arises in response to the distribution of clouds, precipitation, and surface winds also influences the precipitation response.

Taken together the difficulty of representing tropical circulation systems, ranging from the climatology of precipitation to the monsoon, to variability on longer timescales, even when SSTs are prescribed, is indicative of a tropical circulation that is sensitive to the details of its representation. The most important detail being the representation of deep cumulus convection. This, of course, raises the question as to why the tropics should be expected to be so tempered in past and future climates.

4. LEVIATHAN UNLEASHED: CLIMATE SENSITIVITY IN A TROPICAL WORLD

The tropics as tempest, at least for climate models, becomes most apparent when considering the implications of the response of the climate system to forcing in the yet simpler configuration of radiative–convective equilibrium (RCE). RCE has long been the benchmark for our understanding of climate change and is often taken as an analogue for the tropics (Wing et al., 2018). The very first studies of how the climate would respond to warming have been based on one-dimensional RCE models with prescribed distributions of humidity and clouds (Manabe and Wetherald, 1967). These types of calculations have also been used to calibrate estimates of the sensitivity of the climate system to forcing or to interpret the behavior of more comprehensive simulations in the earliest and all subsequent assessments of climate change (Charney et al., 1979). Recently, however, we have begun to explore the solution to RCE using comprehensive climate models (Popke et al., 2013; Wing et al., 2018). To do so, all asymmetries are removed from the model and its forcing. This is equivalent to forcing every point on Earth to be covered by the same depth of water, by the identical distribution of long-lived gases and insolation (which are also kept constant in time thereby eliminating seasonality), and by eliminating rotation. The resulting calculation differs from its one-dimensional counterpart in that, on breaking the symmetry in the initial conditions, large-scale circulations arise and these in turn influence the distribution of clouds, winds, and surface temperatures. The spatial location of particular circulation features is random and varies from realization to realization, so that, assuming ergodicity, the statistics of the stationary state can be derived from averaging over time.

Using a forerunner of the version of our model being developed for the sixth phase of CMIP, coupled to a 25 m slab ocean, we explore the RCE state across a wide range of forcings. These include varying the CO_2 concentrations more than 200-fold, from 2^{-3} to 2^5 times the preindustrial values. Also for CO_2 concentrations fixed at their preindustrial value, we vary the insolation from about 80% to 150% of its present-day (or control) value. The radiative forcing associated with these changes is estimated approximately, as follows: For CO_2 it is assumed that the radiative forcing is logarithmic in the atmospheric CO_2 concentration so that each doubling or halving corresponds to a forcing of $3.7\,\mathrm{W m^{-2}}$ (Ramaswamy et al., 2001). For changes in the insolation, the forcing is assumed to be equal to the change in insolation scaled by the ratio of the insolation absorbed by the system to the total insolation in the control climate.[2] For a tranquil tropics, we would expect that if we plotted the equilibrium surface temperature in RCE versus the estimate of the forcing, all the simulations would lie on a single line, whose slope defines the climate sensitivity parameter, α. Based on past work we expect α to be between about 0.5 and $1.0\,\mathrm{K W^{-1} m^2}$, and if not constant, then slightly increasing as the forcing becomes increasingly positive (Meraner et al., 2013; Popp et al., 2016).

The numerical results are quite different. This is illustrated by plotting the temperature distribution of the last 150 years of the simulations (which range from 250 to 1500 years depending on how long they need to reach stationarity) versus their estimated forcing, in Fig. 9.3. Output from simulations with the standard model is plotted in red; those for the configuration without the convective parameterization are plotted in teal. In this figure, α is defined by the slope of the points that map out the stationary temperature state versus forcing for a given configuration of the model. The expectation of $0.5\,\mathrm{K W^{-1} m^2} < \alpha < 1\,\mathrm{K W^{-1} m^2}$ is roughly met for the case of a negative forcing, but α is considerably larger for the case of CO_2 forcing than for an equivalent solar forcing. For a positive forcing, $\alpha < 0.5\,\mathrm{K W^{-1} m^2}$. Also surprising is the smallness of α in general. Values of less than $0.5\,\mathrm{K W^{-1} m^2}$, which begin to emerge at large ($>10\,\mathrm{W m^{-2}}$) forcing from changes in CO_2 in the model with parameterized convection are often thought difficult to capture in a model by mere virtue of the strength of the water vapor feedback.

As striking as is the state dependence of α, is the change in the character of the equilibrium, as illustrated by the temperature range that is spanned in the stationary state. At higher temperatures the variability about the stationary state becomes large—very large. Whereas for present-day values of

[2] Using the absorbed insolation from the perturbed climate in this ratio does not qualitatively change the results.

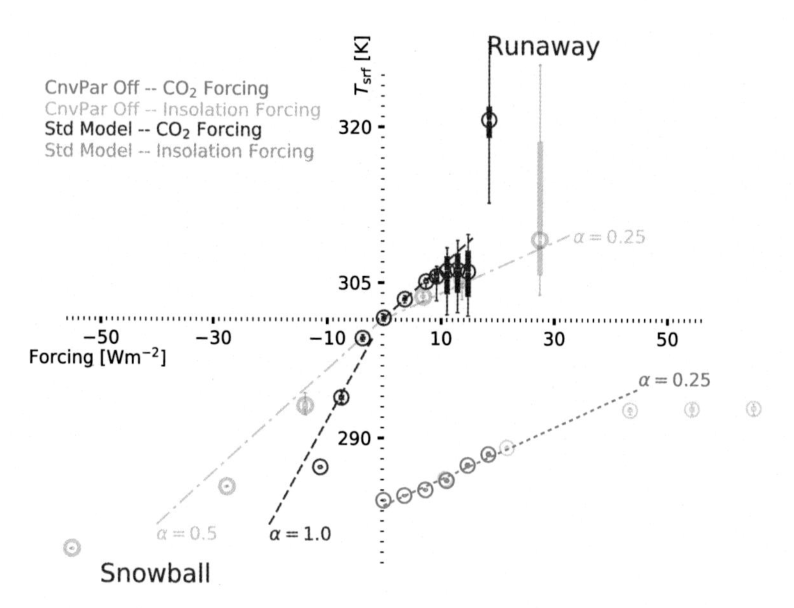

Figure 9.3 Bar and whisker distribution of monthly averaged simulations for RCE simulations in stationarity. The distribution of globally averaged surface temperatures is from the last 600 months of the simulation. The two simulations identified in the "Runaway" part of the parameter space are running away and thus never reach stationarity. For all other simulations, the box denotes the interquartile spread and the whiskers the full range of globally and monthly averaged surface temperatures over the averaging period. The different colors denote different setups as noted in the legend and in the main text.

forcing interannual variability in surface temperature is about 0.1K, with the warmest month in the time span analyzed differing from the coldest month by at most 1K, in a warmer world the fluctuations increase 10-fold. For a world with 16 times as much CO_2, it turns out that there are months that are as cold as the coldest months in a world with preindustrial concentrations of CO_2. For forcings more than about $15\,W\,m^{-2}$, the system runs away to much higher temperatures, and eventually the simulation crashes as it reaches temperatures where subprocesses cease to be defined. At the other extreme, for a forcing less than about $-20\,W\,m^{-2}$, the system becomes metastable whereby sufficiently large monthly temperature perturbations lead to the emergence of sea ice, which pushes the system into a snowball state. A more thorough analysis of the simulations (work in preparation) shows that the reduced sensitivity and very large temperature swings at high forcings is a manifestation of convective self-aggregation (Coppin and Bony, 2016; Bony et al., 2016; Becker et al., 2017).

To test how much of the simulated behavior depends on the representation of convection, simulations were also performed with the convective parameterization turned off (teal points in Fig. 9.3). In this case, the simulations are forced to transport heat from the surface to the atmosphere's interior through resolved-scale circulations. Given the coarse resolution of the model, this is an admittedly extreme assumption. Even so the response is surprising. Without parameterized convection the simulations are much cloudier and colder, and much less sensitive to forcing. Moreover, the response to forcing is rather more consistent as a function of the climate state, and there is relatively little evidence of pronounced fluctuations arising at much warmer temperatures (larger forcing).

The point being that in the very simple RCE analogue to the climate system the response of the system to forcing is complex, and quite sensitive to the assumptions one makes about how convection transports heat into the atmosphere's interior.

5. ROOM FOR SURPRISES

One might be tempted to argue that the seemingly erratic behavior of the modeled climate system in RCE, especially evident at warmer temperatures, is simply an artifact of how convection is parameterized in the model. From this point of view, the tropics as tempest arises from a Leviathan-like representation of convection. But given the role of comprehensive modeling in helping to design effective adaptation and mitigation strategies to climate change, this is not a comforting interpretation. In a more complex system, we expect that some of the instabilities seen in RCE will be tempered by other circulations and by the effects of continents and planetary rotation. Nonetheless, the ability of the tropics, when left to its own devices, to reorganize itself in ways that substantially perturb the planetary energy balance, raises the possibility that large changes in the extratropics might just as well be the climate system's response to a tropical tempest—rather than the other way around. At the very least, these findings raise the possibility that if large surprises are to emerge as the Earth system adjusts to human forcing, then they may well arise in the tropics. Without a deeper understanding of the interplay between convection and circulation, which will be impossible to attain without advancing observations of tropical convection, society may well remain vulnerable to tropical tempests—real or imagined.

REFERENCES

Becker, T., Stevens, B., Hohenegger, C., 2017. Imprint of the convective parameterization and sea-surface temperature on large-scale convective self-aggregation. Journal of Advances in Modeling Earth Systems 9 (2), 1488–1505. http://doi.org/10.1002/2016MS000865.

Blackburn, M., et al., 2013. The aqua-planet experiment (APE): CONTROL SST simulation. JMSJ 91A (0), 17–56.

Bony, S., Stevens, B., Coppin, D., Becker, T., Reed, K.A., Voigt, A., Medeiros, B., 2016. Thermodynamic control of anvil cloud amount. Proc. Natl. Acad. Sci. U.S.A. 201601472.

Charney, J.G., Arakawa, A., Baker, D.J., Bolin, B., 1979. Carbon Dioxide and Climate: A Scientific Assessment. National Research Council.

Christy, J.R., Norris, W.B., Spencer, R.W., Hnilo, J.J., 2007. Tropospheric temperature change since 1979 from tropical radiosonde and satellite measurements. J. Geophys. Res. Atmos. 112 (D6), D12 106–16.

Coppin, D., Bony, S., 2016. Physical mechanisms controlling the initiation of convective self-aggregation in a general circulation model. J. Adv. Model. Earth Syst. 1–50.

Crueger, T., Stevens, B., Brokopf, R., 2013. The Madden–Julian oscillation in ECHAM6 and the introduction of an objective MJO metric. J. Clim. 26 (10), 3241–3257.

Deser, C., Phillips, A.S., Alexander, M.A., 2010. Twentieth century tropical sea surface temperature trends revisited. Geophys. Res. Lett. 37 (10), n/a–n/a.

Emanuel, K.A., 1986. An air-sea interaction theory for tropical cyclones. Part I: steady-state maintenance. J. Atmos. Sci. 43 (6), 585–605.

Fu, Q., Manabe, S., Johanson, C.M., 2011. On the warming in the tropical upper troposphere: models versus observations. Geophys. Res. Lett. 38 (15), n/a–n/a.

Herbert, T.D., Peterson, L.C., Lawrence, K.T., Liu, Z., 2010. Tropical ocean temperatures over the past 3.5 million years. Science 328 (5985), 1530–1534.

Hourdin, F., Găinusă Bogdan, A., Braconnot, P., Dufresne, J.-L., Traore, A.K., Rio, C., 2015. Air moisture control on ocean surface temperature, hidden key to the warm bias enigma. Geophys. Res. Lett. 42 (24), 10885–10893.

Knutti, R., Sedlácek, J., 2012. Robustness and uncertainties in the new CMIP5 climate model projections. Nat. Clim. Change 3 (4), 369–373.

Lindzen, R.S., Nigam, S., 1987. On the role of sea surface temperature gradients in forcing low-level winds and convergence in the tropics. J. Atmos. Sci. 44 (17), 2418–2436.

Manabe, S., Wetherald, R.T., 1967. Thermal equilibrium of the atmosphere with a given distribution of relative humidity. J. Atmos. Sci. 24 (3), 241–259.

Medina-Elizalde, M., Lea, D.W., 2010. Late Pliocene equatorial Pacific. Paleoceanography 25 (2), PA2208. http://doi.org/10.1029/2009PA001780.

Meraner, K., Mauritsen, T., Voigt, A., 2013. Robust increase in equilibrium climate sensitivity under global warming. Geophys. Res. Lett. 40 (22), 5944–5948.

Möbis, B., Stevens, B., 2012. Factors controlling the position of the intertropical convergence zone on an aquaplanet. J. Adv. Model. Earth Syst. 4 (4), 1–16.

Nam, C., Bony, S., Dufresne, J.L., Chepfer, H., 2012. The 'too few, too bright' tropical low-cloud problem in CMIP5 models. Geophys. Res. Lett. 39 (21), n/a–n/a.

Nigam, S., 1997. The annual warm to cold phase transition in the eastern equatorial Pacific: diagnosis of the role of stratus cloud-top cooling. J. Clim. 10 (10), 2447–2467.

O'Brien, C.L., Foster, G.L., Martínez-Botí, M.A., Abell, R., Rae, J.W.B., Pancost, R.D., 2014. High sea surface temperatures in tropical warm pools during the Pliocene. Nat. Geosci. 7 (8), 606–611.

Oueslati, B., Bellon, G., 2013. Convective entrainment and large-scale organization of tropical precipitation: sensitivity of the CNRM-CM5 hierarchy of models. J. Clim. 26 (9), 2931–2946.

Oueslati, B., Bellon, G., 2015. The double ITCZ bias in CMIP5 models: interaction between SST, large-scale circulation and precipitation. Clim. Dyn. 44 (3–4), 585–607.
Palmer, T.N., Owen, J.A., 1986. A possible relationship between some "severe" winters in North America and enhanced convective activity over the tropical west Pacific. Mon. Weather Rev. 114 (3), 648–651.
Popke, D., Stevens, B., Voigt, A., 2013. Climate and climate change in a radiative-convective equilibrium version of ECHAM6. J. Adv. Model. Earth Syst. 5 (1), 1–14.
Popp, M., Schmidt, H., Marotzke, J., 2016. Transition to a moist greenhouse with CO_2 and solar forcing. Nat. Commun. 7, 10 627–10.
Ramaswamy, V., et al., 2001. Radiative forcing of climate. Clim. Change 349–416.
Ravelo, A.C., Lawrence, K.T., Fedorov, A., Ford, H.L., 2014. Comment on "A 12-million-year temperature history of the tropical Pacific Ocean". Science 346 (6216), 1467.
Santer, B.D., et al., 2017. Comparing tropospheric warming in climate models and satellite data. J. Clim. 30 (1), 373–392.
Siongco, A.C., Hohenegger, C., Stevens, B., 2014. The Atlantic ITCZ bias in CMIP5 models. Climate Dynamics. 45 (5–6), 1169–1180. http://doi.org/10.1007/s00382-014-2366-3.
Stevens, B., Bony, S., 2013. What are climate models missing? Science 340 (6136), 1053–1054.
Wara, M.W., Ravelo, A.C., Delaney, M.L., 2005. Permanent El Niño-like conditions during the Pliocene warm period. Science 309 (5735), 758–761.
Williamson, D.L., Blackburn, M., Hoskins, B.J., Nakajima, K., 2012. The APE Atlas. NCAR Tech. Note.
Wing, A.A., Reed, K.A., Satoh, M., Stevens, B., Bony, S., Ohno, T., 2018. Radiative–convective equilibrium model intercomparison project. Geoscientific Model Development. 11 (2), 793–813. http://doi.org/10.5194/gmd-11-793-2018.
Zhang, Y.G., Pagani, M., Liu, Z., 2014. Response to comment on "a 12-million-year temperature history of the tropical Pacific ocean". Science 346 (6216), 1467.

SUBJECT INDEX

Note: 'Page numbers followed by "f" indicate figures, "t" indicate tables and "b" indicate boxes.'

D

E

F

G

H

I

J

L

M

N

O

P

S

T

Z

AUTHOR INDEX

Note: 'Page numbers followed by "f" indicate figures, "t" indicate tables, and "b" indicate boxes.'

A

C

E

F

G

H

I

J

K

L

M

N

O

P

Q

R

S

W

Printed in the United States
By Bookmasters